Differential Games

A Concise Introduction

Differential Games
A Concise Introduction

Jiongmin Yong

University of Central Florida, USA

 **World Scientific**

NEW JERSEY • LONDON • SINGAPORE • BEIJING • SHANGHAI • HONG KONG • TAIPEI • CHENNAI

Published by

World Scientific Publishing Co. Pte. Ltd.

5 Toh Tuck Link, Singapore 596224

USA office: 27 Warren Street, Suite 401-402, Hackensack, NJ 07601

UK office: 57 Shelton Street, Covent Garden, London WC2H 9HE

Library of Congress Cataloging-in-Publication Data
Yong, J. (Jiongmin), 1958–
    Differential games : a concise introduction / by Jiongmin Yong, University of Central Florida,
USA.
        pages cm
    Includes bibliographical references and index.
    ISBN 978-981-4596-22-0 (hardcover : alk. paper)
    1. Differential games. 2. Game theory. I. Title.
    QA272.Y66 2015
    519.3'2--dc23
                                2014038343

British Library Cataloguing-in-Publication Data
A catalogue record for this book is available from the British Library.

Printed in Singapore

In the Memory of Professors
Leonard D. Berkovitz and Xunjing Li

Preface

The study of differential games was initiated by R. Isaacs in the early 1950s, and independently by L. S. Pontryagin a little later in the middle 1950s, both were motivated by pursuit and evasion problems.

In a differential game, as in a classical game which can be regarded as a static counterpart of differential games, there are at least two players involved. Usually, the players have different goals. For example, in a pursuit-evasion situation between two players, the goal of the pursuer is to catch the evader, whereas the goal of the evader is to keep himself/herself from being captured. Another situation is in a gambling of two persons, the gain of one player is the loss of the other player. Therefore, the goals of the players could be completely opposite. On the other hand, sometimes, the two involved players have different goals which might not be completely conflicting one with the other, and it could be better if the players can somehow play cooperatively. Realized the complexity of the problem, Pontryagin and his colleague, instead of attack the differential games, in the middle of 1950s, they first initiate the study of optimal control problems which can be regarded as single-player differential games, formulated in terms of ordinary differential equations.

Mathematically, control theory studies certain interested behaviors of the so-called state trajectory/process subject to some dynamic equations involving a so-called control process. A typical situation is that the state trajectory satisfies an ordinary differential equation containing a control function. By changing the control, the state trajectory will change accordingly. Then one can try to find a control to achieve some specific goals, such as hitting a target by the state, minimizing a cost functional, etc. After having a reasonably good understanding of control theory, people are able to approach differential games.

The purpose of this book is to give a concise introduction to differential games with two players. We will begin with a glance of game theory which is an extension of optimization problems and is the static version of dynamic games. Some useful notions will be introduced there. Then we will briefly present the control theory, including controllability, viability, and optimal control theory. For general optimal control theory, besides the existence theory, there are two main approaches for characterizing optimal controls: variational method which leads to the so-called *Pontryagin maximum principle*, and dynamic programming method based on the so-called *Bellman's optimality principle* which leads to *Hamilton-Jacobi-Bellman* (HJB, for short) *equation* for the value function. Theory of viscosity solutions will be briefly presented. Then time-optimal control problem and viability problem will be treated by dynamic programming method.

Turning to two-person differential games, pursuit and evasion problems are treated as extensions of controllability and viability problems, respectively. For two-person zero-sum differential games, dynamic programming method leads to *Hamilton-Jacobi-Isaacs* (HJI, for short) *equation* for the upper and lower value functions. Then the uniqueness of viscosity solutions together with the *Isaacs condition* leads to the existence of the value function for the differential game. When the controls are unbounded, the above-mentioned procedure becomes much more difficult and technical. We will present results for some of interesting unbounded control cases. A similar theory will be established for the two-person zero-sum differential games with switching controls.

We will also look at the so-called linear-quadratic (LQ, for short) problems, namely, the state equation is a linear ordinary differential equation and the performance functional is quadratic. For such kind of problems, more details about the solutions to the differential games can be obtained. In particular, the open-loop and closed-loop solutions can be clearly described and distinguished.

This book is an expansion of the lecture notes written for the summer school of control theory held at Fudan University, in July 2012. The author would like to take this opportunity to thank Professor Hongwei Lou for organizing the summer school, and inviting me to give a short course on differential games. We assume the readers of the book have basic knowledge of analysis, linear algebra and ordinary differential equations. Several sections, marked by a star, require some more knowledge on functional analysis, etc., and readers can skip them at the first reading.

To conclude this preface, I would like to mention the following. Professor Leonard D. Berkovitz (1924–2009) introduced differential games to me the first time when I was a graduate student at Purdue University in the early 1980s, and later under his supervision, I wrote my doctorial dissertation entitled *"Differential Games of Evasion and Pursuit"*. I would also like to mention that, a little earlier, when I was an undergraduate student at Fudan University, it was Professor Xunjing Li (1935–2003) who taught a course *"Basics of Control Theory"* in which I learned control theory the first time. He also recommended me to go to Purdue University for my graduate study under the supervision of Professor Berkovitz. Moreover, Professor Li had been my mentor, colleague, and collaborator at Fudan University, for 15 years since 1988; from him, I learned a lot in many aspects.

Writing a book on differential games is a dream of mine for a long time. Now the dream comes true, and I sincerely dedicate this book to Professor Berkovitz and Professor Li.

Jiongmin Yong
at Orlando, Florida
October 2014

Contents

Chapter 1

Introduction

1.1 Optimization Problems

Let S be a nonempty set and $\mathbb{R}$ be the set of all real numbers. Let $f : S \to \mathbb{R}$ be a map. We pose the following problem.

Problem (O). Find an $\bar{x} \in S$ such that

$$f(\bar{x}) = \inf_{x \in S} f(x). \tag{1.1}$$

Such a problem is called a *minimization problem*. A point $\bar{x} \in S$ satisfying (1.1) is called a *solution* of Problem (O), or equivalently, a *minimum* of $f(\cdot)$ over S. If such an $\bar{x} \in S$ (uniquely) exists, we say that Problem (O) is (uniquely) *solvable*. Note that we may also pose the following problem.

Problem (O)'. Find an $\bar{x} \in S$ such that

$$f(\bar{x}) = \sup_{x \in S} f(x).$$

We call the above a *maximization problem*. It is clear that by considering $-f(\cdot)$ instead of $f(\cdot)$, one reduces Problem (O)' to Problem (O). Therefore, it suffices to just consider Problem (O). One usually refers to a minimization or maximization problem as an *optimization problem*.

Let us now look at some results on Problem (O). First, let

$$f(S) \equiv \big\{ f(x) \mid x \in S \big\}$$

be a finite set, which is the case, in particular, if S itself is finite. In this case, Problem (O) is solvable, by directly comparing the values in $f(S)$. Theoretically, this is trivial. However, although it will not be pursued further in this book, we point out that, practically, finding a minimum in $f(S)$ might be quite a nontrivial job if the number of elements in set $f(S)$ is very large.

Next, we let $f(S)$ be infinite. Then the existence of a minimum is not guaranteed. To ensure the existence of a minimum, one needs some topology on the set S and certain continuity on the map $f(\cdot)$. The following is such a result.

Proposition 1.1.1. (i) *Let S be a compact metric space and let $f : S \to \mathbb{R}$ be a lower semi-continuous function which is bounded from below, i.e.,*

$$f(x) \leqslant \varliminf_{y \to x} f(y), \qquad \forall x \in S,$$

and for some $M \geqslant 0$,

$$f(x) \geqslant -M, \qquad \forall x \in S.$$

Then Problem (O) admits a solution.

(ii) *Let S be a complete metric space (not necessarily compact) having the following property: There exists a sequence of compact subspaces S_k with*

$$S_1 \subseteq S_2 \subseteq S_3 \subseteq \cdots \subseteq S, \quad \bigcup_{k \geq 1} S_k = S.$$

Let $f : S \to \mathbb{R}$ be lower semi-continuous, bounded from below, and for some $x_0 \in S$,

$$\lim_{k \to \infty} \inf_{x \in S \setminus S_k} f(x) > f(x_0),$$

Then $f(\cdot)$ admits a minimum over S.

The proof of the above proposition is straightforward and is left to the readers. An example of the above (i) is: $S = [a, b]$, a closed interval, and $f : [a, b] \to \mathbb{R}$ is a continuous function. An example of the above (ii) is $S = \mathbb{R}$ and $f(x) = xe^{-x^2}$. Another important example for (ii) is $S = \mathbb{R}$ and $f : \mathbb{R} \to \mathbb{R}$ is continuous and *coercive*, i.e.,

$$\lim_{|x| \to \infty} f(x) = \infty,$$

for example, $f(x) = x^2$.

When S is an open set in $\mathbb{R}^n \equiv \{(x_1, \cdots, x_n) \mid x_1, \cdots, x_n \in \mathbb{R}\}$, we have more interesting results concerning Problem (O). The following is a collection of standard results from calculus.

Proposition 1.1.2. *Let $S \subseteq \mathbb{R}^n$ be an open set and $f : S \to \mathbb{R}$ be a continuously differentiable function. Suppose $f(\cdot)$ attains a local minimum at $x_0 \in S$, i.e., there exists a $\delta > 0$ such that*

$$B_\delta(x_0) = \{x \in \mathbb{R}^n \mid |x - x_0| < \delta\} \subseteq S,$$

and

$$f(x_0) \leqslant f(x), \qquad x \in B_\delta(x_0).$$

Then

$$f_x(x_0) = 0. \tag{1.2}$$

In addition, if $f(\cdot)$ is twice continuously differentiable, then

$$f_{xx}(x_0) \geqslant 0,$$

i.e., the Hessian matrix $f_{xx}(x_0)$ is positive semi-definite. Conversely, if (1.2) *holds and*

$$f_{xx}(x_0) > 0,$$

i.e., the Hessian $f_{xx}(x_0)$ is positive definite, then $f(\cdot)$ attains a local minimum at x_0.

The necessary condition (1.2) is called the *Fermat's theorem*. There are many other interesting results relevant to Problem (O). We omit them here.

1.2 Game Theory — A Brief Glance

We now let S_1 and S_2 be two sets and for $i = 1, 2$, let $f_i : S_i \to \mathbb{R}$ be a given map. Let us vaguely describe the following problem.

Problem (G). There are two persons involved, called *Players* 1 and 2, respectively. For $i = 1, 2$, Player i tries to find an $\bar{x}_i \in S_i$ so that the function $f_i(\cdot, \cdot)$ is minimized.

The above Problem (G) is called a *two-person game*. In the above, $f_i(\cdot, \cdot)$ is called the *cost function* of Player i. Any 4-tuple $\{S_1, S_2, f_1(\cdot, \cdot), f_2(\cdot, \cdot)\}$ determines a two-person game. If S_2 is a singleton, Problem (G) is reduced to Problem (O). Thus, Problem (G) is a natural extension of Problem (O). Or, we can say that optimization problems are *single-player* games.

Next, we note that since $f_1(x_1, x_2)$ depends on x_2, if Player 1 finds a minimum $\bar{x}_1$ of the map $x_1 \mapsto f_1(x_1, x_2)$, it must depend on x_2. Thus, we should denote it by $\bar{x}_1 = \varphi_1(x_2)$. Likewise, if Player 2 finds a minimum $\bar{x}_2$ of the map $x_2 \mapsto f_2(x_1, x_2)$, it must depend on x_1. Thus, we should write $\bar{x}_2 = \varphi_2(x_1)$. Hence, a pair $(\bar{x}_1, \bar{x}_2)$ is satisfactory to both players if the following hold:

$$\bar{x}_1 = \varphi_1(\varphi_2(\bar{x}_1)), \quad \bar{x}_2 = \varphi_2(\varphi_1(\bar{x}_2)). \tag{1.3}$$

However, in general, the above is not necessarily true. Here is a simple example.

Example 1.2.1. Let $S_1 = S_2 = [1, \infty)$ and

$$f_1(x_1, x_2) = (x_1 - x_2)^2, \qquad f_2(x_1, x_2) = (2x_1 - x_2)^2.$$

Then $x_1 \mapsto f_1(x_1, x_2)$ attains its minimum value 0 at

$$\bar{x}_1 = \varphi_1(x_2) = x_2, \qquad \forall x_2 \in S_2,$$

and $x_2 \mapsto f_2(x_1, x_2)$ attains its minimum value 0 at

$$\bar{x}_2 = \varphi_2(x_1) = 2x_1, \qquad \forall x_1 \in S_1.$$

Clearly,

$$\varphi_1(\varphi_2(\bar{x}_1)) = \varphi_1(2\bar{x}_1) = 2\bar{x}_1 \neq \bar{x}_1,$$

and

$$\varphi_2(\varphi_1(\bar{x}_2)) = \varphi_2(\bar{x}_2) = 2\bar{x}_2 \neq \bar{x}_2.$$

Hence, both relations in (1.3) fail. This means that in general two players might not get a pair $(\bar{x}_1, \bar{x}_2) \in S_1 \times S_2$ for which both are satisfactory.

From the above simple example, we get some taste of games, which is quite different from optimization problems.

1.2.1 *Pareto optimum and Nash equilibrium*

For Problem (G), let us introduce some concepts.

Definition 1.2.2. (i) A pair $(\bar{x}_1, \bar{x}_2) \in S_1 \times S_2$ is called a *Pareto optimum* of Problem (G) if there exists no other pair $(x_1, x_2) \in S_1 \times S_2$ such that

$$f_i(x_1, x_2) \leqslant f_i(\bar{x}_1, \bar{x}_2), \qquad i = 1, 2,$$

and at least one of the inequalities is strict. In this case, $(f_1(\bar{x}_1, \bar{x}_2), f_2(\bar{x}_1, \bar{x}_2))$ is called a *Pareto optimal value vector* of Problem (G).

(ii) A pair $(\bar{x}_1, \bar{x}_2) \in S_1 \times S_2$ is called a *Nash equilibrium* of Problem (G) if

$$\begin{cases} f_1(\bar{x}_1, \bar{x}_2) \leqslant f_1(x_1, \bar{x}_2), & \forall x_1 \in S_1, \\ f_2(\bar{x}_1, \bar{x}_2) \leqslant f_2(\bar{x}_1, x_2), & \forall x_2 \in S_2. \end{cases}$$

Roughly speaking, $(\bar{x}_1, \bar{x}_2)$ is a Pareto optimum if there are no strictly "better pairs". Whereas, $(\bar{x}_1, \bar{x}_2)$ is a Nash equilibrium if one player is deviating from the point, the cost function of this player would get larger; and there is no information given if both players are deviating from the Nash equilibrium point. Before going further, let us introduce the following assumption.

(G1) For $i = 1, 2$, let S_i be a compact metric space, and $f_i : S_1 \times S_2 \to \mathbb{R}$ be a continuous function.

The following result is concerned with the existence of Pareto optima.

Proposition 1.2.3. *For any* $\lambda_1, \lambda_2 > 0$, *if* $(\bar{x}_1, \bar{x}_2) \in S_1 \times S_2$ *is a minimum of* $\lambda_1 f_1(\cdot, \cdot) + \lambda_2 f_2(\cdot, \cdot)$, *then it is a Pareto optimum of Problem* (G). *Consequently, under* (G1), *Problem* (G) *always admits a Pareto optimum.*

Proof. We prove the conclusion by contradiction. Let $(\bar{x}_1, \bar{x}_2) \in S_1 \times S_2$ be a minimum of $\lambda_1 f_1(\cdot, \cdot) + \lambda_2 f_2(\cdot, \cdot)$. Suppose there exists a pair $(x_1, x_2) \in S_1 \times S_2$ such that, say,

$$f_1(x_1, x_2) < f_1(\bar{x}_1, \bar{x}_2) \qquad \text{and} \qquad f_2(x_1, x_2) \leqslant f_2(\bar{x}_1, \bar{x}_2).$$

Then

$$\lambda_1 f_1(x_1, x_2) + \lambda_2 f_2(x_1, x_2) < \lambda_1 f_1(\bar{x}_1, \bar{x}_2) + \lambda_2 f_2(\bar{x}_1, \bar{x}_2)$$
$$\leqslant \lambda_1 f_1(x_1, x_2) + \lambda_2 f_2(x_1, x_2),$$

which is a contradiction.

Now, under (G1), $(x_1, x_2) \mapsto \lambda_1 f_1(x_1, x_2) + \lambda_2 f_2(x_1, x_2)$ is continuous on the compact metric space $S_1 \times S_2$. Thus, a minimum exists, which gives a Pareto optimum of Problem (G). $\qquad\square$

It is clear that by choosing different $\lambda_1, \lambda_2 > 0$, the minimum of $\lambda_1 f_1(\cdot, \cdot) + \lambda_2 f_2(\cdot, \cdot)$ might be different in general. Hence, from the above result, one can expect that Pareto optima are not unique in general. We will see such kind of examples below.

Let us look at the Pareto optima from a little different angle. Consider the set

$$\mathcal{D} = \left\{ \left(f_1(x_1, x_2), f_2(x_1, x_2) \right) \mid (x_1, x_2) \in S_1 \times S_2 \right\} \subseteq \mathbb{R}^2.$$

Then a point $(\bar{y}_1, \bar{y}_2) \in \mathcal{D}$ is a pair of Pareto optimal value vector of Problem (G) if and only if there is no other pair $(y_1, y_2) \in \mathcal{D}$ such that

$$y_1 \leqslant \bar{y}_1, \quad y_2 \leqslant \bar{y}_2, \quad y_1 + y_2 < \bar{y}_1 + \bar{y}_2,$$

which means that in the first two inequalities, at least one of them must be strict, which is denoted by the following:

$$(y_1, y_2) < (\bar{y}_1, \bar{y}_2).$$

In Proposition 1.2.3, we have constructed Pareto optima of Problem (G) by means of minimizing map $(x_1, x_2) \mapsto \lambda_1 f_1(x_1, x_2) + \lambda_2 f_2(x_1, x_2)$. Let us now point out that Problem (G) may have some other Pareto optima that cannot be constructed in such a way. Here is an example.

Example 1.2.4. Let $S_1 = S_2 = \{a, b\}$ and

$$
\begin{cases}
f_1(a, a) = f_2(b, b) = 1, & f_1(b, b) = f_2(a, a) = 0, \\
f_1(a, b) = f_1(b, a) = f_2(a, b) = f_2(b, a) = \dfrac{3}{4}.
\end{cases}
$$

Therefore, in the current case,

$$
\mathcal{D} = \left\{ (1, 0), (0, 1), \left(\frac{3}{4}, \frac{3}{4} \right) \right\}.
$$

Clearly, all of these three are Pareto optimal values, i.e., all the points $(a, a), (a, b), (b, a), (b, b)$ are Pareto optima. However, for any $\lambda_1, \lambda_2 > 0$, if we define

$$
\psi^{\lambda_1, \lambda_2}(x_1, x_2) = \lambda_1 f_1(x_1, x_2) + \lambda_2 f_2(x_1, x_2),
$$

then

$$
\begin{cases}
\psi^{\lambda_1, \lambda_2}(a, a) = \lambda_1, & \psi^{\lambda_1, \lambda_2}(b, b) = \lambda_2, \\
\psi^{\lambda_1, \lambda_2}(a, b) = \psi^{\lambda_1, \lambda_2}(b, a) = \dfrac{3}{4}(\lambda_1 + \lambda_2).
\end{cases}
$$

We claim that for any $\lambda_1, \lambda_2 > 0$, (a, b) and (b, a) are not minimum of $\psi^{\lambda_1, \lambda_2}(\cdot, \cdot)$. In fact, if, say, (a, b) is a minimum, then

$$
\frac{3}{4}(\lambda_1 + \lambda_2) \leqslant \lambda_1, \lambda_2,
$$

which implies

$$
3\lambda_2 \leqslant \lambda_1, \qquad 3\lambda_1 \leqslant \lambda_2.
$$

These lead to a contradiction. Hence, the Pareto optima (a, b) and (b, a) cannot be characterized by the minimum of $(x_1, x_2) \mapsto \psi^{\lambda_1, \lambda_2}(x_1, x_2)$, for any $\lambda_1, \lambda_2 > 0$.

Next, for Nash equilibria, we claim that all the following situations can happen:

• Nash equilibria do not necessarily exist. This implies that a Pareto optimum might not be a Nash equilibrium, since the latter always exists.

• The Nash equilibrium might not be unique, and in this case, different Nash equilibria can yield different costs to each player.

• A Nash equilibrium might not be a Pareto optimum.

To illustrate the above, let us present some examples.

Example 1.2.5. (Nonexistence of Nash equilibria) Let $S_1 = S_2 = \{0, 1\}$ and let

$$\begin{cases} f_1(0,0) = f_1(1,1) = 1, & f_1(0,1) = f_1(1,0) = 0, \\ f_2(0,1) = f_2(1,0) = 1, & f_2(0,0) = f_2(1,1) = 0. \end{cases}$$

We claim that this game does not have Nash equilibria. In fact, suppose it has a Nash equilibrium $(\bar{x}_1, \bar{x}_2) \in S_1 \times S_2$. If $\bar{x}_1 = \bar{x}_2$, then for $x_1 \neq \bar{x}_2$,

$$1 = f_1(\bar{x}_1, \bar{x}_2) \leqslant f_1(x_1, \bar{x}_2) = 0,$$

which is a contradiction. If $\bar{x}_1 \neq \bar{x}_2$, then for $x_2 = \bar{x}_1$,

$$1 = f_2(\bar{x}_1, \bar{x}_2) \leqslant f_2(\bar{x}_1, x_2) = 0,$$

which is also a contradiction. Now, we claim that every pair $(x_1, x_2) \in S_1 \times S_2$ is a Pareto optimum. In fact, by letting

$$\psi(x_1, x_2) = f_1(x_1, x_2) + f_2(x_1, x_2), \qquad (x_1, x_2) \in S_1 \times S_2,$$

we have

$$\psi(0,0) = \psi(1,1) = \psi(0,1) = \psi(1,0) = 1.$$

Thus, any $(\bar{x}_1, \bar{x}_2) \in S_1 \times S_2$ is a minimum of $\psi(\cdot, \cdot)$. Hence, by Proposition 1.2.3, every point is a Pareto optimum.

Example 1.2.6. (Prisoners' dilemma) Let c and d represent "confess" and "denial", respectively. Let $S_1 = S_2 = \{c, d\}$. The total number of years in jail is given by $f_i(x_1, x_2)$ for the i-th prisoner when the strategy $(x_1, x_2) \in S_1 \times S_2$ is taken by these two prisoners. Suppose

$$\begin{cases} f_1(c,c) = f_2(c,c) = 3, & f_1(d,d) = f_2(d,d) = 1, \\ f_1(c,d) = f_2(d,c) = 0, & f_1(d,c) = f_2(c,d) = 10. \end{cases}$$

We claim the following:

• (c, c) is the Nash equilibrium, but it is not a Pareto optimum;

• (c, d), (d, c), and (d, d) are Pareto optima, but none of them is a Nash equilibrium.

In fact,

$$f_1(c,c) = 3 < 10 = f_1(d,c), \qquad f_2(c,c) = 3 < 10 = f_2(c,d).$$

Thus, (c, c) is a Nash equilibrium. On the other hand,

$$f_i(d, d) = 1 < 3 = f_i(c, c), \qquad i = 1, 2.$$

This implies that (c, c) is not a Pareto optimum.

Next, for any $\lambda_1, \lambda_2 > 0$, if we let

$$\psi^{\lambda_1, \lambda_2}(x_1, x_2) = \lambda_1 f_1(x_1, x_2) + \lambda_2 f_2(x_1, x_2), \qquad (x_1, x_2) \in S_1 \times S_2,$$

then

$$\begin{cases} \psi^{\lambda_1, \lambda_2}(c, c) = 3(\lambda_1 + \lambda_2), \quad \psi^{\lambda_1, \lambda_2}(d, d) = \lambda_1 + \lambda_2, \\ \psi^{\lambda_1, \lambda_2}(c, d) = 10\lambda_2, \quad \psi^{\lambda_1, \lambda_2}(d, c) = 10\lambda_1. \end{cases}$$

Clearly,

$$2 = \psi^{1,1}(d, d) = \min_{(x_1, x_2) \in S_1 \times S_2} \psi^{1,1}(x_1, x_2) = \min\{6, 2, 10, 10\},$$

$$10 = \psi^{19,1}(c, d) = \min_{(x_1, x_2) \in S_1 \times S_2} \psi^{19,1}(x_1, x_2) = \min\{60, 20, 10, 190\},$$

and

$$10 = \psi^{1,19}(d, c) = \min_{(x_1, x_2) \in S_1 \times S_2} \psi^{1,19}(x_1, x_2) = \min\{60, 20, 190, 10\}.$$

Hence, (d, d), (c, d), and (d, c) are Pareto optima. We now show that (d, d), (c, d), and (d, c) are not Nash equilibria. In fact,

$$f_1(d, d) = 1 > 0 = f_1(c, d), \qquad f_2(d, d) = 1 > 0 = f_2(d, c),$$

$$f_1(d, c) = 10 > 3 = f_1(c, c), \qquad f_2(c, d) = 10 > 3 = f_2(c, c).$$

Any one of the first two shows that (d, d) is not a Nash equilibrium, and the last two show (d, c) and (c, d) are not Nash equilibria, respectively.

Example 1.2.7. (Non-uniqueness of Nash equilibria) Let $S_1 = S_2 = \{1, 2\}$. Define

$$\begin{cases} f_1(1, 1) = f_1(2, 2) = f_2(1, 1) = f_2(2, 2) = 3, \\ f_1(1, 2) = 0, \qquad f_2(1, 2) = 1, \qquad f_1(2, 1) = f_2(2, 1) = 2. \end{cases}$$

We claim that $(1, 2)$ and $(2, 1)$ are Nash equilibria. In fact,

$$f_1(1, 2) = 0 < 3 = f_1(2, 2), \qquad f_2(1, 2) = 1 < 3 = f_2(1, 1),$$

and

$$f_1(2, 1) = 2 < 3 = f_1(1, 1), \qquad f_2(2, 1) = 2 < 3 = f_2(2, 2).$$

Thus, Nash equilibria are not unique and the costs at different Nash equilibria could be different.

For the existence of Nash equilibria, we need the following result.

Lemma 1.2.8. (Kakutani's Fixed Point Theorem) *Let $S \subseteq \mathbb{R}^n$ be non-empty, convex and compact. Let $\varphi : S \to 2^S$ (2^S stands for the set of all subsets of S) be a set-valued function with the following properties:*

(i) *For each $x \in S$, $\varphi(x)$ is non-empty and convex;*

(ii) *The graph $\mathcal{G}(\varphi)$ of $\varphi(\cdot)$ defined by*

$$\mathcal{G}(\varphi) = \{(x,y) \mid x \in S, \ y \in \varphi(x)\}$$

is closed in $S \times S$.

Then there exists an x^* such that $x^* \in \varphi(x^*)$.

Theorem 1.2.9. *Let* (G1) *hold and both S_1 and S_2 be convex sets in Euclidean spaces. Suppose*

$$\begin{cases} x_1 \mapsto f_1(x_1, x_2) & \text{is convex}, \quad \forall x_2 \in S_2, \\ x_2 \mapsto f_2(x_1, x_2) & \text{is convex}, \quad \forall x_1 \in S_1. \end{cases}$$

Then Problem (G) *admits a Nash equilibrium.*

Proof. Under (G1), both $f_1(\cdot, \cdot)$ and $f_2(\cdot, \cdot)$ are uniformly continuous on $S_1 \times S_2$. Hence,

$$\varphi_1(x_2) = \inf_{x_1 \in S_1} f_1(x_1, x_2), \qquad x_2 \in S_2,$$
$$\varphi_2(x_1) = \inf_{x_2 \in S_2} f_2(x_1, x_2), \qquad x_1 \in S_1,$$

are well-defined and continuous. Let

$$\begin{cases} F_1(x_2) = \arg\min f_1(\cdot, x_2) \\ \qquad \equiv \Big\{ x_1 \in S_1 \mid f_1(x_1, x_2) = \inf_{\widetilde{x}_1 \in S_1} f_1(\widetilde{x}_1, x_2) \equiv \varphi_1(x_2) \Big\}, \\ F_2(x_1) = \arg\min f_2(x_1, \cdot) \\ \qquad \equiv \Big\{ x_2 \in S_2 \mid f_2(x_1, x_2) = \inf_{\widetilde{x}_2 \in S_2} f_2(x_1, \widetilde{x}_2) \equiv \varphi_2(x_1) \Big\}. \end{cases}$$

Then $F_1 : S_2 \to 2^{S_1}$ and $F_2 : S_1 \to 2^{S_2}$. Note that the graph $\mathcal{G}(F_1)$ of F_1 is given by

$$\mathcal{G}(F_1) = \Big\{ (x_2, x_1) \in S_2 \times S_1 \mid x_1 \in F_1(x_2) \Big\}$$
$$= \Big\{ (x_2, x_1) \mid f_1(x_1, x_2) = \varphi_1(x_2) \Big\},$$

which is closed (due to the continuity of $f_1(\cdot,\cdot)$ and $\varphi_1(\cdot)$). Likewise, the graph $\mathcal{G}(F_2)$ of F_2 is also closed. Next, we claim that for each $x_2 \in S_2$, $F_1(x_2)$ is convex. In fact, for any $x_1, \bar{x}_1 \in F_1(x_2)$, we have

$$f_1(x_1, x_2) = f_1(\bar{x}_1, x_2) = \inf_{\tilde{x}_1 \in S_1} f_1(\tilde{x}_1, x_2) = \varphi_1(x_2).$$

Thus, for any $\lambda \in (0,1)$, by the convexity of $x_1 \mapsto f_1(x_1, x_2)$, we have

$$f_1(\lambda x_1 + (1-\lambda)\bar{x}_1, x_2) \leqslant \lambda f_1(x_1, x_2) + (1-\lambda)f_1(\bar{x}_1, x_2)$$
$$= \varphi_1(x_2) \leqslant f_1(\lambda x_1 + (1-\lambda)\bar{x}_1, x_2).$$

Hence,

$$\lambda x_1 + (1-\lambda)\bar{x}_1 \in F_1(x_2),$$

proving the convexity of $F_1(x_2)$. Similarly, $F_2(x_1)$ is convex for any $x_1 \in S_1$.

We now consider the following map

$$\Phi(x_1, x_2) = F_1(x_2) \times F_2(x_1).$$

Clearly, $\Phi : S_1 \times S_2 \to S_1 \times S_2$ takes compact and convex set values. The graph of Φ is topologically equivalent to $\mathcal{G}(F_1) \times \mathcal{G}(F_2)$:

$$\mathcal{G}(\Phi) = \Big\{ (x_1, x_2, y_1, y_2) \mid y_1 \in F_1(x_2), y_2 \in F_2(x_1) \Big\}$$
$$= \Big\{ (x_1, x_2, y_1, y_2) \mid f_1(y_1, x_2) = \varphi_1(x_2), \ f_2(x_1, y_2) = \varphi_2(x_1) \Big\}.$$

Thus, $\mathcal{G}(\Phi)$ is also closed. Therefore, by Kakutani's fixed point theorem, there exists a pair $(\bar{x}_1, \bar{x}_2) \in S_1 \times S_2$ such that

$$(\bar{x}_1, \bar{x}_2) \in \Phi(\bar{x}_1, \bar{x}_2).$$

That is

$$\bar{x}_1 \in F_1(\bar{x}_2), \quad \bar{x}_2 \in F_2(\bar{x}_1),$$

which means

$$f_1(\bar{x}_1, \bar{x}_2) = \varphi_1(\bar{x}_2) = \inf_{x_1 \in S_1} f_1(x_1, \bar{x}_2) \leqslant f_1(x_1, \bar{x}_2), \qquad \forall x_1 \in S_1,$$

and

$$f_2(\bar{x}_1, \bar{x}_2) = \varphi_2(\bar{x}_1) = \inf_{x_2 \in S_2} f_2(\bar{x}_1, x_2) \leqslant f_2(\bar{x}_1, x_2), \qquad \forall x_2 \in S_2.$$

Hence, $(\bar{x}_1, \bar{x}_2) \in S_1 \times S_2$ is a Nash equilibrium. $\square$

Let us look at a simple example for which the above is applicable.

Example 1.2.10. (**Matrix Game**) Let $S_i \subseteq \mathbb{R}^{n_i}$ be convex and compact, $i = 1, 2$. Let

$$f_i(x_1, x_2) = \langle A_i x_1, x_2 \rangle, \qquad \forall x_i \in S_i, \quad i = 1, 2,$$

where $A_i \in \mathbb{R}^{n_2 \times n_1}$, the set of all $(n_2 \times n_1)$ real matrices. Then the associated two-person game admits a Nash equilibrium. In fact, all the conditions assumed in Theorem 1.2.9 are satisfied. It is not hard to see that if we define

$$f_i(x_1, x_2) = \langle Q_i x_i, x_i \rangle + \langle A_i x_1, x_2 \rangle, \qquad \forall x_i \in S_i, \quad i = 1, 2,$$

with $Q_i \in \mathbb{R}^{n_i \times n_i}$ being positive definite, and $S_i = \mathbb{R}^{n_i}$. Then the associated two-person game also admits a Nash equilibrium. We leave the proof to interested readers.

1.2.2 *Two-person zero-sum game*

In this subsection, we consider an important special case which is described in the following definition.

Definition 1.2.11. Problem (G) is called a *two-person zero-sum game* if

$$f_1(x_1, x_2) + f_2(x_1, x_2) = 0, \qquad \forall (x_1, x_2) \in S_1 \times S_2. \tag{1.4}$$

If the above is not satisfied, we call the game a *two-person non-zero-sum game*.

In the case of zero-sum, we let

$$f(x_1, x_2) = f_1(x_1, x_2) = -f_2(x_1, x_2), \qquad \forall (x_1, x_2) \in S_1 \times S_2,$$

and call it the *performance index* of the game. Then Player 1 is a *minimizer* and Player 2 is a *maximizer*. To distinguish from the general Problem (G), hereafter, we denote the two-person zero-sum game associated with $\{S_1, S_2, f(\cdot, \cdot)\}$ by Problem (G^0).

We now introduce the following definition.

Definition 1.2.12. (i) The *upper value* V^+ and the *lower value* V^- of Problem (G^0) are defined by the following:

$$\begin{cases} V^+ = \inf_{x_1 \in S_1} \sup_{x_2 \in S_2} f(x_1, x_2), \\ V^- = \sup_{x_2 \in S_2} \inf_{x_1 \in S_1} f(x_1, x_2), \end{cases}$$

for which the following is automatically true:

$$V^- \leqslant V^+.$$

In the case that

$$V^+ = V^- \equiv V, \tag{1.5}$$

we call V the *value* of Problem (G^0).

(ii) The upper value V^+ is (uniquely) *achievable* if there is a (unique) $\mu_2 : S_1 \to S_2$ and a (unique) $\bar{x}_1 \in S_1$ such that

$$\begin{cases} f(x_1, \mu_2(x_1)) = \sup_{x_2 \in S_2} f(x_1, x_2), & \forall x_1 \in S_1, \\ f(\bar{x}_1, \mu_2(\bar{x}_1)) = \inf_{x_1 \in S_1} f(x_1, \mu_2(x_1)) = \inf_{x_1 \in S_1} \sup_{x_2 \in S_2} f(x_1, x_2) = V^+. \end{cases} \tag{1.6}$$

In the above case, we say that $(\bar{x}_1, \mu_2(\cdot))$ (uniquely) achieves V^+. Similarly, we can define (unique) achievability of V^-.

(iii) A pair $(\bar{x}_1, \bar{x}_2) \in S_1 \times S_2$ is called a *saddle point* of Problem (G^0) if

$$f(\bar{x}_1, x_2) \leqslant f(\bar{x}_1, \bar{x}_2) \leqslant f(x_1, \bar{x}_2), \qquad \forall (x_1, x_2) \in S_1 \times S_2. \tag{1.7}$$

Note that a pair $(\bar{x}_1, \bar{x}_2) \in S_1 \times S_2$ is a saddle point of Problem (G^0) if and only if it is a Nash equilibrium of Problem (G) with (1.4).

For convenience, we rewrite the assumption (G1) corresponding to the current zero-sum case as follows.

(G2) The sets S_1 and S_2 are compact metric spaces, and the function $f : S_1 \times S_2 \to \mathbb{R}$ is continuous.

We now present the following result.

Proposition 1.2.13. (i) *If* $(\bar{x}_1, \bar{x}_2) \in S_1 \times S_2$ *is a saddle point of Problem* (G^0), *then the game has a value V and*

$$V = f(\bar{x}_1, \bar{x}_2).$$

(ii) *If Problem* (G^0) *has a value V. Moreover, $V = V^+$ is achieved by* $(\bar{x}_1, \mu_2(\cdot))$ *and $V = V^-$ is achieved by* $(\mu_1(\cdot), \bar{x}_2)$, *respectively, then* $(\bar{x}_1, \bar{x}_2)$ *is a saddle point of the game.*

(iii) *If (G2) holds and Problem* (G^0) *has a value V, then the game admits a saddle point.*

(iv) *Let S_1 and S_2 be convex compact sets in some linear spaces, let* $f(x_1, x_2)$ *be convex in x_1 and concave in x_2. Then Problem* (G^0) *admits a saddle point.*

Proof. (i) From (1.7), we have

$$V^+ \leqslant \sup_{x_2 \in S_2} f(\bar{x}_1, x_2) \leqslant f(\bar{x}_1, \bar{x}_2) \leqslant \inf_{x_1 \in S_1} f(x_1, \bar{x}_2) \leqslant V^- \leqslant V^+,$$

which implies the existence of the value.

(ii) Let $V = V^+ = V^-$. Let $\mu_2 : S_1 \to S_2$ and $\bar{x}_1$ such that (1.6) holds, and let $\mu_1 : S_2 \to S_1$ and $\bar{x}_2$ such that

$$
\begin{cases}
f(\mu_1(x_2), x_2) = \inf_{x_1 \in S_1} f(x_1, x_2), & \forall x_2 \in S_2, \\
f(\mu_1(\bar{x}_2), \bar{x}_2) = \sup_{x_2 \in S_2} f(\mu_1(x_2), x_2) \\
\qquad = \sup_{x_2 \in S_2} \inf_{x_1 \in S_1} f(x_1, x_2) = V^-.
\end{cases}
$$

We claim that $(\bar{x}_1, \bar{x}_2)$ is a saddle point of the game. In fact, from the above, we see that

$$ f(\mu_1(x_2), x_2) \leqslant f(x_1, x_2), \qquad \forall (x_1, x_2) \in S_1 \times S_2. $$

Taking $x_2 = \bar{x}_2$ in the above, we get

$$ V^- = f(\mu_1(\bar{x}_2), \bar{x}_2) \leqslant f(x_1, \bar{x}_2), \qquad \forall x_1 \in S_1. $$

Similarly,

$$ f(x_1, \mu_2(x_1)) \geqslant f(x_1, x_2), \qquad \forall (x_1, x_2) \in S_1 \times S_2. $$

Thus, by taking $x_1 = \bar{x}_1$ in the above, one has

$$ V^+ = f(\bar{x}_1, \mu_2(\bar{x}_1)) \geqslant f(\bar{x}_1, x_2), \qquad \forall x_2 \in S_2. $$

Consequently, if (1.5) holds, then

$$ f(\bar{x}_1, x_2) \leqslant V^+ = V = V^- \leqslant f(x_1, \bar{x}_2), \qquad \forall (x_1, x_2) \in S_1 \times S_2. $$

This yields

$$ f(\bar{x}_1, \bar{x}_2) \leqslant \sup_{x_2 \in S_2} f(\bar{x}_1, x_2) \leqslant V \leqslant \inf_{x_1 \in S_1} f(x_1, \bar{x}_2) \leqslant f(\bar{x}_1, \bar{x}_2). $$

Hence, all the equalities in the above must hold, which implies

$$ f(\bar{x}_1, x_2) \leqslant f(\bar{x}_1, \bar{x}_2) \leqslant f(x_1, \bar{x}_2), \qquad \forall (x_1, x_2) \in S_1 \times S_2. $$

Therefore, $(\bar{x}_1, \bar{x}_2)$ is a saddle point of the game.

(iii) Suppose $V^+ = V^- = V$. Let

$$
\begin{cases}
\varphi_1(x_2) = \inf_{x_1 \in S_1} f(x_1, x_2), & \forall x_2 \in S_2, \\
\varphi_2(x_1) = \sup_{x_2 \in S_2} f(x_1, x_2), & \forall x_1 \in S_1.
\end{cases}
$$

Then under (G2), $\varphi_1(\cdot)$ and $\varphi_2(\cdot)$ are continuous, and

$$ \sup_{x_2 \in S_2} \varphi_1(x_2) = V^- = V^+ = \inf_{x_1 \in S_1} \varphi_2(x_1). $$

Thus, for any $\varepsilon > 0$, we can find a pair $(x_1^\varepsilon, x_2^\varepsilon) \in S_1 \times S_2$ such that

$$V^+ + \varepsilon > \varphi_2(x_1^\varepsilon) \geqslant V^+, \tag{1.8}$$

$$V^- - \varepsilon < \varphi_1(x_2^\varepsilon) \leqslant V^-. \tag{1.9}$$

Since both S_1 and S_2 are compact, we may assume that

$$(x_1^\varepsilon, x_2^\varepsilon) \to (\bar{x}_1, \bar{x}_2), \qquad \varepsilon \to 0.$$

Then by the continuity of $\varphi_1(\cdot)$ and $\varphi_2(\cdot)$, we obtain from (1.8)–(1.9) that

$$\varphi_1(\bar{x}_2) = \varphi_2(\bar{x}_1) = V^\pm = V.$$

Hence,

$$\begin{aligned}
f(\bar{x}_1, \bar{x}_2) &\leqslant \sup_{x_2 \in S_2} f(\bar{x}_1, x_2) = \varphi_2(\bar{x}_1) \\
&= \varphi_1(\bar{x}_2) = \inf_{x_1 \in S_1} f(x_1, \bar{x}_2) \leqslant f(\bar{x}_1, \bar{x}_2).
\end{aligned}$$

Clearly, all the equalities must hold in the above. Consequently, for any $(x_1, x_2) \in S_1 \times S_2$,

$$\begin{aligned}
f(\bar{x}_1, x_2) &\leqslant \sup_{x_2 \in S_2} f(\bar{x}_1, x_2) = \varphi_2(\bar{x}_1) = f(\bar{x}_1, \bar{x}_2) \\
&= \varphi_1(\bar{x}_2) = \inf_{x_1 \in S_1} f(x_1, \bar{x}_2) \leqslant f(x_1, \bar{x}_2),
\end{aligned}$$

which shows that $(\bar{x}_1, \bar{x}_2)$ is a saddle point.

(iv) Applying Theorem 1.2.9 to the current situation, we see that Problem (G^0) admits a saddle point. $\qquad\square$

We emphasize that part (ii) of the above proposition does not need the compactness of S_1 and S_2, and it provides an effective way of finding a saddle point, via optimization problems. Part (iv) of the above result is essentially the *von Neumann's minimax theorem*.

1.3   Control and Differential Game Problems

In this section, we briefly look at some typical problems in control theory and differential games.

1.3.1 *Control problems*

In what follows, we denote $\mathbb{R}_+ = [0, \infty)$. For any *initial pair* $(t, x) \in \mathbb{R}_+ \times \mathbb{R}^n$, consider the following ordinary differential equation:

$$\begin{cases} \dot{X}(s) = f(s, X(s), u(s)), & s \in [t, \infty), \\ X(t) = x, \end{cases} \tag{1.10}$$

where $f : \mathbb{R}_+ \times \mathbb{R}^n \times U \to \mathbb{R}^n$ is a given map. In the above, $X(\cdot)$ is called the *state trajectory*, taking values in $\mathbb{R}^n$ and $u(\cdot)$ is called the *control*, taking values in some metric space U. We call (1.10) a *control system*. Note that in the case that $f(\cdot, \cdot, \cdot)$ is only defined on $[0, T] \times \mathbb{R}^n \times U$ for some $T \in (0, \infty)$, we may extend it to $\mathbb{R}_+ \times \mathbb{R}^n \times U$ in some natural ways. For any $0 \leqslant t < T < \infty$, we introduce the following:

$$\mathcal{U}[t, T] = \Big\{ u : [t, T] \to U \mid u(\cdot) \text{ is measurable} \Big\},$$

and

$$\mathcal{U}[t, \infty) = \Big\{ u : [t, \infty) \to U \mid u(\cdot) \text{ is measurable} \Big\}.$$

Any $u(\cdot) \in \mathcal{U}[t, T]$ (resp. $u(\cdot) \in \mathcal{U}[t, \infty)$) is called a *feasible control* on $[t, T]$ (resp. on $[t, \infty)$).

Under proper conditions, for any initial pair $(t, x) \in \mathbb{R}_+ \times \mathbb{R}^n$, and feasible control $u(\cdot) \in \mathcal{U}[t, \infty)$, (1.10) admits a unique solution $X(\cdot) = X(\cdot\,; t, x, u(\cdot))$ defined on $[t, \infty)$. Clearly, different choices of $u(\cdot)$ will result in different state trajectories $X(\cdot)$. We refer to $(X(\cdot), u(\cdot))$ as a *state-control pair* of the control system (1.10).

Next, we recall that $2^{\mathbb{R}^n}$ is the set of all subsets of $\mathbb{R}^n$. Any map $M : \mathbb{R}_+ \to 2^{\mathbb{R}^n}$ is called a *moving target* in $\mathbb{R}^n$ if for any $t \in \mathbb{R}_+$, $M(t)$ is a measurable set in $\mathbb{R}^n$. We allow $M(t)$ to be empty for some or all t, which will give us some flexibility below. In most situations, for any $t \in \mathbb{R}_+$, $M(t)$ is assumed to be closed or open. Let us look at some simple examples of moving targets.

- Let $b : \mathbb{R}_+ \to \mathbb{R}^n$ be a continuous function and let

$$M(t) = \{ x \in \mathbb{R}^n \mid |x - b(t)| \leqslant 1 \} \equiv \bar{B}_1(b(t)), \qquad t \in \mathbb{R}_+,$$

where we recall that $B_r(x)$ stands for the open ball centered at x with radius r, and $\bar{B}_r(x)$ is its closure. This $M(\cdot)$ is a moving closed unit ball with the center moving along the path $b(\cdot)$.

- Let $M \subseteq \mathbb{R}^n$ be a fixed subset which could be $\mathbb{R}^n$ and ϕ. Let

$$M(t) = M, \qquad \forall t \in \mathbb{R}_+.$$

In this case, we simply call M a (fixed) *target*. For convenience, in the case $M = \phi$, we call it an *empty fixed target set* and in the case $M \neq \phi$, we call it a *nonempty fixed target set*.

- For some $T \in (0, \infty)$, let

$$\mathbb{R}^n_T(t) = \phi I_{[0,T)}(t) + \mathbb{R}^n I_{[T,\infty)}(t) \equiv \begin{cases} \phi, & t \in [0, T), \\ \mathbb{R}^n, & t \in [T, \infty), \end{cases}$$

hereafter, for any $\Omega \subset \mathbb{R}$, $I_\Omega(\cdot)$, called the *characteristic function* of Ω, defined by

$$I_\Omega(t) = \begin{cases} 1, & t \in \Omega, \\ 0, & t \notin \Omega. \end{cases}$$

We now formulate the following problem.

Problem (C). Let $M(\cdot)$ be a moving target set in $\mathbb{R}^n$. For given $(t, x) \in \mathbb{R}_+ \times \mathbb{R}^n$, find a control $u(\cdot) \in \mathcal{U}[t, \infty)$ such that for some $\tau \geqslant t$,

$$X(\tau; t, x, u(\cdot)) \in M(\tau).$$

The above is called a *controllability problem* for system (1.10) with the moving target set $M(\cdot)$. For a moving target set $M(\cdot)$ in $\mathbb{R}^n$, and $T \in (0, \infty)$, we define

$$\mathcal{U}^{M(\cdot)}_x[t, T] = \Big\{ u(\cdot) \in \mathcal{U}[t, \infty) \mid X(\tau; t, x, u(\cdot)) \in M(\tau), \text{ for some } \tau \in [t, T] \Big\},$$
$$(t, x) \in [0, T] \times \mathbb{R}^n,$$

and

$$\mathcal{U}^{M(\cdot)}_x[t, \infty) = \bigcup_{T \geqslant t} \mathcal{U}^{M(\cdot)}_x[t, T], \qquad \forall (t, x) \in \mathbb{R}_+ \times \mathbb{R}^n.$$

Note that if $u(\cdot) \in \mathcal{U}^{M(\cdot)}_x[t, T]$, then under $u(\cdot)$, the state starting from (t, x) will hit the moving target $M(\cdot)$ at some $\tau \in [t, T]$. Let us look at some simple special cases.

(i) $M(t) = \mathbb{R}^n_T(t) \equiv \phi I_{[0,T)}(t) + \mathbb{R}^n I_{[T,\infty)}(t)$, for any $t \in \mathbb{R}_+$, with some given $T \in (0, \infty)$. For this case, one has

$$\mathcal{U}^{\mathbb{R}^n_T(\cdot)}_x[t, T] = \bigcup_{S \geqslant T} \mathcal{U}[t, S], \qquad \forall (t, x) \in [0, T] \times \mathbb{R}^n. \tag{1.11}$$

In this case, the target is hit by the state at $t = T$. Such a case is referred to as a *fixed-duration case*.

(ii) $M(t) = \mathbb{R}^n$, for any $t \in \mathbb{R}_+$. For this case, one has

$$X(s; t, x, u(\cdot)) \in M(t), \quad \forall u(\cdot) \in \mathcal{U}[t, \infty), s \in [t, \infty), \forall (t, x) \in \mathbb{R}_+ \times \mathbb{R}^n.$$

Thus,

$$\mathcal{U}_x^{\mathbb{R}^n}[t, T] = \bigcup_{S \geqslant t} \mathcal{U}[t, S], \quad \forall (t, x) \in \mathbb{R}_+ \times \mathbb{R}^n, \ T \in (t, \infty).$$

Such a case is referred to as the *no-constraint case*.

(iii) $M(t) = \phi$, for any $t \in \mathbb{R}_+$. For this case, one has

$$X(s; t, x, u(\cdot)) \notin M(t), \quad \forall u(\cdot) \in \mathcal{U}[t, \infty), s \in [t, \infty), \forall (t, x) \in \mathbb{R}_+ \times \mathbb{R}^n.$$

Thus,

$$\mathcal{U}_x^{\phi}[t, T] = \phi, \quad \forall (t, x) \in \mathbb{R}_+ \times \mathbb{R}^n, \ T \in (t, \infty).$$

As a convention, for this case, we let

$$\mathcal{U}_x^{\phi}[t, \infty) = \mathcal{U}[t, \infty), \quad \forall (t, x) \in \mathbb{R}_+ \times \mathbb{R}^n.$$

This amounts to saying that we regard the empty target formally as

$$M(t) = \mathbb{R}_{\infty}^n(t) \equiv \phi I_{[0, \infty)}(t) + \mathbb{R}^n I_{\{\infty\}}(t), \quad t \in [0, \infty] \equiv \bar{\mathbb{R}}_+,$$

which can be regarded as the limiting case of (i) with $T \to \infty$ (see (1.11)).

Relevant to the above, for any moving target $M(\cdot)$ and initial pair $(t, x) \in \mathbb{R}_+ \times \mathbb{R}^n$, we define

$$\mathcal{T}_{M(\cdot)}(t, x; u(\cdot)) = \inf \Big\{ s \in [t, \infty) \ \big| \ X(s; t, x, u(\cdot)) \in M(s) \Big\},$$

with the convention that $\inf \phi = \infty$. We call $\mathcal{T}_{M(\cdot)}(t, x; u(\cdot))$ the *first hitting time* of the system to the target $M(\cdot)$ from the initial pair (t, x). Trivially, one has the following:

$$\mathcal{T}_{\mathbb{R}^n}(t, x; u(\cdot)) = t, \quad \forall (t, x) \in \mathbb{R}_+ \times \mathbb{R}^n, \ u(\cdot) \in \mathcal{U}[t, \infty),$$
$$\mathcal{T}_{\mathbb{R}_T^n(\cdot)}(t, x; u(\cdot)) = T, \quad \forall (t, x) \in \mathbb{R}_+ \times \mathbb{R}^n, u(\cdot) \in \mathcal{U}[t, \infty), \ T \geqslant t,$$

and, by the convention $\inf \phi = \infty$,

$$\mathcal{T}_{\phi}(t, x; u(\cdot)) = \infty, \quad \forall (t, x) \in \mathbb{R}_+ \times \mathbb{R}^n, \ u(\cdot) \in \mathcal{U}[t, \infty).$$

In all the three examples above, the first hitting times are independent of the choice of controls. However, in general, for given $(t, x) \in \mathbb{R}_+ \times \mathbb{R}^n$, the first hitting time $\mathcal{T}_{M(\cdot)}(t, x; u(\cdot))$ depends on the control $u(\cdot)$. Therefore, the following problem is meaningful.

Problem (T). For $(t, x) \in \mathbb{R}_+ \times \mathbb{R}^n$, find a $\bar{u}(\cdot) \in \mathcal{U}_x^{M(\cdot)}[t, \infty)$ such that

$$\mathcal{T}_{M(\cdot)}(t, x; \bar{u}(\cdot)) = \inf_{u(\cdot) \in \mathcal{U}_x^{M(\cdot)}[t,\infty)} \mathcal{T}_{M(\cdot)}(t, x; u(\cdot)) \equiv \mathcal{T}_{M(\cdot)}(t, x). \qquad (1.12)$$

The above is called a *time optimal control problem* for system (1.10) with the moving target set $M(\cdot)$. The control $\bar{u}(\cdot)$ satisfying (1.12) is called a *time optimal control* of Problem (T), and $\mathcal{T}_{M(\cdot)}(t, x)$ is called the *minimum time* of Problem (T) for the initial pair (t, x). Clearly,

$$\mathcal{T}_{M(\cdot)}(t, x) < +\infty \quad \Longleftrightarrow \quad \mathcal{U}_x^{M(\cdot)}[t, \infty) \neq \phi.$$

Next, in the case that $\mathcal{U}_x^{M(\cdot)}[t, \infty) \neq \phi$, to measure the performance of the controls, we may introduce the following cost functional

$$J(t, x; u(\cdot)) = \int_t^{\mathcal{T}_{M(\cdot)}(t,x;u(\cdot))} g(s, X(s), u(s)) ds \qquad (1.13)$$
$$+ h\big(\mathcal{T}_{M(\cdot)}(t, x; u(\cdot)), X(\mathcal{T}_{M(\cdot)}(t, x; u(\cdot)))\big),$$

for some maps $g(\cdot)$ and $h(\cdot)$. According to our convention, for the case of empty target set, the corresponding cost functional could read

$$J(t, x; u(\cdot)) = \int_t^{\infty} g(s, X(s), u(s)) ds + \overline{\lim_{s \to \infty}} \, h\big(s, X(s)\big).$$

As a convention, if for some $(t, x, u(\cdot))$, $J(t, x; u(\cdot))$ is not defined or not finite, we let

$$J(t, x; u(\cdot)) = \infty.$$

Let us now pose the following problem.

Problem (OC). For any initial pair $(t, x) \in \mathbb{R}_+ \times \mathbb{R}^n$ with $\mathcal{U}_x^{M(\cdot)}[t, \infty) \neq \phi$, find a $\bar{u}(\cdot) \in \mathcal{U}_x^{M(\cdot)}[t, \infty)$ such that

$$J(t, x; u(\cdot)) = \inf_{u(\cdot) \in \mathcal{U}_x^{M(\cdot)}[t,\infty)} J(t, x; u(\cdot)) \equiv V(t, x).$$

The above is called an *optimal control problem*, with *terminal state constraint* and with a *non-fixed duration*. The function $V(\cdot, \cdot)$ is called the *value function* of Problem (OC).

By our convention, the above Problem (OC) is to find a $\bar{u}(\cdot)$ such that $J(t, x; \bar{u}(\cdot))$ is defined, and it achieves the minimum among all the $u(\cdot)$ such that $J(t, x; u(\cdot))$ is well-defined.

Let us look at several special cases of the above Problem (OC).

Let $M \subseteq \mathbb{R}^n$ be non-empty and closed, $T \in (0, \infty)$, and let

$$M(t) = \phi I_{\mathbb{R}_+ \setminus \{T\}}(t) + M I_{\{T\}}(t), \qquad t \in \mathbb{R}_+.$$

Then, for any $(t, x) \in [0, T) \times \mathbb{R}^n$,

$$\mathcal{U}_x^{M(\cdot)}[t, T] = \left\{ u(\cdot) \in \mathcal{U}[t, T] \mid X(T; t, x, u(\cdot)) \in M \right\} \equiv \widetilde{\mathcal{U}}_x^M[t, T].$$

The cost functional (1.13) becomes

$$J(t, x; u(\cdot)) = \int_t^T g(s, X(s), u(s)) ds + h(T, X(T)) \equiv J^T(t, x; u(\cdot)).$$

In this case, Problem (OC) is posed on a fixed time interval $[0, T]$ and the terminal state $X(T)$ is constrained in M. We may restate Problem (OC) as follows:

Problem (OC)^T. For $(t, x) \in [0, T) \times \mathbb{R}^n$ with $\widetilde{\mathcal{U}}_x^M[t, T] \neq \phi$, find a $\bar{u}(\cdot) \in \widetilde{\mathcal{U}}_x^M[t, T]$ such that

$$J^T(t, x; \bar{u}(\cdot)) = \inf_{u(\cdot) \in \widetilde{\mathcal{U}}_x^M[t, T]} J^T(t, x; u(\cdot)).$$

The above is called an *optimal control problem*, with a *fixed terminal time* and a *terminal state constraint*.

In the case that $M = \mathbb{R}^n$, namely, $M(t) = \phi I_{\mathbb{R}_+ \setminus \{T\}}(t) + \mathbb{R}^n I_{\{T\}}(t)$, the above problem is referred to as an *optimal control problem* with a *fixed terminal time* and *free terminal state*. From this, we see that by allowing $M(t)$ to be the empty set for some t, optimal control problems with fixed duration are included in the general formulation of Problem (OC).

Next, for the case that $M(t) = \phi I_{[0,\infty)}(t) + \mathbb{R}^n I_{\{\infty\}}(t)$, and $h(t, x) = 0$, the cost functional becomes

$$J(t, x; u(\cdot)) = \int_t^\infty g(s, X(s), u(s)) ds \equiv J^\infty(t, x; u(\cdot)).$$

In this case, Problem (OC) becomes the following.

Problem (OC)$^\infty$. For any initial pair $(t, x) \in \mathbb{R}_+ \times \mathbb{R}^n$, find a $\bar{u}(\cdot) \in \mathcal{U}[t, \infty)$ such that

$$J^\infty(t, x; u(\cdot)) = \inf_{u(\cdot) \in \mathcal{U}[t,\infty)} J^\infty(t, x; u(\cdot)) \equiv V^\infty(t, x).$$

The above problem is called an *infinite horizon optimal control problem*.

Now, we look at two more specific cases of Problem (OC).

(i) Let

$$g(s, x, u) = 1, \quad h(s, x) = 0, \qquad \forall (s, x, u) \in \mathbb{R}_+ \times \mathbb{R}^n \times U.$$

Then
$$J(t, x; u(\cdot)) = \mathcal{T}_{M(\cdot)}(t, x; u(\cdot)) - t, \qquad \forall u(\cdot) \in \mathcal{U}[t, \infty).$$

From this, we see that Problem (T) can be regarded as a special case of Problem (OC).

(ii) Let
$$g(s, x, u) = -1, \quad h(s, x) = 0, \qquad \forall (s, x, u) \in \mathbb{R}_+ \times \mathbb{R}^n \times U.$$

Then
$$J(t, x; u(\cdot)) = t - \mathcal{T}_{M(\cdot)}(t, x; u(\cdot)), \qquad u(\cdot) \in \mathcal{U}[t, \infty).$$

Thus, minimizing such a cost functional amounts to maximizing the first hitting time. In other words, the controller tries to avoid touching the moving target set $M(\cdot)$, or tries to stay inside of $M(\cdot)^c \equiv \mathbb{R}^n \setminus M(\cdot) \equiv \Omega(\cdot)$ which is called a *moving survival set*. This leads to the following problem.

Problem (V). Let $\Omega : \mathbb{R}_+ \to 2^{\mathbb{R}^n}$ be a moving survival set. For any $(t, x) \in \mathbb{R}_+ \times \mathbb{R}^n$ with $x \in \Omega(t)$, find a control $u(\cdot) \in \mathcal{U}[t, \infty)$ such that
$$X(s, t, x, u(\cdot)) \in \Omega(s), \qquad \forall s \in [t, \infty).$$

The above is called a *viability problem* for system (1.10) associated with the moving survival set $\Omega(\cdot)$. If we let $M(\cdot) = \Omega(\cdot)^c \equiv \mathbb{R}^n \setminus \Omega(\cdot)$, then Problem (V) is an opposite problem of Problem (C).

1.3.2 *Differential game problems*

We now look at the state equation (1.10) with $U = U_1 \times U_2$ and $u(\cdot) = (u_1(\cdot), u_2(\cdot))$. For $i = 1, 2$, U_i is a metric space, and $u_i(\cdot) \in \mathcal{U}_i[t, \infty)$ with
$$\mathcal{U}_i[t, \infty) = \{u_i : [t, \infty) \to U_i \mid u_i(\cdot) \text{ is measurable}\}, \quad i = 1, 2.$$

The state equation now reads
$$\begin{cases} \dot{X}(s) = f(s, X(s), u_1(s), u_2(s)), & s \in [t, \infty), \\ X(t) = x. \end{cases} \tag{1.14}$$

As $u_1(\cdot)$ and $u_2(\cdot)$ varying in $\mathcal{U}_1[t, \infty)$ and $\mathcal{U}_2[t, \infty)$, the state trajectory $X(\cdot)$ changes. It is clear that as far as the state equation is concerned, (1.14) is a special case of (1.10). If we further let
$$X(s) = \begin{pmatrix} X_1(s) \\ X_2(s) \end{pmatrix}, \qquad f(s, x, u) = \begin{pmatrix} f_1(s, x_1, u_1) \\ f_2(s, x_2, u_2) \end{pmatrix},$$

then (1.14) becomes the following two systems: for $i = 1, 2$,

$$\begin{cases} \dot{X}_i(s) = f_i(s, X_i(s), u_i(s)), & s \in [t, \infty), \\ X_i(t) = x_i. \end{cases}$$

One may use the above to describe two systems controlled by two persons whose objectives may be different. A typical example is the following. Two aircrafts are fighting each other in the air, whose positions are denoted by $X_1(s)$ and $X_2(s)$ at time s, respectively, and they are following the above dynamics. Suppose that $X_1(\cdot)$ is pursuing $X_2(\cdot)$. Then $X_1(\cdot)$ and $X_2(\cdot)$ are in a *dynamic* game situation.

With the above description, we now return to (1.14). We say that (1.14) together with $\mathcal{U}_1[t, \infty)$ and $\mathcal{U}_2[t, \infty)$ form a *two-person differential game*, for which Player i takes control $u_i(\cdot)$ from $\mathcal{U}_i[t, \infty)$. We now pose several problems that are comparable with those for control problems.

Suppose Player 1 is trying to catch Player 2, and Player 2 is trying to escape from Player 1. This is a pursuit-evasion situation, and we generally call it a *two-person differential pursuit and evasion game*. More precisely, we may roughly pose the following two problems.

Problem (P). Let $M(\cdot)$ be a moving target set. For any $(t, x) \in \mathbb{R}_+ \times \mathbb{R}^n$, and any $u_2(\cdot) \in \mathcal{U}_2[t, \infty)$, find a $u_1(\cdot) \in \mathcal{U}_1[t, \infty)$ such that for some $\tau \in [t, \infty)$,

$$X(\tau; t, x, u_1(\cdot), u_2(\cdot)) \in M(\tau).$$

The above is referred to as a (two-person) *differential pursuit game*. In a symmetric way, we may pose the following problem.

Problem (E). Let $\Omega(\cdot)$ be a moving survival set. For any $(t, x) \in \mathbb{R}_+ \times \mathbb{R}^n$ with $x \in \Omega(t)$, and any $u_1(\cdot) \in \mathcal{U}_1[t, \infty)$, find a $u_2(\cdot) \in \mathcal{U}_2[t, \infty)$ such that

$$X(s; t, x, u_1(\cdot), u_2(\cdot)) \in \Omega(s), \qquad s \in [t, \infty).$$

The above problem is referred to as a (two-person) *differential evasion game*.

Note that if U_2 is a singleton, Problem (P) is reduced to Problem (C); and if U_1 is a singleton, Problem (E) is reduced to Problem (V).

Next, let $M(\cdot)$ be a moving target set. For any $(t, x) \in \mathbb{R}_+ \times \mathbb{R}^n$ and $(u_1(\cdot), u_2(\cdot)) \in \mathcal{U}_1[t, \infty) \times \mathcal{U}_2[t, \infty)$, let

$$\mathcal{T}_{M(\cdot)}(t, x; u_1(\cdot), u_2(\cdot)) = \inf\Big\{ s \geqslant t \mid X(s; t, x, u_1(\cdot), u_2(\cdot)) \in M(s) \Big\},$$

with the convention inf $\phi = +\infty$ again. We introduce the following cost functionals: for $i = 1, 2$,

$$J_i(t, x; u_1(\cdot), u_2(\cdot)) = \int_t^{\mathcal{T}_{M(\cdot)}} g_i(s, X(s), u_1(s), u_2(s))ds + h_i\big(\mathcal{T}_{M(\cdot)}, X(\mathcal{T}_{M(\cdot)})\big),$$

with $\mathcal{T}_{M(\cdot)} = \mathcal{T}_{M(\cdot)}(t, x; u_1(\cdot), u_2(\cdot))$. Then

$$\Big\{ \mathcal{U}_i[t, \infty), J_i(t, x; u_1(\cdot), u_2(\cdot)), i = 1, 2 \Big\}$$

form a two-person game in the sense of classical game theory (see Section 1.2), in which Player i wants to minimize the cost functional $J_i(t, x; u_1(\cdot), u_2(\cdot))$, by selecting $u_i(\cdot) \in \mathcal{U}_i[t, \infty)$. We refer to such a game as a *two-person differential game in a non-fixed duration*, denoted by Problem (DG). Further, if

$$g_1(t, x, u_1, u_2) + g_2(t, x, u_1, u_2) = 0, \quad h_1(t, x) + h_2(t, x) = 0,$$
$$\forall (t, x, u_1, u_2) \in \mathbb{R}_+ \times \mathbb{R}^n \times U_1 \times U_2,$$

then

$$J_1(t, x; u_1(\cdot), u_2(\cdot)) + J_2(t, x; u_1(\cdot), u_2(\cdot)) = 0,$$
$$\forall (t, x) \in \mathbb{R}_+ \times \mathbb{R}^n, \ u_i(\cdot) \in \mathcal{U}_i[t, \infty).$$

In this case, Problem (DG) is called a *two-person zero-sum differential game*, denoted by Problem (Z).

Next, for a fixed $T \in (0, \infty)$, let $M(\cdot) = \mathbb{R}^n_T(\cdot) \equiv \phi I_{\mathbb{R}_+ \setminus \{T\}}(\cdot) + \mathbb{R}^n I_{\{T\}}(\cdot)$. Then the corresponding cost functionals become

$$J_i^T(t, x; u_1(\cdot), u_2(\cdot)) = \int_t^T g_i(s, X(s), u_1(s), u_2(s))ds + h_i(T, X(T)).$$

The corresponding Problems (DG) and (Z) are denoted by Problem $(DG)^T$ and Problem $(Z)^T$, respectively, and are called *two-person differential game*, and *two-person zero-sum differential game, in a fixed duration*, respectively.

We point out an important issue of differential games: The selected controls of Players 1 and 2 should be *non-anticipating*. More precisely, each player does not know the future values of the other player's control. Because of this, some more sophisticated discussion will be necessary for the controls in differential games. We will carefully investigate this in a later chapter.

1.4 Some Mathematical Preparations

In this section, we present some results which will be used in the following chapters.

Let us introduce some spaces. For any $0 \leqslant t < T < \infty$ and $1 \leqslant p < \infty$, define

$$C([t,T];\mathbb{R}^n) = \Big\{\varphi : [t,T] \to \mathbb{R}^n \mid \varphi(\cdot) \text{ is continuous}\Big\},$$

$$L^\infty(t,T;\mathbb{R}^n) = \Big\{\varphi : [t,T] \to \mathbb{R}^n \mid \varphi(\cdot) \text{ measurable}, \ \operatorname*{esssup}_{s\in[t,T]}|\varphi(s)| < \infty\Big\},$$

$$L^p(t,T;\mathbb{R}^n) = \Big\{\varphi : [t,T] \to \mathbb{R}^n \mid \varphi(\cdot) \text{ measurable}, \ \int_t^T |\varphi(s)|^p ds < \infty\Big\},$$

which are Banach spaces under the following norms, respectively,

$$\|\varphi(\cdot)\|_{C([t,T];\mathbb{R}^n)} = \sup_{s\in[t,T]}|\varphi(s)|, \quad \forall \varphi(\cdot) \in C([t,T];\mathbb{R}^n),$$

$$\|\varphi(\cdot)\|_{L^\infty(t,T;\mathbb{R}^n)} = \operatorname*{esssup}_{s\in[t,T]}|\varphi(s)|, \quad \forall \varphi(\cdot) \in L^\infty(t,T;\mathbb{R}^n),$$

$$\|\varphi(\cdot)\|_{L^p(t,T;\mathbb{R}^n)} = \Big(\int_t^T |\varphi(s)|^p ds\Big)^{\frac{1}{p}}, \quad \forall \varphi(\cdot) \in L^p(t,T;\mathbb{R}^n).$$

We now present some standard results.

Theorem 1.4.1. (Contraction Mapping Theorem) *Let* $\mathbb{X}$ *be a Banach space, and* $\mathcal{S} : \mathbb{X} \to \mathbb{X}$ *be a map satisfying*

$$\|\mathcal{S}(x) - \mathcal{S}(y)\| \leqslant \alpha\|x - y\|, \qquad \forall x,y \in \mathbb{X}, \tag{1.15}$$

with $\alpha \in (0,1)$. *Then there exists a unique* $\bar{x} \in \mathbb{X}$ *such that*

$$\mathcal{S}(\bar{x}) = \bar{x}.$$

Proof. First of all, by (1.15), $\mathcal{S}$ is continuous. Pick any $x_0 \in M$. Define

$$x_k = \mathcal{S}^k(x_0), \qquad k \geqslant 1.$$

Then for any $k, \ell \geqslant 1$,

$$\|x_{k+\ell} - x_k\| \leqslant \Big\|\sum_{i=k+1}^{k+\ell}(x_i - x_{i-1})\Big\| \leqslant \sum_{i=k+1}^{k+\ell}\alpha^k\|x_1 - x_0\|.$$

Thus, $\{x_k\}_{k\geq 0}$ is a Cauchy sequence. Consequently, there exists a unique $\bar{x} \in \mathbb{X}$ such that

$$\lim_{k\to\infty}\|x_k - \bar{x}\| = 0.$$

Then by the continuity of $\mathcal{S}$, we obtain

$$\bar{x} = \lim_{k \to \infty} x_k = \lim_{k \to \infty} \mathcal{S}(x_{k-1}) = \mathcal{S}(\bar{x}).$$

This means that $\bar{x}$ is a fixed point of $\mathcal{S}$. Finally, if $\bar{x}$ and $\widetilde{x}$ are two fixed points. Then

$$\|\bar{x} - \widetilde{x}\| = \mathcal{S}(\bar{x}) - \mathcal{S}(\widetilde{x})\| \leqslant \alpha \|\bar{x} - \widetilde{x}\|.$$

Hence, $\bar{x} = \widetilde{x}$, proving the uniqueness.   □

Theorem 1.4.2. (Arzela–Ascoli) *Let $\mathcal{Z} \subseteq C([t, T]; \mathbb{R}^n)$ be an infinite set which is uniformly bounded and equi-continuous, i.e.,*

$$\sup_{\varphi(\cdot) \in \mathcal{Z}} \|\varphi(\cdot)\|_{C([t,T];\mathbb{R}^n)} < \infty,$$

and for any $\varepsilon > 0$, there exists a $\delta > 0$ such that

$$|\varphi(t) - \varphi(s)| < \varepsilon, \quad \forall |t - s| < \delta, \quad \forall \varphi(\cdot) \in \mathcal{Z}.$$

Then there exists a sequence $\varphi_k(\cdot) \in \mathcal{Z}$ such that

$$\lim_{k \to \infty} \|\varphi_k(\cdot) - \bar{\varphi}(\cdot)\|_{C([t,T];\mathbb{R}^n)} = 0,$$

for some $\bar{\varphi}(\cdot) \in C([t, T]; \mathbb{R}^n)$.

Proof. Let $\mathcal{T} \stackrel{\Delta}{=} \{t_k\}_{k \geq 1}$ be a dense set of $[t, T]$. For any $k \geq 1$, the set $\{\varphi(t_1) \mid \varphi(\cdot) \in \mathcal{Z}\}$ is bounded. Thus, there exists a sequence denoted by $\{\varphi_{\sigma_1(i)}(t_1)\}$ converging some point in $\mathbb{R}^n$, denoted by $\bar{\varphi}(t_k)$. Next, the set $\{\varphi_{\sigma_1(i)}(t_2)\}$ is bounded. Thus, we may let $\{\varphi_{\sigma_2(i)}(t_2)\}$ be a subsequence of $\{\varphi_{\sigma_1(i)}(t_2)\}$, which is convergent to some point in $\mathbb{R}^n$, denoted by $\bar{\varphi}(t_2)$. Continue this process, we obtain a function $\bar{\varphi} : \mathcal{T} \to \mathbb{R}$. By letting

$$\bar{\varphi}_i(\cdot) = \varphi_{\sigma_i(i)}(\cdot), \quad i \geqslant 1,$$

we have

$$\lim_{i \to \infty} \bar{\varphi}_i(s) = \bar{\varphi}(s), \quad \forall s \in \mathcal{T}.$$

By the equi-continuity of the sequence $\{\varphi_k(\cdot)\}$, we see that for any $\varepsilon > 0$, there exists a $\delta = \delta(\varepsilon) > 0$, independent of $i \geqslant 1$ such that

$$|\bar{\varphi}_i(s_1) - \bar{\varphi}_i(s_2)| < \varepsilon, \quad \forall s_1, s_2 \in \mathcal{T}, \ |s_1 - s_2| < \delta. \tag{1.16}$$

Then letting $i \to \infty$, we obtain

$$|\bar{\varphi}(s_1) - \bar{\varphi}(s_2)| \leqslant \varepsilon, \quad \forall s_1, s_2 \in \mathcal{T}, \ |s_1 - s_2| < \delta.$$

This means that $\bar{\varphi} : \mathcal{T} \to \mathbb{R}^n$ is uniformly continuous on $\mathcal{T}$. Consequently, we may extend $\bar{\varphi}(\cdot)$ on $\overline{\mathcal{T}} = [t, T]$ which is still continuous. Finally, for any

$\varepsilon > 0$, let $\delta = \delta(\varepsilon) > 0$ be such that (1.16) holds and let $\mathcal{S}_\varepsilon = \{s_j, \ 1 \leqslant j \leqslant M\} \subseteq \mathcal{T}$ with $M > 1$ depending on $\varepsilon > 0$ such that

$$\bigcup_{j=1}^{M} (s_j - \delta, s_j + \delta) \supseteq [t, T].$$

Next, we may let $i_0 > 1$ such that

$$|\bar{\varphi}_i(s_j) - \bar{\varphi}(s_j)| < \varepsilon, \qquad i \geqslant i_0, \quad 1 \leqslant j \leqslant M.$$

Then for any $s \in [t, T]$, there is an $s_j \in \mathcal{S}_\varepsilon$ such that $|s - s_j| < \delta$. Consequently,

$$|\bar{\varphi}_i(s) - \bar{\varphi}(s)| \leqslant |\bar{\varphi}_i(s) - \bar{\varphi}_i(s_j)| + |\bar{\varphi}_i(s_j) - \bar{\varphi}(s_j)| + |\bar{\varphi}(s_j) - \bar{\varphi}(s)|$$

$$\leqslant 3\varepsilon.$$

This shows that $\bar{\varphi}_i(\cdot)$ converges to $\bar{\varphi}(\cdot)$ uniformly in $s \in [t, T]$. $\qquad \square$

Theorem 1.4.3. (Banach–Saks) *Let $\varphi_k(\cdot) \in L^2(a, b; \mathbb{R}^n)$ be a sequence which is weakly convergent to $\bar{\varphi}(\cdot) \in L^2(a, b; \mathbb{R}^n)$, i.e.,*

$$\lim_{k \to \infty} \int_a^b \langle \varphi_k(s) - \bar{\varphi}(s), \eta(s) \rangle \, ds = 0, \qquad \forall \eta(\cdot) \in L^2(a, b; \mathbb{R}^n).$$

Then there is a subsequence $\{\varphi_{k_j}(\cdot)\}$ such that

$$\lim_{N \to \infty} \left\| \frac{1}{N} \sum_{j=1}^{N} \varphi_{k_j}(\cdot) - \bar{\varphi}(\cdot) \right\|_{L^2(a, b; \mathbb{R}^n)} = 0.$$

Proof. Without loss of generality, we may assume that $\bar{\varphi}(\cdot) = 0$ (Why?). Let $k_1 = 1$. By the weak convergence of $\varphi_k(\cdot)$, we may find $k_1 < k_2 < k_3 < \cdots < k_N$ such that

$$\left| \int_a^b \langle f_{k_i}(s), f_{k_j}(s) \rangle \, ds \right| < \frac{1}{N}, \qquad 1 \leqslant i < j \leqslant N.$$

Observe

$$\left\| \frac{1}{N} \sum_{i=1}^{N} f_{k_i}(\cdot) \right\|_{L^2(a, b; \mathbb{R}^n)}^2 = \frac{1}{N^2} \int_a^b \left| \sum_{i=1}^{N} f_{k_i}(s) \right|^2 ds$$

$$= \frac{1}{N^2} \int_a^b \sum_{i,j=1}^{N} \langle f_{k_i}(s), f_{k_j}(s) \rangle \, ds$$

$$= \frac{1}{N^2} \sum_{i=1}^{N} \| f_{k_i}(\cdot) \|_{L^2(a, b; \mathbb{R}^n)}^2 + \frac{2}{N^2} \sum_{1 \leqslant i < j \leqslant N} \int_a^b \langle f_{k_i}(s), f_{k_j}(s) \rangle \, ds$$

$$\leqslant \frac{1}{N} \sup_{i \geqslant 1} \| f_{k_i}(\cdot) \|_{L^2(a, b; \mathbb{R}^n)}^2 + \frac{2}{N^3} \frac{N(N-1)}{2}$$

$$\leqslant \frac{1}{N} \sup_{i \geqslant 1} \| f_{k_i}(\cdot) \|_{L^2(a, b; \mathbb{R}^n)}^2 + \frac{1}{N} \to 0.$$

This completes the proof. □

Lemma 1.4.4. (Filippov) *Let U be a complete separable metric space whose metric is denoted by $d(\cdot\,,\cdot)$. Let $g : [0,T] \times U \to \mathbb{R}^n$ be a map which is measurable in $t \in [0,T]$ and*

$$|g(t,u) - g(t,v)| \leq \omega(d(u,v)), \qquad \forall u, v \in U, t \in [0,T],$$

for some continuous and increasing function $\omega : \mathbb{R}_+ \to \mathbb{R}_+$ with $\omega(0) = 0$, called a modulus of continuity. Moreover,

$$0 \in g(t,U), \qquad \text{a.e. } t \in [0,T].$$

Then there exists a measurable map $u : [0,T] \to U$, such that

$$g(t, u(t)) = 0, \qquad \text{a.e. } t \in [0,T]. \tag{1.17}$$

Proof. Define

$$\bar{d}(u,v) \stackrel{\Delta}{=} \frac{d(u,v)}{1 + d(u,v)} < 1, \qquad \forall u, v \in U.$$

Then $\bar{d}$ is a metric on U under which U is still complete and separable. Hence, without loss of generality, we assume that the original metric $d(\cdot\,,\cdot)$ already satisfies

$$d(u,v) < 1, \qquad \forall u, v \in U.$$

Next, we define

$$\Gamma(t) \stackrel{\Delta}{=} \{u \in U \mid g(t,u) = 0\}, \quad t \in [0,T].$$

Without loss of generality, we may assume that $\Gamma(t) \neq \phi$ for all $t \in [0,T]$ (Why?). Let $U_0 \stackrel{\Delta}{=} \{v_k \mid k \geq 1\}$ be a countable dense subset of U. We claim that for any $u \in U$ and $0 \leq c < 1$,

$$\left\{t \in [0,T] \mid d(u, \Gamma(t)) \leq c\right\}$$
$$= \bigcap_{i=1}^{\infty} \bigcup_{j=i}^{\infty} \left\{t \in [0,T] \mid d(u, v_j) \leq c + \frac{1}{i}, \ |g(t, v_j)| \leq \frac{1}{i}\right\}, \tag{1.18}$$

where

$$d(u, \Gamma(t)) \stackrel{\Delta}{=} \inf_{v \in \Gamma(t)} d(u,v).$$

To show (1.18), we note that $t \in [0,T]$, with $d(u, \Gamma(t)) \leq c$ if and only if there exists a sequence $u_k \in \Gamma(t)$, i.e., $g(t, u_k) = 0$, such that

$$d(u, u_k) \leq c + \frac{1}{k}.$$

Since $\overline{U_0} = U$, there exists a sequence $v_{j_k} \in U_0$ such that

$$d(u_k, v_{j_k}) < \frac{1}{k}.$$

Hence,

$$d(u, v_{j_k}) \leqslant d(u, u_k) + d(u_k, v_{j_k}) \leqslant c + \frac{2}{k}.$$

Next, by the uniform continuity of $u \mapsto g(t, u)$, we have

$$|g(t, v_{j_k})| \leqslant |g(t, v_{j_k}) - g(t, u_k)| \leqslant \omega(d(v_{j_k}, u_k)) \leqslant \omega\left(\frac{1}{k}\right).$$

Hence, one has

$$\begin{cases} \lim_{k \to \infty} d(u, v_{j_k}) \leqslant c, \\ \lim_{k \to \infty} g(t, v_{j_k}) = 0. \end{cases}$$

Thus, $d(u, \Gamma(t)) \leq c$ if and only if for any $i \geqslant 1$, there exists a $j \geqslant i$, such that

$$\begin{cases} \bar{d}(u, v_j) \leqslant c + \frac{1}{i}, \\ |g(t, v_j)| \leqslant \frac{1}{i}. \end{cases}$$

This proves (1.18). Since the right-hand side of (1.18) is measurable, so is the left-hand side. On the other hand,

$$\begin{cases} \{t \in [0, T] \mid d(u, \Gamma(t)) \leqslant c\} = [0, T], & \forall c \geq 1, \\ \{t \in [0, T] \mid d(u, \Gamma(t)) \leqslant c\} = \phi, & \forall c < 0. \end{cases}$$

Hence, the function $t \mapsto d(u, \Gamma(t))$ is measurable. Now, we define

$$u_0(t) \overset{\Delta}{=} v_1, \qquad \forall t \in [0, T].$$

Clearly, $u_0(\cdot)$ is measurable and

$$d(u_0(t), \Gamma(t)) < 1, \qquad \forall t \in [0, T].$$

Suppose that we have defined $u_{k-1}(\cdot)$ such that

$$\begin{cases} d(u_{k-1}(t), \Gamma(t)) < 2^{1-k}, \\ d(u_{k-1}(t), u_{k-2}(t)) < 2^{2-k}, \end{cases} \qquad t \in [0, T]. \tag{1.19}$$

We define sets

$$\begin{cases} C_i^k \overset{\Delta}{=} \{t \in [0, T] \mid d(v_i, \Gamma(t)) < 2^{-k}\}, \\ D_i^k \overset{\Delta}{=} \{t \in [0, T] \mid d(v_i, u_{k-1}(t)) < 2^{1-k}\}. \end{cases}$$

Since $t \mapsto d(v_i, \Gamma(t))$ is measurable, C_i^k is measurable. Likewise, D_i^k is also measurable. Set

$$A_i^k = C_i^k \bigcap D_i^k, \quad k, i \geqslant 1.$$

Then A_i^k is measurable as well. We claim that

$$[0, T] = \bigcup_{i=1}^{\infty} A_i^k, \quad \forall k \geqslant 1. \tag{1.20}$$

In fact, for any $t \in [0, T]$, by (1.19), there exists a $u \in \Gamma(t)$ such that

$$d(u_{k-1}(t), u) < 2^{1-k}.$$

By the density of U_0 in U, there exists an $i \geqslant 1$ such that

$$d(v_i, \Gamma(t)) \leqslant \begin{cases} d(v_i, u) < 2^{-k}, \\ d(v_i, u_{k-1}(t)) < 2^{1-k}, \end{cases}$$

which means $t \in A_i^k$, proving (1.20). Now we define $u_k(\cdot) : [0, T] \to U_0 \subseteq U$ as follows:

$$u_k(t) = v_i, \quad \forall t \in A_i^k \setminus \bigcup_{j=1}^{i-1} A_j^k.$$

By $t \in C_i^k$, we have

$$d(u_k(t), \Gamma(t)) < 2^{-k},$$

and by $t \in D_i^k$, we have

$$d(u_k(t), u_{k-1}(t)) < 2^{1-k}.$$

This completes the construction of the sequence $\{u_k(\cdot)\}$ inductively. Clearly, (1.19) holds for all $k \geqslant 1$. This also implies that for each $t \in [0, T]$, $\{u_k(t)\}$ is Cauchy in U. By completeness of U, we obtain

$$\lim_{k \to \infty} u_k(t) = u(t), \quad t \in [0, T].$$

Of course, $u(\cdot)$ is measurable, and moreover, by the closeness of $\Gamma(t)$, we have

$$u(t) \in \Gamma(t), \quad \forall t \in [0, T].$$

This means (1.17) holds. $\qquad\qquad\qquad\qquad\qquad\qquad\qquad\qquad\qquad\qquad\qquad$ $\square$

Theorem 1.4.5. (Ekeland Variational Principle) *Let (U, d) be a complete metric space and $F : U \to (0, \infty)$ be a continuous function. Suppose $u^* \in U$ such that*

$$F(u^*) < \inf_{u \in U} F(u) + \varepsilon.$$

Then there exists a $u^\varepsilon \in U$ such that

$$F(u^\varepsilon) \leqslant F(u^*), \qquad d(u^\varepsilon, u^*) \leqslant \sqrt{\varepsilon}, \tag{1.21}$$

$$F(u^\varepsilon) \leqslant F(u) + \sqrt{\varepsilon}\, d(u, u^\varepsilon), \qquad \forall u \in U. \tag{1.22}$$

Proof. Define

$$G(v) = \{w \in U \mid F(w) + \sqrt{\varepsilon}d(w, v) \leqslant F(v)\}.$$

For any $v \in U$, $G(v)$ is a closed set in U since F is continuous. Also,

$$v \in G(v), \qquad \forall v \in U. \tag{1.23}$$

We claim that

$$w \in G(v) \quad \Rightarrow \quad G(w) \subseteq G(v). \tag{1.24}$$

In fact, $w \in G(v)$ implies

$$F(w) + \sqrt{\varepsilon}d(w, v) \leqslant F(v), \tag{1.25}$$

and for any $u \in G(w)$, we have

$$F(u) + \sqrt{\varepsilon}d(u, w) \leqslant F(w). \tag{1.26}$$

Combining (1.25) and (1.26), one has

$$F(u) + \sqrt{\varepsilon}d(u, v) \leqslant F(u) + \sqrt{\varepsilon}d(u, w) + \sqrt{\varepsilon}d(w, v)$$
$$\leqslant F(w) + \sqrt{\varepsilon}d(w, v) \leqslant F(v).$$

Thus, (1.24) holds. Next, we define

$$f(v) = \inf_{w \in G(v)} F(w), \qquad \forall v \in U.$$

Then, for any $w \in G(v)$,

$$f(v) \leqslant F(w) \leqslant F(v) - \sqrt{\varepsilon}d(w, v).$$

Hence,

$$\sqrt{\varepsilon}d(w, v) \leqslant F(v) - f(v).$$

This implies that the diameter $\operatorname{diam} G(v)$ of the set $G(v)$ satisfies

$$\operatorname{diam} G(v) \equiv \sup_{w, u \in G(v)} d(w, u) \leqslant \frac{2}{\sqrt{\varepsilon}}\big(F(v) - f(v)\big). \tag{1.27}$$

Now, we define a sequence $\{v_n\}_{n \geqslant 0}$ in the following way:

$$\begin{cases} v_0 = u^*, \\ v_{n+1} \in G(v_n), \quad F(v_{n+1}) \leqslant f(v_n) + \dfrac{1}{2^n}, \quad n \geqslant 0. \end{cases} \tag{1.28}$$

By (1.24), we know that $G(v_{n+1}) \subseteq G(v_n)$. Thus,

$$f(v_n) \leqslant f(v_{n+1}), \qquad n \geqslant 0. \tag{1.29}$$

On the other hand, $f(w) \leqslant F(w)$ since $w \in G(w)$ (see (1.23)). Thus, together with (1.28) and (1.29), we obtain

$$0 \leqslant F(v_{n+1}) - f(v_{n+1}) \leqslant f(v_n) + \frac{1}{2^n} - f(v_{n+1}) \leqslant \frac{1}{2^n}.$$

Then, by (1.27), we see that the diameter of $G(v_n)$ goes to 0 as $n \to \infty$. Since $G(v_n)$ is a sequence of nested closed sets in U (i.e., $G(v_{n+1}) \subseteq G(v_n)$, $\forall n \geq 0$) and U is complete, we must have a unique point, denoted by $u^\varepsilon \in U$, such that

$$\bigcap_{n \geq 0} G(v_n) = \{u^c\}.$$

In particular, $u^\varepsilon \in G(u^*)$, which gives

$$F(u^\varepsilon) \leqslant F(u^\varepsilon) + \sqrt{\varepsilon} d(u^\varepsilon, u^*) \leqslant F(u^*) < \inf_{u \in U} F(u) + \varepsilon \leqslant F(u^\varepsilon) + \varepsilon.$$

Thus, (1.21) follows. Further, from (1.24), we have

$$G(u^\varepsilon) \subseteq \bigcap_{n \geq 0} G(v_n) = \{u^\varepsilon\}.$$

This implies that $G(u^\varepsilon) = \{u^\varepsilon\}$. Hence, for any $u \neq u^\varepsilon$, we have $u \notin G(u^\varepsilon)$, which means

$$F(u) + \sqrt{\varepsilon} d(u, u^\varepsilon) > F(u^\varepsilon),$$

proving (1.22).   □

The following is called a *spike-variation lemma*.

Lemma 1.4.6. Suppose $f(\cdot) \in L^1(0, T; \mathbb{R}^n)$ and for $\delta > 0$, let

$$\mathbb{E}_\delta = \Big\{ E_\delta \subseteq [0, T] \mid |E_\delta| = \delta T \Big\},$$

where $|E_\delta|$ is the Lebesgue measure of E_δ. Then

$$\inf_{E_\delta \in \mathbb{E}_\delta} \Big\| \int_0^{\cdot} \Big(1 - \frac{1}{\delta} I_{E_\delta}\Big) f(s) ds \Big\|_{C([0,T];\mathbb{R}^n)} = 0.$$

Proof. For any $f(\cdot) \in L^1(0, T; \mathbb{R}^n)$, and any $\varepsilon > 0$, there exists an $f_\varepsilon(\cdot) \in C([0, T]; \mathbb{R}^n)$ such that

$$\int_0^T |f(r) - f_\varepsilon(r)| dr < \varepsilon.$$

Next, we can find a partition $0 = t_0 < t_1 < \cdots < t_k = T$ of $[0, T]$ such that

$$\|f_\varepsilon(\cdot)\|_{C([0,T];\mathbb{R}^n)} \max_{1 \leqslant i \leqslant k} (t_i - t_{i-1}) < \varepsilon,$$

and by defining step function

$$\bar{f}_\varepsilon(r) = \sum_{i=1}^{k} f_\varepsilon(t_i) I_{(t_{i-1}, t_i]}(r), \qquad r \in [0, T],$$

one has

$$\int_0^T |f_\varepsilon(r) - \bar{f}_\varepsilon(r)| dr < \varepsilon.$$

Now, let

$$E_\delta = \bigcup_{i=1}^{k} [t_{i-1} + \delta(t_i - t_{i-1})].$$

Then

$$|E_\delta| = \sum_{i=1}^{k} \delta(t_i - t_{i-1}) = \delta T.$$

For any $s \in (0, T]$, there exists a j such that $t_{j-1} < s \leqslant t_j$. For $1 \leqslant i < j$, we have

$$\int_{t_{i-1}}^{t_i} \left(1 - \frac{1}{\delta} I_{E_\delta}(r)\right) \bar{f}_\varepsilon(r) ds$$

$$= f_\varepsilon(t_i) \int_{t_{i-1}}^{t_i} \left(1 - \frac{1}{\delta} I_{(t_{i-1}, t_{i-1}+\delta(t_i-t_{i-1})]}(r)\right) dr$$

$$= f_\varepsilon(t_i) \left[(t_i - t_{i-1}) - \frac{1}{\delta} \delta(t_i - t_{i-1})\right] = 0,$$

and

$$\left| \int_s^{t_j} \left(1 - \frac{1}{\delta} I_{E_\delta}(r)\right) \bar{f}_\varepsilon(r) ds \right|$$

$$= |f_\varepsilon(t_j)| \left| \int_{t_{j-1}}^s \left(1 - \frac{1}{\delta} I_{(t_{j-1}, t_{j-1}+\delta(t_j-t_{j-1})]}(r)\right) dr \right|$$

$$= |f_\varepsilon(t_j)| \left| (s - t_{j-1}) - \frac{1}{\delta} \{(s - t_{j-1}) \wedge [\delta(t_j - t_{j-1})]\} \right|$$

$$\leqslant |f_\varepsilon(t_j)| (t_j - t_{j-1}) < \varepsilon.$$

Consequently,

$$\left| \int_0^s \left(1 - \frac{1}{\delta} I_{E_\delta}(r) \right) f(r) dr \right| \leqslant \left| \int_0^s \left(1 - \frac{1}{\delta} I_{E_\delta}(r) \right) \left(f(r) - f_\varepsilon(r) \right) dr \right|$$

$$+ \left| \int_0^s \left(1 - \frac{1}{\delta} I_{E_\delta}(r) \right) \left(f_\varepsilon(r) - \bar{f}_\varepsilon(r) \right) dr \right| + \left| \int_0^s \left(1 - \frac{1}{\delta} I_{E_\delta}(r) \right) \bar{f}_\varepsilon(r) dr \right|$$

$$\leqslant \frac{2(1+\delta)\varepsilon}{\delta} + \left| f_\varepsilon(t_j) \int_{t_{j-1}}^s \left(1 - \frac{1}{\delta} I_{E_\delta}(r) \right) dr \right| \leqslant \frac{2(1+\delta)\varepsilon}{\delta} + \varepsilon.$$

Since $\varepsilon > 0$ is arbitrary, we obtain our conclusion. $\qquad\square$

Proposition 1.4.7. (Gronwall's Inequality) *Let* $\theta : [a,b] \to \mathbb{R}_+$ *be continuous and satisfy*

$$\theta(s) \leqslant \alpha(s) + \int_a^s \beta(r)\theta(r) dr, \qquad s \in [a,b],$$

for some $\alpha(\cdot), \beta(\cdot) \in L^1(a,b;\mathbb{R}_+)$. *Then*

$$\theta(s) \leqslant \alpha(s) + \int_a^s \alpha(\tau)\beta(\tau) e^{\int_\tau^s \beta(r) dr} d\tau, \qquad s \in [a,b]. \tag{1.30}$$

In particular, if $\alpha(\cdot) = \alpha$ *is a constant, then*

$$\theta(s) \leqslant \alpha e^{\int_a^s \beta(r) dr}, \qquad s \in [a,b]. \tag{1.31}$$

Proof. Let

$$\varphi(s) = \int_a^s \beta(r)\theta(r) dr.$$

Then

$$\dot{\varphi}(s) = \beta(s)\theta(s) \leqslant \beta(s)\Big[\alpha(s) + \varphi(s)\Big].$$

This leads to

$$\left[\varphi(s) e^{-\int_a^s \beta(r) dr} \right]' \leqslant \alpha(s)\beta(s) e^{-\int_a^s \beta(r) dr}.$$

Consequently,

$$\varphi(s) e^{-\int_a^s \beta(r) dr} \leqslant \int_a^s \alpha(\tau)\beta(\tau) e^{-\int_a^\tau \beta(r) dr} d\tau.$$

Hence, (1.30) and (1.31) follow. $\qquad\square$

Proposition 1.4.8. *Let* $M \subseteq \mathbb{R}^n$ *be a non-empty closed convex set. Then there exists a map* $P_M : \mathbb{R}^n \to M$ *such that*

$$|x - P_M(x)| = \inf_{y \in M} |x - y| \equiv d(x, M),$$

and

$$|P_M(x_1) - P_M(x_2)| \leqslant |x_1 - x_2|, \qquad \forall x_1, x_2 \in \mathbb{R}^n. \tag{1.32}$$

Moreover, for $z \in M$, $z = P_M(x)$ if and only if

$$\langle x - z, y - z \rangle \leqslant 0, \qquad \forall y \in M. \tag{1.33}$$

Proof. First of all, for any $x \in \mathbb{R}^n$, let $z_k \in M$ such that

$$\lim_{k \to \infty} |x - z_k| = d(z, M).$$

Clearly, $\{z_k\}$ is bounded. Thus, we may assume that $z_k \to \bar{z}$ (if necessary, we may take a subsequence). Then

$$|x - \bar{z}| = d(x, M).$$

Now, suppose $\bar{y} \in M$ also satisfies

$$|x - \bar{y}| = d(x, M).$$

By the convexity of M, $\frac{\bar{y} + \bar{z}}{2} \in M$, which implies

$$\begin{aligned}
d(x, M)^2 &\leqslant \left| x - \frac{\bar{y} + \bar{z}}{2} \right|^2 = \frac{1}{4} |x - \bar{y} + x - \bar{z}|^2 \\
&= \frac{1}{4} \Big(|(x - \bar{y}) + (x - \bar{z})|^2 + |(x - \bar{y}) - (x - \bar{z})|^2 - |\bar{y} - \bar{z}|^2 \Big) \\
&= \frac{1}{4} \Big(2|x - \bar{y}|^2 + 2|x - \bar{z}|^2 - |\bar{y} - \bar{z}|^2 \Big) \\
&= d(x, M)^2 - \frac{1}{4} |\bar{y} - \bar{z}|^2.
\end{aligned}$$

Thus, $\bar{y} = \bar{z}$. Consequently, $P_M : \mathbb{R}^n \to M$ is a well-defined map, which is called the *projection* onto the convex set M.

Next, by the definition of $P_M(x)$, for any $y \in M$ and $\alpha \in (0, 1)$, we have

$$P_M(x) + \alpha [y - P_M(x)] = (1 - \alpha) P_M(x) + \alpha y \in M.$$

Hence,

$$\begin{aligned}
0 &\leqslant |P_M(x) + \alpha[y - P_M(x)] - x|^2 - |P_M(x) - x|^2 \\
&= 2\alpha \langle P_M(x) - x, y - P_M(x) \rangle + \alpha^2 |y - P_M(x)|^2.
\end{aligned}$$

Dividing α and sending $\alpha \to 0$, we obtain

$$\langle y - P_M(x), x - P_M(x) \rangle \leqslant 0, \qquad \forall y \in M. \tag{1.34}$$

Conversely, if $z \in M$ satisfies (1.33), then for any $y \in M$

$$|y - x|^2 - |z - x|^2 = |y - z|^2 + 2 \langle y - z, z - x \rangle \geq 0.$$

Hence, $z = P_M(x)$.

Finally, from (1.34), for any $x_1, x_2 \in \mathbb{R}^n$, we have

$$0 \geqslant \langle\, P_M(x_1) - P_M(x_2), x_2 - P_M(x_2) \,\rangle,$$

and

$$0 \geqslant \langle\, P_M(x_2) - P_M(x_1), x_1 - P_M(x_1) \,\rangle$$
$$= \langle\, P_M(x_1) - P_M(x_2), P_M(x_1) - x_1 \,\rangle.$$

Adding the above together, we have

$$0 \geqslant \langle\, P_M(x_1) - P_M(x_2), P_M(x_1) - P_M(x_2) - (x_1 - x_2) \,\rangle,$$

which implies

$$|P_M(x_1) - P_M(x_2)|^2 \leqslant \langle\, P_M(x_1) - P_M(x_2), x_1 - x_2 \,\rangle$$
$$\leqslant |P_M(x_1) - P_M(x_2)|\,|x_1 - x_2|.$$

Thus (1.32) follows.   □

Next, for any matrix $A \in \mathbb{R}^{n \times m}$ there exists a unique pseudo-inverse, denoted by $A^\dagger \in \mathbb{R}^{m \times n}$ such that

$$AA^\dagger A = A, \qquad A^\dagger A A^\dagger = A^\dagger,$$

and $AA^\dagger : \mathbb{R}^n \to \mathcal{R}(A)$ and $A^\dagger A : \mathbb{R}^m \to \mathcal{R}(A^\dagger) \subseteq \mathbb{R}^m$ are the *orthogonal projections*. To construct $A^\dagger$, we suppose

$$\operatorname{rank} A = r \leq m \wedge n.$$

Then A can be decomposed as follows:

$$A = BC, \qquad B \in \mathbb{R}^{n \times r}, \qquad C \in \mathbb{R}^{r \times m}, \qquad \operatorname{rank} B = \operatorname{rank} C = r.$$

With such a decomposition, we have

$$A^\dagger = C^T (CC^T)^{-1} (B^T B)^{-1} B^T.$$

In the case that $A \in \mathbb{S}^n$, the set of all $(n \times n)$ symmetric matrices, we have an orthogonal matrix Q such that

$$A = Q^T \begin{pmatrix} \Lambda & 0 \\ 0 & 0 \end{pmatrix} Q,$$

with

$$\Lambda = \begin{pmatrix} \lambda_1 & 0 & \cdots & 0 \\ 0 & \lambda_2 & \cdots & 0 \\ \vdots & \vdots & \ddots & \vdots \\ 0 & 0 & \cdots & \lambda_r \end{pmatrix}, \qquad \lambda_k \neq 0, \qquad 1 \leq k \leq r = \operatorname{rank} M.$$

Then
$$A^\dagger = Q^T \begin{pmatrix} \Lambda^{-1} & 0 \\ 0 & 0 \end{pmatrix} Q,$$

Consequently,
$$A^\dagger A = A A^\dagger = Q^T \begin{pmatrix} I_r & 0 \\ 0 & 0 \end{pmatrix} Q,$$

and
$$I - A^\dagger A = I - A A^\dagger = Q^T \begin{pmatrix} 0 & 0 \\ 0 & I_{n-r} \end{pmatrix} Q.$$

Now, if A is a *self-adjoint operator* in some Hilbert space, then by *spectral decomposition*, we have
$$A = \int_{\sigma(A)} \lambda dE_\lambda,$$
where $\sigma(A) \subseteq \mathbb{R}$ is the spectrum of A and $\{E_\lambda, \lambda \in \sigma(A)\}$ is the family of projection-valued measures. Then one has
$$A^\dagger = \int_{\sigma(A)\backslash\{0\}} \lambda^{-1} dE_\lambda.$$

Finally, we look at the following integral operator $\mathbf{F} : L^2(a,b;\mathbb{R}^n) \to L^2(a,b;\mathbb{R}^n)$ defined by
$$\mathbf{F}y(s) = y(s) - \int_a^b K(s,r)y(r)dr, \qquad s \in [a,b], \tag{1.35}$$
where $K(\cdot,\cdot) \in L^2([a,b]^2; \mathbb{R}^{n \times n})$. We have the following result, called the *Fredholm Alternative*.

Proposition 1.4.9. *For any $z(\cdot) \in L^2(a,b;\mathbb{R}^n)$ either there exists a $y(\cdot) \in L^2(a,b;\mathbb{R}^n)$ such that*
$$z(\cdot) = \mathbf{F}y(\cdot),$$
or there exists a $y(\cdot) \in L^2(a,b;\mathbb{R}^n)$ such that
$$\mathbf{F}y(\cdot) = 0, \quad \langle y(\cdot), z(\cdot) \rangle \neq 0.$$
This implies
$$\mathcal{R}(\mathbf{F}) = \mathcal{N}(\mathbf{F}^*)^\perp = \overline{\mathcal{R}(\mathbf{F})}.$$
Consequently, the range $\mathcal{R}(\mathbf{F})$ of $\mathbf{F}$ is closed.

Note that for any matrix $A \in \mathbb{R}^{n \times m}$, one has
$$\mathcal{R}(A) = \mathcal{N}(A^T)^\perp, \quad \mathcal{N}(A) = \mathcal{R}(A^T)^\perp.$$
The first relation above means that $\mathcal{R}(A)$ is closed. Whereas, for general bounded operator A in a Hilbert space, one only has
$$\overline{\mathcal{R}(A)} = \mathcal{N}(A^T)^\perp, \quad \mathcal{N}(A) = \mathcal{R}(A^T)^\perp.$$
Thus, $\mathcal{R}(A)$ might not be closed. Operator $\mathbf{F}$ of form (1.35) is a typical example of the so-called *Fredholm operator*.

1.5   Brief Historic Remarks

The idea of game should exist since people realized the existence of competition among human being: Better action/strategy would lead to a better outcome. The first known minimax mixed strategy equilibrium for a two-person game was provided in a letter of Francis Waldegrave to Pierre-Remond de Montmort dated on November 13, 1713. In 1838, Cournot published his book [28], in which he discussed a special case of duopoly and used a solution concept which is a kind of Nash equilibrium. In 1913, Zermelo established the first formal theorem of game theory [130] (see [108] for further comments). During 1921–1927, Borel published several notes ([21], [22], [23]) in which he gave the first modern formulation of a mixed strategy and found the minimax solution for two-person games in some special cases. See more comments in [48], [49] by Fréchet, and [116] by von Neumann. In 1928, von Neumann proved the minimax theorem for two-person zero-sum games under mixed strategies ([115]). The book by von Neumann and Morganstain ([117]) is a milestone in the history of game theory. They developed a general theory for two-person games and introduced an axiomatic utility theory, leading to wide range applications of game theory in economics. In the early 1950s, Nash introduced n-person non-cooperative games ([79], [80], [81], [82]) for which some significant applications were found in economics. As a result, Nash became a winner of Nobel memorial prize in economic sciences in 1994.

For remarks on control theory and differential games, see later chapters.

Chapter 2

Control Theory — Single-Player Differential Games

In this chapter, we will concisely present a control theory which can be regarded as a theory for single-player differential games.

2.1 Control Systems

We assume that U is a non-empty closed subset in $\mathbb{R}^m$ (it could be more generally a separable complete metric space). For any initial pair $(t, x) \in \mathbb{R}_+ \times \mathbb{R}^n$, we rewrite the control system here:

$$\begin{cases} \dot{X}(s) = f(s, X(s), u(s)), & s \in [t, \infty), \\ X(t) = x. \end{cases} \tag{2.1}$$

Let us begin with the following assumption.

(C1) The map $f : \mathbb{R}_+ \times \mathbb{R}^n \times U \to \mathbb{R}^n$ is measurable and there exists a constant $L > 0$ such that

$$\begin{cases} |f(t, x_1, u) - f(t, x_2, u)| \leqslant L|x_1 - x_2|, & (t, u) \in \mathbb{R}_+ \times U, \ x_1, x_2 \in \mathbb{R}^n, \\ |f(t, 0, u)| \leqslant L, & \forall (t, u) \in \mathbb{R}_+ \times U. \end{cases}$$

Note that conditions in (C1) imply

$$|f(t, x, u)| \leqslant L\big(1 + |x|\big), \qquad (t, x, u) \in \mathbb{R}_+ \times \mathbb{R}^n \times U.$$

A key feature of the above is that the bound of $|f(t, x, u)|$, depending on $|x|$, is uniform in u. We will discuss the case that the bound also depends on u in a later chapter. The following result is concerned with the state trajectory of (2.1).

Proposition 2.1.1. *Let* (C1) *hold. Then, for any* $(t, x) \in \mathbb{R}_+ \times \mathbb{R}^n$, *and* $u(\cdot) \in \mathcal{U}[t, \infty)$, *there exists a unique solution* $X(\cdot) \equiv X(\cdot\,; t, x, u(\cdot))$ *to* the state equation (2.1). Moreover, the following estimates hold:

$$\begin{cases} |X(s;t,x,u(\cdot))| \leqslant e^{L(s-t)}(1+|x|) - 1, \\ |X(s;t,x,u(\cdot)) - x| \leqslant \left[e^{L(s-t)} - 1\right](1+|x|), \\ \qquad\qquad\qquad s \in [t,\infty), \ u(\cdot) \in \mathcal{U}[t,\infty). \end{cases} \tag{2.2}$$

Further, for any $t \in \mathbb{R}_+$, $x_1, x_2 \in \mathbb{R}^n$, and $u(\cdot) \in \mathcal{U}[t,\infty)$,

$$|X(s;t,x_1,u(\cdot)) - X(s;t,x_2,u(\cdot))| \leqslant e^{L(s-t)}|x_1 - x_2|, \tag{2.3}$$
$$\forall s \in [t,\infty).$$

Proof. It suffices to prove our conclusion on any $[t,T]$ with $0 \leqslant t < T < \infty$. For any $X(\cdot) \in C([t,T];\mathbb{R}^n)$, we define

$$[\mathcal{S}X(\cdot)](s) = x + \int_t^s f(r,X(r),u(r))dr, \qquad s \in [t,T].$$

Then for any $X_1(\cdot), X_2(\cdot) \in C([t,T];\mathbb{R}^n)$, we have

$$\left\|[\mathcal{S}X_1(\cdot)] - [\mathcal{S}X_2(\cdot)]\right\|_{C([t,t+\delta];\mathbb{R}^n)} \leqslant L\delta\|X_1(\cdot) - X_2(\cdot)\|_{C([t,t+\delta];\mathbb{R}^n)}.$$

Therefore, by choosing $\delta < \frac{1}{L}$, we see that $\mathcal{S} : C([t,t+\delta];\mathbb{R}^n) \to C([t,t+\delta];\mathbb{R}^n)$ is contractive. Hence, by Contraction Mapping Theorem, (2.1) admits a unique solution on $[t,t+\delta]$. Repeating the argument, we can obtain that (2.1) admits a unique solution on $[t,T]$.

Next, for the unique solution $X(\cdot)$ of (2.1), we have

$$|X(s)| \leqslant |x| + L\int_t^s \big(1+|X(r)|\big)dr, \qquad s \in [t,T].$$

If we denote the right-hand side of the above by $\theta(s)$, then

$$\dot{\theta}(s) = L + L|X(s)| \leqslant L + L\theta(s),$$

which leads to

$$\theta(s) \leqslant e^{L(s-t)}|x| + L\int_t^s e^{L(s-r)}dr = e^{L(s-t)}|x| + e^{L(s-t)} - 1.$$

This gives the first estimate in (2.2). Next,

$$|X(s) - x| \leqslant \int_t^s |f(r,X(r),u(r))|dr \leqslant L\int_t^s \big(1+|X(r)|\big)dr$$

$$\leqslant L\int_t^s e^{L(r-t)}(1+|x|)dr = (1+|x|)\big[e^{L(s-t)} - 1\big].$$

This proves the second estimate in (2.2). Finally, for $x_1, x_2 \in \mathbb{R}^n$, let us denote $X_i(\cdot) = X(\cdot;t,x_i,u(\cdot))$. Then

$$|X_1(s) - X_2(s)| \leqslant L\int_t^s |X_1(r) - X_2(r)|dr.$$

By Gronwall's inequality, we obtain (2.3). □

We point out that the estimates in (2.2) are uniform in $u(\cdot) \in \mathcal{U}[t,T]$. This will play an interesting role later.

2.2 Optimal Control — Existence Theory

Let $T \in (0, \infty)$ be fixed. Consider control system

$$\begin{cases} \dot{X}(s) = f(s, X(s), u(s)), & s \in [t, T], \\ X(t) = x, \end{cases}$$

with *terminal state constraint*:

$$X(T; t, x, u(\cdot)) \in M,$$

where $M \subseteq \mathbb{R}^n$ is fixed, and with cost functional

$$J^T(t, x; u(\cdot)) = \int_t^T g(s, X(s), u(s))ds + h(X(T)).$$

We recall (from Section 1.3.1)

$$\widetilde{\mathcal{U}}_x^M[t, T] = \left\{ u(\cdot) \in \mathcal{U}[t, T] \mid X(T; t, x, u(\cdot)) \in M \right\},$$

and recall the following optimal control problem:

Problem (OC)^T. For given $(t, x) \in [0, T) \times \mathbb{R}^n$ with $\widetilde{\mathcal{U}}_x^M[t, T] \neq \emptyset$, find a $\bar{u}(\cdot) \in \widetilde{\mathcal{U}}_x^M[t, T]$ such that

$$J^T(t, x; \bar{u}(\cdot)) = \inf_{u(\cdot) \in \widetilde{\mathcal{U}}_x^M[t, T]} J^T(t, x; u(\cdot)) \equiv V(t, x). \qquad (2.4)$$

Any $\bar{u}(\cdot) \in \widetilde{\mathcal{U}}_x^M[t, T]$ satisfying (2.4) is called an *optimal control*, the corresponding $\bar{X}(\cdot) \equiv X(\cdot; t, x, \bar{u}(\cdot))$ is called an *optimal state trajectory*, and $(\bar{X}(\cdot), \bar{u}(\cdot))$ is called an *optimal pair*. We call $V(\cdot, \cdot)$ the *value function* of Problem (OC)^T.

We introduce the following assumption for the cost functional.

(C2) The maps $g : \mathbb{R}_+ \times \mathbb{R}^n \times U \to \mathbb{R}$ and $h : \mathbb{R}^n \to \mathbb{R}$ are measurable and there exists a continuous function $\omega : \mathbb{R}_+ \times \mathbb{R}_+ \to \mathbb{R}_+$, called a *local modulus of continuity*, which is increasing in each argument, and $\omega(r, 0) = 0$, for all $r \geqslant 0$, such that

$$|g(s, x_1, u) - g(s, x_2, u)| + |h(x_1) - h(x_2)|$$
$$\leqslant \omega(|x_1| \vee |x_2|, |x_1 - x_2|), \qquad \forall (s, u) \in \mathbb{R}_+ \times U, \ x_1, x_2 \in \mathbb{R}^n,$$

where $|x_1| \vee |x_2| = \max\{|x_1|, |x_2|\}$, and

$$\sup_{(s, u) \in \mathbb{R}_+ \times U} |g(s, 0, u)| \equiv g_0 < \infty.$$

In what follows, $\omega(\cdot, \cdot)$ will stand for a generic local modulus of continuity which can be different from line to line.

For any $(t, x) \in [0, T] \times \mathbb{R}^n$, let us introduce the following set

$$\mathbb{E}(t, x) = \left\{ (z^0, z) \in \mathbb{R} \times \mathbb{R}^n \mid z^0 \geqslant g(t, x, u), \right.$$
$$\left. z = f(t, x, u), u \in U \right\}.$$

The following assumption gives some compatibility between the control system and the cost functional.

(C3) For almost all $t \in [0, T]$, the following *Cesari property* holds at any $x \in \mathbb{R}^n$:

$$\bigcap_{\delta > 0} \overline{\text{co}}\, \mathbb{E}(t, B_\delta(x)) = \mathbb{E}(t, x),$$

where, we recall that $B_\delta(x)$ is the open ball centered at x with radius $\delta > 0$, and $\overline{\text{co}}(E)$ stands for the *closed convex hull* of the set E (the smallest closed convex set containing E).

It is clear that if $\mathbb{E}(t, x)$ has the Cesari property at x, then $\mathbb{E}(t, x)$ is convex and closed.

Theorem 2.2.1. *Let* (C1)–(C3) *hold, Let* $M \subseteq \mathbb{R}^n$ *be a non-empty closed set. Let* $(t, x) \in [0, T] \times \mathbb{R}^n$ *be given and* $\widetilde{\mathcal{U}}_x^M[t, T] \neq \emptyset$. *Then Problem* $(OC)^T$ *admits at least one optimal pair.*

Proof. Let $u_k(\cdot) \in \widetilde{\mathcal{U}}_x^M[t, T]$ be a minimizing sequence. By Proposition 2.1.1,

$$|X_k(s)| \leqslant e^{L(s-t)}(1 + |x|) - 1, \qquad s \in [t, T], \qquad k \geqslant 1, \qquad (2.5)$$

and for any $t \leqslant \tau < s \leqslant T$,

$$|X_k(s) - X_k(\tau)| = |X(s; t, x, u_k(\cdot)) - X(\tau; t, x, u_k(\cdot))|$$
$$\leqslant \left[e^{L(s-\tau)} - 1 \right] \left[1 + |X(\tau; t, x, u_k(\cdot))| \right]$$
$$\leqslant \left[e^{L(s-\tau)} - 1 \right] e^{L(\tau-t)}(1 + |x|).$$

Hence, the sequence $\{X_k(\cdot)\}$ is uniformly bounded and equi-continuous. Therefore, by Arzela–Ascoli Theorem, we may assume that the sequence is convergent to some $\bar{X}(\cdot)$ in $C([t, T]; \mathbb{R}^n)$. On the other hand,

$$|f(s, X_k(s), u_k(s))| \leqslant L(1 + |X_k(s)|) \leqslant L e^{L(s-t)}(1 + |x|).$$

Also, by (2.5) and (C2), we have

$$|g(s, X_k(s), u_k(s))| \leqslant |g(s, 0, u_k(s))| + |g(s, X_k(s), u_k(s)) - g(s, 0, u_k(s))|$$
$$\leqslant g_0 + \omega\big(|X_k(s)|, |X_k(s)|\big) \leqslant g_0 + \omega\big(e^{LT}(1 + |x|), e^{LT}(1 + |x|)\big) \leqslant K,$$
$$s \in [t, T], \qquad k \geqslant 1.$$

Hereafter, $K > 0$ stands for a generic constant which could be different from line to line. Hence, by extracting a subsequence if necessary, we may assume that

$$\begin{cases} g(\cdot, X_k(\cdot), u_k(\cdot)) \to \bar{g}(\cdot), & \text{weakly in } L^2([t,T]; \mathbb{R}), \\ f(\cdot, X_k(\cdot), u_k(\cdot)) \to \bar{f}(\cdot), & \text{weakly in } L^2([t,T]; \mathbb{R}^n), \end{cases}$$

for some $\bar{g}(\cdot)$ and $\bar{f}(\cdot)$. Then by Banach–Saks Theorem, we have

$$\begin{cases} \widetilde{g}_k(\cdot) \stackrel{\Delta}{=} \dfrac{1}{k} \sum_{i=1}^{k} g(\cdot, X_i(\cdot), u_i(\cdot)) \to \bar{g}(\cdot), & \text{strongly in } L^2(t,T; \mathbb{R}), \\ \widetilde{f}_k(\cdot) \stackrel{\Delta}{=} \dfrac{1}{k} \sum_{i=1}^{k} f(\cdot, X_i(\cdot), u_i(\cdot)) \to \bar{f}(\cdot), & \text{strongly in } L^2(t,T; \mathbb{R}^n). \end{cases} \tag{2.6}$$

On the other hand, by (C1) and the convergence of $X_k(\cdot) \to \bar{X}(\cdot)$ in $C([t,T]; \mathbb{R}^n)$, we have

$$\Big| \widetilde{f}_k(s) - \frac{1}{k} \sum_{i=1}^{k} f(s, \bar{X}(s), u_i(s)) \Big|$$

$$\leqslant \frac{1}{k} \sum_{i=1}^{k} \big| f(s, X_i(s), u_i(s)) - f(s, \bar{X}(s), u_i(s)) \big|$$

$$\leqslant \frac{L}{k} \sum_{i=1}^{k} |X_i(s) - \bar{X}(s)| \to 0, \quad (k \to \infty),$$

uniformly in $s \in [t,T]$. Similarly, by (C2),

$$\Big| \widetilde{g}_k(s) - \frac{1}{k} \sum_{i=1}^{k} g(s, \bar{X}(s), u_i(s)) \Big|$$

$$\leqslant \frac{1}{k} \sum_{i=1}^{k} \big| g(s, X_i(s), u_i(s)) - g(s, \bar{X}(s), u_i(s)) \big|$$

$$\leqslant \frac{1}{k} \sum_{i=1}^{k} \omega \big(|X_i(s)| \vee |\bar{X}(s)|, |X_i(s) - \bar{X}(s)| \big) \to 0, \quad (k \to \infty)$$

uniformly in $s \in [t,T]$. Next, by the definition of $\mathbb{E}(s,x)$, we have

$$\begin{pmatrix} g(s, X_i(s), u_i(s)) \\ f(s, X_i(s), u_i(s)) \end{pmatrix} \in \mathbb{E}(s, X_i(s)), \quad i \geqslant 1, \ s \in [t,T].$$

Hence, for any $\delta > 0$, there exists a $k_\delta > 0$ such that

$$\begin{pmatrix} \widetilde{g}_k(s) \\ \widetilde{f}_k(s) \end{pmatrix} \in \overline{co}\, \mathbb{E}\big(s, B_\delta(\bar{X}(s))\big), \quad k \geqslant k_\delta, \ s \in [t,T]. \tag{2.7}$$

Combining (2.6) and (2.7), using (C3), we obtain

$$\begin{pmatrix} \bar{g}(s) \\ \bar{f}(s) \end{pmatrix} = \lim_{k \to \infty} \begin{pmatrix} \widetilde{g}_k(s) \\ \widetilde{f}_k(s) \end{pmatrix} \in \bigcap_{\delta > 0} \overline{\mathrm{co}}\,\mathbb{E}\big(s, B_\delta(\bar{X}(s))\big) = \mathbb{E}(s, \bar{X}(s)).$$

Then by Filippov's Lemma, there exists a $\bar{u}(\cdot) \in \mathcal{U}[t, T]$ such that

$$\bar{g}(s) \geqslant g(s, \bar{X}(s), \bar{u}(s)), \quad \bar{f}(s) = f(s, \bar{X}(s), \bar{u}(s)), \qquad s \in [t, T].$$

This means $\bar{X}(\cdot) = X(\cdot\,; t, x, \bar{u}(\cdot))$. On the other hand, since

$$X_k(T) \equiv X(T; t, x, u_k(\cdot)) \in M, \qquad k \geqslant 1,$$

one has

$$\bar{X}(T) \equiv X(T; t, x, \bar{u}(\cdot)) \in M,$$

which means that $\bar{u}(\cdot) \in \widetilde{\mathcal{U}}_x^M[t, T]$. Finally, by Fatou's Lemma

$$J^T(t, x; \bar{u}(\cdot)) \leqslant \int_t^T \bar{g}(s)ds + h(\bar{X}(T)) \leqslant \varliminf_{k \to \infty} \Big[\int_t^T \widetilde{g}_k(s)ds + h(X_k(T)) \Big]$$

$$= \lim_{k \to \infty} \frac{1}{k} \sum_{i=1}^k J^T(t, x; u_i(\cdot)) = \lim_{k \to \infty} J^T(t, x; u_k(\cdot))$$

$$= \inf_{u(\cdot) \in \widetilde{\mathcal{U}}_x^M[t,T]} J^T(t, x; u(\cdot)).$$

This means that $(\bar{X}(\cdot), \bar{u}(\cdot))$ is an optimal pair. $\qquad\square$

For Problem (OC), i.e., the optimal control problem with moving target set and with non-fixed duration, one can similarly establish an existence theorem of optimal controls. We leave the details to the readers.

2.3 Pontryagin Maximum Principle — A Variational Method

In this subsection, we are going to derive a necessary condition for optimal controls of Problem $(OC)^T$ by means of *variational method*. To this end, we introduce the following assumption.

(C4) In addition to (C1)–(C2), the map

$$x \mapsto (f(t, x, u), g(t, x, u), h(x))$$

is differentiable and the map

$$(x, u) \mapsto (f(t, x, u), f_x(t, x, u), g(t, x, u), g_x(t, x, u), h_x(x))$$

is continuous.

We have the following necessary conditions for optimal controls.

Theorem 2.3.1. (Pontryagin Maximum Principle) *Let* (C4) *hold. Let M be a non-empty closed convex set. Suppose $(\bar{X}(\cdot), \bar{u}(\cdot))$ is an optimal pair of Problem* (OC)^T *for the initial pair (t, x). Then the following maximum condition holds:*

$$H(s, \bar{X}(s), \bar{u}(s), \psi^0, \psi(s)) = \max_{u \in U} H(s, \bar{X}(s), u, \psi^0, \psi(s)),$$

$$\text{a.e. } s \in [t, T], \tag{2.8}$$

where $H(s, x, u, \psi^0, \psi)$ is called the Hamiltonian which is defined by

$$H(s, x, u, \psi^0, \psi) = \psi^0 g(s, x, u) + \langle \psi, f(s, x, u) \rangle,$$

$$(s, x, u, \psi^0, \psi) \in [t, T] \times \mathbb{R}^n \times U \times \mathbb{R} \times \mathbb{R}^n,$$

and $\psi(\cdot)$ is the solution to the adjoint equation:

$$\dot{\psi}(s) = -f_x(s, \bar{X}(s), \bar{u}(s))^T \psi(s) + \psi^0 g_x(s, \bar{X}(s), \bar{u}(s))^T, \quad s \in [t, T], \tag{2.9}$$

with

$$\begin{cases} \psi^0 \leqslant 0, \\ |\psi^0|^2 + |\psi(T) - \psi^0 h_x(\bar{X}(T))^T|^2 = 1, \end{cases} \tag{2.10}$$

and the following transversality condition holds:

$$\langle \psi(T) - \psi^0 h_x(\bar{X}(T))^T, y - \bar{X}(T) \rangle \geq 0, \qquad \forall y \in M. \tag{2.11}$$

Note that in the case $M = \mathbb{R}^n$, the transversality condition becomes

$$\psi(T) = \psi^0 h_x(\bar{X}(T))^T.$$

Then (2.10) *implies $\psi^0 \neq 0$. By rescaling if necessary, we may assume that $\psi^0 = -1$. Hence, in this case, the adjoint equation reads*

$$\begin{cases} \dot{\psi}(s) = -f_x(s, \bar{X}(s), \bar{u}(s))^T \psi(s) - g_x(s, \bar{X}(s), \bar{u}(s))^T, \quad s \in [t, T], \\ \psi(T) = -h_x(\bar{X}(T))^T. \end{cases}$$

Proof. First, we introduce

$$\rho(u(\cdot), v(\cdot)) = \big|\{s \in [t, T] \mid u(s) = v(s)\}\big|, \qquad \forall u(\cdot), v(\cdot) \in \mathcal{U}[t, T].$$

Recall that $|A|$ stands for the Lebesgue measure of the set $A \subseteq \mathbb{R}$. We can show that $\rho(\cdot, \cdot)$ is a metric under which $\mathcal{U}[t, T]$ is a complete metric space. Now, if necessary, by adding a constant to $h(\cdot)$, we may assume that

$$J^T(t, x; \bar{u}(\cdot)) = 0.$$

For any $\varepsilon > 0$, define

$$J_\varepsilon^T(t, x; u(\cdot)) = \left\{ \left[\left(J^T(t, x; u(\cdot)) + \varepsilon \right)^+ \right]^2 + d_M \left(X(T; t, x, u(\cdot)) \right)^2 \right\}^{\frac{1}{2}},$$

where

$$d_M(x) = \inf_{y \in M} |x - y|, \qquad \forall x \in \mathbb{R}^n.$$

Then

$$J_\varepsilon^T(t, x; \bar{u}(\cdot)) = \varepsilon \leqslant \inf_{u(\cdot) \in \tilde{\mathcal{U}}_x^M[t,T]} J^T(t, x; u(\cdot)) + \varepsilon.$$

By Ekeland Variational Principle (Theorem 1.4.5), there exists a $u^\varepsilon(\cdot) \in \mathcal{U}[t, T]$ such that

$$\rho(u^\varepsilon(\cdot), \bar{u}(\cdot)) \leqslant \sqrt{\varepsilon},$$

and

$$J_\varepsilon^T(t, x; u^\varepsilon(\cdot)) \leqslant J_\varepsilon^T(t, x; u(\cdot)) + \sqrt{\varepsilon}\, \rho(u(\cdot), u^\varepsilon(\cdot)), \quad \forall u(\cdot) \in \mathcal{U}[t, T].$$

This means that $u^\varepsilon(\cdot)$ is a minimum of the map

$$u(\cdot) \mapsto J_\varepsilon^T(t, x; u(\cdot)) + \sqrt{\varepsilon}\, \rho(u(\cdot), u^\varepsilon(\cdot)).$$

Let $X^\varepsilon(\cdot) = X(\cdot; t, x, u^\varepsilon(\cdot))$. Now, making use of Lemma 1.4.6, for any $u(\cdot) \in \mathcal{U}[t, T]$ and $\delta \in (0, 1)$, we define $u_\delta^\varepsilon(\cdot) \in \mathcal{U}[t, T]$ as follows:

$$u_\delta^\varepsilon(s) = \begin{cases} u^\varepsilon(s), & s \in [t, T] \setminus E_\delta^\varepsilon, \\ u(s), & s \in E_\delta^\varepsilon, \end{cases}$$

where $E_\delta^\varepsilon \subseteq [t, T]$ can be chosen so that

$$|E_\delta^\varepsilon| = \delta(T - t),$$

and for $s \in [t, T]$,

$$\delta \int_t^s \begin{pmatrix} g(\tau, X^\varepsilon(\tau), u(\tau)) - g(\tau, X^\varepsilon(\tau), u^\varepsilon(\tau)) \\ f(\tau, X^\varepsilon(\tau), u(\tau)) - f(\tau, X^\varepsilon(\tau), u^\varepsilon(\tau)) \end{pmatrix} d\tau$$
$$= \int_{E_\delta^\varepsilon \cap [t, s]} \begin{pmatrix} g(\tau, X^\varepsilon(\tau), u(\tau)) - g(\tau, X^\varepsilon(\tau), u^\varepsilon(\tau)) \\ f(\tau, X^\varepsilon(\tau), u(\tau)) - f(\tau, X^\varepsilon(\tau), u^\varepsilon(\tau)) \end{pmatrix} d\tau + \begin{pmatrix} r_\delta^{0,\varepsilon}(s) \\ r_\delta^\varepsilon(s) \end{pmatrix},$$

with

$$|r_\delta^{0,\varepsilon}(s)| + |r_\delta^\varepsilon(s)| \leqslant \delta^2, \qquad s \in [t, T].$$

Let $X_\delta^\varepsilon(\cdot) = X(\cdot; t, x, u_\delta^\varepsilon(\cdot))$, and define

$$Y_\delta^\varepsilon(s) = \frac{X_\delta^\varepsilon(s) - X^\varepsilon(s)}{\delta}, \qquad s \in [t, T].$$

Then

$$
\begin{aligned}
Y_\delta^\varepsilon(s) &= \frac{1}{\delta} \int_t^s \Big[f(\tau, X_\delta^\varepsilon(\tau), u_\delta^\varepsilon(\tau)) - f(\tau, X^\varepsilon(\tau), u^\varepsilon(\tau)) \Big] d\tau \\
&= \frac{1}{\delta} \int_t^s \Big[f(\tau, X_\delta^\varepsilon(\tau), u_\delta^\varepsilon(\tau)) - f(\tau, X^\varepsilon(\tau), u_\delta^\varepsilon(\tau)) \Big] d\tau \\
&\quad + \frac{1}{\delta} \int_{E_\delta^\varepsilon \cap [t,s]} \Big[f(\tau, X^\varepsilon(\tau), u(\tau)) - f(\tau, X^\varepsilon(\tau), u^\varepsilon(\tau)) \Big] d\tau \\
&= \int_t^s \int_0^1 f_x(\tau, X^\varepsilon(\tau) + \theta[X_\delta^\varepsilon(\tau) - X^\varepsilon(\tau)], u_\delta^\varepsilon(\tau)) Y_\delta^\varepsilon(\tau) d\theta d\tau \\
&\quad + \int_t^s \Big[f(\tau, X^\varepsilon(\tau), u(\tau)) - f(\tau, X^\varepsilon(\tau), u^\varepsilon(\tau)) \Big] d\tau + \int_t^s \frac{r_\delta^\varepsilon(\tau)}{\delta} d\tau.
\end{aligned}
$$

Next, we let $Y^\varepsilon(\cdot)$ be the solution to the following:

$$
\begin{cases}
\dot{Y}^\varepsilon(s) = f_x(s, X^\varepsilon(s), u^\varepsilon(s)) Y^\varepsilon(s) \\
\qquad\qquad + f(s, X^\varepsilon(s), u(s)) - f(s, X^\varepsilon(s), u^\varepsilon(s)), \qquad s \in [t, T], \\
Y^\varepsilon(t) = 0.
\end{cases}
$$

It is not hard to show that

$$
\lim_{\delta \to 0} \| Y_\delta^\varepsilon(\cdot) - Y^\varepsilon(\cdot) \|_{C([t,T];\mathbb{R}^n)} = 0.
$$

On the other hand, by the optimality of $u^\varepsilon(\cdot)$, we have

$$
\begin{aligned}
-\sqrt{\varepsilon}(T - t) &= -\sqrt{\varepsilon} \frac{|E_\delta^\varepsilon|}{\delta} \leqslant -\sqrt{\varepsilon} \frac{\rho(u_\delta^\varepsilon(\cdot), u^\varepsilon(\cdot))}{\delta} \\
&\leqslant \frac{1}{\delta} \big[J_\varepsilon^T(t, x; u_\delta^\varepsilon(\cdot)) - J_\varepsilon^T(t, x; u^\varepsilon(\cdot)) \big] \\
&= \frac{\big\{ \big[J^T(t, x; u_\delta^\varepsilon(\cdot)) + \varepsilon \big]^+ \big\}^2 - \big\{ \big[J^T(t, x; u^\varepsilon(\cdot)) + \varepsilon \big]^+ \big\}^2}{\delta \big[J_\varepsilon^T(t, x; u_\delta^\varepsilon(\cdot)) + J_\varepsilon^T(t, x; u^\varepsilon(\cdot)) \big]} \\
&\quad + \frac{d_M(X_\delta^\varepsilon(T))^2 - d_M(X^\varepsilon(T))^2}{\delta \big[J_\varepsilon^T(t, x; u_\delta^\varepsilon(\cdot)) + J_\varepsilon^T(t, x; u^\varepsilon(\cdot)) \big]} \\
&= \bar{\psi}_\delta^{0,\varepsilon} \Big\{ \frac{1}{\delta} \int_t^T \big[g(s, X_\delta^\varepsilon(s), u_\delta^\varepsilon(s)) - g(t, X^\varepsilon(s), u^\varepsilon(s)) \big] ds \\
&\quad + \frac{h(X_\delta^\varepsilon(T)) - h(X^\varepsilon(T))}{\delta} \Big\} + \bar{\psi}_\delta^\varepsilon Y_\delta^\varepsilon(T),
\end{aligned}
\tag{2.12}
$$

where

$$
\begin{cases}
\bar{\psi}_\delta^{0,\varepsilon} = \dfrac{\big[J^T(t, x; u_\delta^\varepsilon(\cdot)) + \varepsilon \big]^+ + \big[J^T(t, x; u^\varepsilon(\cdot)) + \varepsilon \big]^+}{J_\varepsilon^T(t, x; u_\delta^\varepsilon(\cdot)) + J_\varepsilon^T(t, x; u^\varepsilon(\cdot))}, \\[2ex]
\bar{\psi}_\delta^\varepsilon = \dfrac{\int_0^1 \nabla(d_M^2)(X^\varepsilon(T) + \theta[X_\delta^\varepsilon(T) - X^\varepsilon(T)]) d\theta}{J_\varepsilon^T(t, x; u_\delta^\varepsilon(\cdot)) + J_\varepsilon^T(t, x; u^\varepsilon(\cdot))}.
\end{cases}
$$

Then

$$
\begin{cases}
\lim_{\delta \to 0} \bar{\psi}_\delta^{0,\varepsilon} = \bar{\psi}^{0,\varepsilon} \equiv \dfrac{\left[J^T(t,x;u^\varepsilon(\cdot)) + \varepsilon\right]^+}{J_\varepsilon^T(t,x;u^\varepsilon(\cdot))}, \\[3mm]
\lim_{\delta \to 0} \bar{\psi}_\delta^{\varepsilon} = \bar{\psi}^{\varepsilon} \equiv \dfrac{d_M(X^\varepsilon(T))\partial d_M(X^\varepsilon(T))}{J_\varepsilon^T(t,x;u^\varepsilon(\cdot))},
\end{cases}
$$

where

$$
\partial d_M(X^\varepsilon(T)) =
\begin{cases}
\dfrac{X^\varepsilon(T) - P_M(X^\varepsilon(T))}{|X^\varepsilon(T) - P_M(X^\varepsilon(T))|}, & X^\varepsilon(T) \notin M, \\[3mm]
\nu(X^\varepsilon(T)), & X^\varepsilon(T) \in \partial M, \\[1mm]
0, & X^\varepsilon(T) \in M \setminus \partial M.
\end{cases}
$$

In the above, recall that $P_M(X^\varepsilon(T))$ is the projection of $X^\varepsilon(T)$ onto the convex set M (when $X^\varepsilon(T) \notin M$), $\nu(X^\varepsilon(T))$ is a *unit outward normal* of M at $X^\varepsilon(T) \in \partial M$, defined to be a unit vector satisfying

$$
\langle \nu(X^\varepsilon(T)), y - X^\varepsilon(T) \rangle \leqslant 0, \qquad \forall y \in M.
$$

It is clear that

$$
\begin{cases}
|\bar{\psi}^{0,\varepsilon}|^2 + |\bar{\psi}^\varepsilon|^2 = 1, & \forall \varepsilon > 0, \\[1mm]
\langle \bar{\psi}^\varepsilon, y - X^\varepsilon(T) \rangle \leqslant 0, & \forall y \in M.
\end{cases}
\tag{2.13}
$$

Further,

$$
\frac{1}{\delta} \int_t^T \Big[g(\tau, X_\delta^\varepsilon(\tau), u_\delta^\varepsilon(\tau)) - g(\tau, X^\varepsilon(\tau), u^\varepsilon(\tau)) \Big] d\tau
$$

$$
= \frac{1}{\delta} \int_t^T \Big[g(\tau, X_\delta^\varepsilon(\tau), u_\delta^\varepsilon(\tau)) - g(\tau, X^\varepsilon(\tau), u_\delta^\varepsilon(\tau)) \Big] d\tau
$$

$$
+ \frac{1}{\delta} \int_{E_\delta^\varepsilon \cap [t,s]} \Big[g(\tau, X^\varepsilon(\tau), u(\tau)) - g(\tau, X^\varepsilon(\tau), u^\varepsilon(\tau)) \Big] d\tau
$$

$$
= \int_t^T \int_0^1 g_x(\tau, X^\varepsilon(\tau) + \theta[X_\delta^\varepsilon(\tau) - X^\varepsilon(\tau)], u_\delta^\varepsilon(\tau)) Y_\delta^\varepsilon(\tau) d\theta d\tau
$$

$$
+ \int_t^T \Big[g(\tau, X^\varepsilon(\tau), u(\tau)) - g(\tau, X^\varepsilon(\tau), u^\varepsilon(\tau)) \Big] d\tau + \int_t^s \frac{r_\delta^{0,\varepsilon}(\tau)}{\delta} d\tau
$$

$$
\to \int_t^T \Big[g_x(\tau, X^\varepsilon(\tau), u^\varepsilon(\tau)) Y^\varepsilon(\tau) + g(\tau, X^\varepsilon(\tau), u(\tau)) - g(\tau, X^\varepsilon(\tau), u^\varepsilon(\tau)) \Big] d\tau.
$$

Consequently, sending $\delta \to 0$ in (2.12), one has

$$
-\sqrt{\varepsilon}\,(T-t) \leqslant \bar{\psi}^{0,\varepsilon} \int_t^T \Big[g_x(\tau, X^\varepsilon(\tau), u^\varepsilon(\tau)) Y^\varepsilon(\tau) + g(\tau, X^\varepsilon(\tau), u(\tau))
$$

$$
- g(\tau, X^\varepsilon(\tau), u^\varepsilon(\tau)) \Big] d\tau + \Big[\bar{\psi}^{0,\varepsilon} h_x(X^\varepsilon(T)) + \bar{\psi}^\varepsilon \Big] Y^\varepsilon(T).
\tag{2.14}
$$

By (2.13), we may assume that $(\psi^{0,\varepsilon}, \psi^\varepsilon)$ is convergent, as $\varepsilon \to 0$ (if necessary, we may take a subsequence). Denote

$$\lim_{\varepsilon \to 0} (\psi^{0,\varepsilon}, \psi^\varepsilon) = -(\psi^0, \psi),$$

with

$$\begin{cases} |\psi^0|^2 + |\psi|^2 = 1, & \psi^0 \leqslant 0, \\ \langle \psi, y - \bar{X}(T) \rangle \geqslant 0. \end{cases}$$

Also,

$$\lim_{\varepsilon \to 0} \|Y^\varepsilon(\cdot) - Y(\cdot)\|_{C([t,T];\mathbb{R}^n)} = 0,$$

where

$$\begin{cases} \dot{Y}(s) = f_x(s, \bar{X}(s), \bar{u}(s))Y(s) + f(s, \bar{X}(s), u(s)) \\ \qquad\qquad - f(s, \bar{X}(s), \bar{u}(s)), & s \in [t, T], \\ Y(t) = 0. \end{cases}$$

Let $\psi(\cdot)$ be the solution of the adjoint equation (2.9) with

$$\psi(T) = \psi + \psi^0 h_x(\bar{X}(T)).$$

Then (2.10)–(2.11) hold. Now, let $\varepsilon \to 0$ in (2.14), we obtain

$$\begin{aligned}
0 \geqslant & \int_t^T \psi^0 \Big[g_x(s, \bar{X}(s), u(s))Y(s) + g(s, \bar{X}(s), u(s)) \\
& \qquad - g(s, \bar{X}(s), \bar{u}(s)) \Big] ds + \Big[\psi^0 h_x(X(T)) + \psi \Big] Y(T) \\
= & \int_t^T \Big\{ \psi^0 \Big[g_x(s, \bar{X}(s), u(s))Y(s) + g(s, \bar{X}(s), u(s)) - g(s, \bar{X}(s), \bar{u}(s)) \Big] \\
& \quad + \langle \dot{\psi}(s), Y(s) \rangle + \langle \psi(s), f_x(s, \bar{X}(s), \bar{u}(s))Y(s) \rangle \\
& \quad + \langle \psi(s), f(s, \bar{X}(s), u(s)) - f(s, \bar{X}(s), \bar{u}(s)) \rangle \Big\} ds \\
= & \int_t^T \Big\{ \psi^0 \Big[g(s, \bar{X}(s), u(s)) - g(s, \bar{X}(s), \bar{u}(s)) \Big] \\
& \quad + \langle \psi(s), f(s, \bar{X}(s), u(s)) - f(s, \bar{X}(s), \bar{u}(s)) \rangle \Big\} ds \\
= & \int_t^T \Big(H(s, \bar{X}(s), u(s), \psi^0, \psi(s)) - H(s, \bar{X}(s), \bar{u}(s), \psi^0, \psi(s)) \Big) ds.
\end{aligned}$$

This implies (2.8). $\qquad\qquad\qquad\qquad\qquad\qquad\qquad\qquad\qquad\qquad\qquad$ □

2.4 Dynamic Programming and HJB Equation

In this section, we consider the case that $M = \mathbb{R}^n$. Thus,

$$\widetilde{\mathcal{U}}_x^M[t,T] = \widetilde{\mathcal{U}}_x^{\mathbb{R}^n}[t,T] = \mathcal{U}[t,T], \qquad \forall (t,x) \in [0,T] \times \mathbb{R}^n.$$

Now, we introduce the following assumption.

(C5) In addition to (C1)–(C2), we assume that functions $f(\cdot,\cdot,\cdot)$ and $g(\cdot,\cdot,\cdot)$ are continuous.

Note that in (C5), no differentiability condition is assumed for the functions $f(\cdot,\cdot,\cdot)$ and $g(\cdot,\cdot,\cdot)$. Recall that $V : [0,T] \times \mathbb{R}^n \to \mathbb{R}$ defined by (2.4) is the *value function* of Problem (OC)^T. The following result is concerned with the function $V(\cdot,\cdot)$.

Proposition 2.4.1. *Let* (C5) *hold. Then the value function* $V(\cdot,\cdot)$ *is continuous on* $[0,T] \times \mathbb{R}^n$.

Proof. For any $t \in [0,T]$, $x_1, x_2 \in \mathbb{R}^n$, and $u(\cdot) \in \mathcal{U}[t,T]$, let $X_i(\cdot) = X(\cdot\,; t, x_i, u(\cdot))$. Then

$$|J^T(t, x_1; u(\cdot)) - J^T(t, x_2; u(\cdot))|$$

$$\leqslant \int_t^T |g(s, X_1(s), u(s)) - g(s, X_2(s), u(s))| ds + |h(X_1(T)) - h(X_2(T))|$$

$$\leqslant \int_t^T \omega\big(|X_1(s)| \vee |X_2(s)|, |X_1(s) - X_2(s)|\big) ds$$

$$\qquad + \omega\big(|X_1(T)| \vee |X_2(T)|, |X_1(T) - X_2(T)|\big)$$

$$\leqslant (T+1)\omega\big(e^{LT}(1 + |x_1| \vee |x_2|), e^{LT}|x_1 - x_2|\big).$$

Hence, for some local modulus of continuity $\bar{\omega}(\cdot,\cdot)$,

$$|V(t, x_1) - V(t, x_2)| \leqslant \bar{\omega}\big(|x_1| \vee |x_2|, |x_1 - x_2|\big),$$
$$\forall t \in [0,T], x_1, x_2 \in \mathbb{R}^n.$$

Next, for any $0 \leqslant t < \tau \leqslant T$, $x \in \mathbb{R}^n$ and $u(\cdot) \in \mathcal{U}[t,T]$, we have

$$J^T(t, x; u(\cdot)) = \int_t^\tau g(s, X(s), u(s)) ds + J\big(\tau, X(\tau); u(\cdot)|_{[\tau,T]}\big)$$

$$\geqslant -\int_t^\tau \Big[g_0 + \omega\big(|X(s)|, |X(s)|\big)\Big] ds + V(\tau, X(\tau))$$

$$\geqslant -(\tau - t)\Big[g_0 + \omega\big(e^{LT}(1 + |x|), e^{LT}(1 + |x|)\big)\Big]$$

$$\qquad + V(\tau, x) + \omega\big(|x| \vee |X(\tau)|, |X(\tau) - x|\big)$$

$$\geqslant V(\tau, x) - (\tau - t)\Big[g_0 + \omega\big(e^{LT}(1 + |x|), e^{LT}(1 + |x|)\big)\Big]$$

$$\qquad - \omega\big(e^{LT}(1 + |x|), (e^{L(\tau - t)} - 1)(1 + |x|)\big).$$

Hence,

$$V(t,x) - V(\tau,x) \geqslant -(\tau - t)\Big[g_0 + \omega\big(e^{LT}(1+|x|), e^{LT}(1+|x|)\big)\Big]$$
$$-\omega\big(e^{LT}(1+|x|), L(\tau - t)e^{L(\tau - t)}(1+|x|)\big).$$

On the other hand,

$$V(t,x) \leqslant J^T(t,x;u(\cdot)) = \int_t^\tau g(s, X(s), u(s))ds + J^T\big(\tau, X(\tau); u(\cdot)\big|_{[t,\tau]}(\cdot)\big)$$
$$\leqslant \int_t^\tau \Big[g_0 + \omega\big(|X(s)|, |X(s)|\big)\Big]ds + J\big(\tau, x; u(\cdot)\big|_{[t,\tau]}\big)$$
$$+(T+1)\omega\big(e^{LT}(1 + |x| \vee |X(\tau)|), e^{LT}|X(\tau) - x|\big)$$
$$\leqslant (\tau - t)\Big[g_0 + \omega\big(e^{LT}(1+|x|), e^{LT}(1+|x|)\big)\Big] + J\big(\tau, x; u(\cdot)\big|_{[\tau,T]}\big)$$
$$+(T+1)\omega\big(e^{LT}(1 + e^{LT}(1+|x|)), e^{LT}(e^{L(\tau - t)} - 1)(1+|x|)\big).$$

Consequently,

$$V(t,x) - V(\tau,x) \leqslant \bar\omega(|x|, \tau - t), \qquad \forall x \in \mathbb{R}^n, 0 \leqslant t < \tau \leqslant T.$$

This proves our result. $\qquad\qquad\square$

The following result gives a dynamic property of the value function.

Theorem 2.4.2. (Bellman Optimality Principle) *Let* (C5) *hold. Then for any* $(t,x) \in [0,T) \times \mathbb{R}^n$, *and* $\tau \in (t,T)$, *the following holds:*

$$V(t,x) = \inf_{u(\cdot) \in \mathcal{U}[t,\tau]} \Big\{ \int_t^\tau g(s, X(s), u(s))ds + V(\tau, X(\tau)) \Big\}. \qquad (2.15)$$

Proof. Let us denote the right-hand side by $\widehat{V}(t,x)$. For any $u(\cdot) \in \mathcal{U}[t,T]$, by the definition of $V(\cdot,\cdot)$,

$$V(t,x) \leqslant J^T(t,x;u(\cdot)) = \int_t^\tau g(s, X(s), u(s))ds + J^T\big(\tau, X(\tau); u(\cdot)\big|_{[\tau,T]}\big).$$

Taking infimum over $u(\cdot)\big|_{[\tau,T]} \in \mathcal{U}[\tau,T]$, we have

$$V(t,x) \leqslant \int_t^\tau g(s, X(s), u(s))ds + V(\tau, X(\tau)).$$

This leads to

$$V(t,x) \leqslant \widehat{V}(t,x).$$

On the other hand, for any $\varepsilon > 0$, there exists a $u^\varepsilon(\cdot) \in \mathcal{U}[t, T]$ such that (with $X^\varepsilon(\cdot) = X(\cdot\,; t, x, u^\varepsilon(\cdot))$)

$$V(t, x) + \varepsilon > J^T(t, x; u^\varepsilon(\cdot))$$

$$= \int_t^\tau g(s, X^\varepsilon(s), u^\varepsilon(s))ds + J^T(\tau, X^\varepsilon(\tau); u^\varepsilon(\cdot)|_{[\tau,T]})$$

$$\geqslant \int_t^\tau g(s, X^\varepsilon(s), u^\varepsilon(s))ds + V(\tau, X^\varepsilon(\tau)) \geqslant \widehat{V}(t, x).$$

Since $\varepsilon > 0$ is arbitrary, we obtain our conclusion. $\qquad\qquad\square$

Let us make an observation on the above principle. Suppose $(\bar{X}(\cdot), \bar{u}(\cdot))$ is an optimal pair for the initial pair (t, x). Then by (2.15), we have

$$V(t, x) = J^T(t, x; \bar{u}(\cdot))$$

$$= \int_t^\tau g(s, \bar{X}(s), \bar{u}(s))ds + \int_\tau^T g(s, \bar{X}(s), \bar{u}(s))ds + h(\bar{X}(T))$$

$$= \int_t^\tau g(s, \bar{X}(s), \bar{u}(s))ds + J^T\left(\tau, \bar{X}(\tau); \bar{u}(\cdot)|_{[\tau,T]}\right)$$

$$\geqslant \int_t^\tau g(s, \bar{X}(s), \bar{u}(s))ds + V(\tau, \bar{X}(\tau))$$

$$\geqslant \inf_{u(\cdot)\in[t,\tau]}\left\{\int_t^\tau g(s, X(s), u(s))ds + V(\tau, X(\tau))\right\} = V(t, x).$$

Hence, all the equalities have to be true. This leads to the following:

$$J^T\left(\tau, \bar{X}(\tau); \bar{u}(\cdot)|_{[t,\tau]}\right) = V(\tau, \bar{X}(\tau)) = \inf_{u(\cdot)\in\mathcal{U}[\tau,T]} J^T(\tau, \bar{X}(\tau); u(\cdot)),$$

which means

The restriction $\bar{u}(\cdot)|_{[\tau,T]}$ of the optimal control $\bar{u}(\cdot)$

for the initial pair (t, x) on a later interval $[\tau, T]$

is optimal for the resulting initial pair $(\tau, \bar{X}(\tau))$.

Such a property is called the *time-consistency* of Problem $(OC)^T$. This also tells us that "*Global optimality implies local optimality*".

Next, we note that (2.15) is a functional equation for the value function $V(\cdot, \cdot)$ which is difficult to solve. The following gives a localization of the above.

Proposition 2.4.3. (HJB Equation) *Let* (C5) *hold. Suppose the value function $V(\cdot, \cdot)$ is C^1. Then it is a solution to the following Hamilton–Jacobi–Bellman (HJB, for short) equation:*

$$\begin{cases} V_t(t, x) + H(t, x, V_x(t, x)) = 0, & (t, x) \in [0, T] \times \mathbb{R}^n, \\ V(T, x) = h(x), \end{cases} \qquad (2.16)$$

where

$$H(t,x,p) = \inf_{u \in U} \Big[\langle p, f(t,x,u) \rangle + g(t,x,u) \Big],$$

$$(t,x,p) \in [0,T] \times \mathbb{R}^n \times \mathbb{R}^n. \tag{2.17}$$

Rigorously speaking (similar to the previous section), the gradient $V_x(t,x)$ of a scalar function $V(t,x)$ is a row vector. Therefore, according to (2.17), instead of $H(t,x,V_x(t,x))$ in (2.16), we should have $H(t,x,V_x(t,x)^T)$ which looks not as good. On the other hand, we may change $\langle p, f(t,x,u) \rangle$ to $pf(t,x,u)$ with $p \in \mathbb{R}^{1 \times n}$ in (2.17). But this will lead to many other bad-looking notations later on. Hence, as a trade-off, we identify $V_x(t,x)$ with $V_x(t,x)^T$ hereafter.

Proof. From Theorem 2.4.2, we know that for any $(t,x) \in [0,T) \times \mathbb{R}^n$ and $\delta > 0$ with $t + \delta \in [0,T]$,

$$0 = \inf_{u(\cdot) \in \mathcal{U}[t,t+\delta]} \Big\{ \int_t^{t+\delta} g(s,X(s),u(s))ds + V\big(t+\delta, X(t+\delta)\big) - V(t,x) \Big\}.$$

Thus, for any fixed $u \in U$, one has

$$0 \leqslant \int_t^{t+\delta} g(s,X(s),u)ds + V(t+\delta, X(t+\delta)) - V(t,x)$$

$$= \int_t^{t+\delta} \Big[g(s,X(s),u) + V_t(s,X(s)) + \langle V_x(s,X(s)), f(s,X(s),u) \rangle \Big] ds$$

$$\leqslant g(t,x,u) + V_t(t,x) + \langle V_x(t,x), f(t,x,u) \rangle + R(\delta),$$

where (noting that all the involved functions are continuous)

$$R(\delta) = \frac{1}{\delta} \int_t^{t+\delta} \omega\big(|X(s)| \vee |x|, |X(s) - x|\big)ds$$

$$\leqslant \frac{1}{\delta} \int_t^{t+\delta} \omega\big(e^{LT}(1+|x|), (e^{L\delta}-1)(1+|x|)\big)ds = o(1),$$

with $\omega(\cdot, \cdot)$ being some local modulus of continuity. Hence, letting $\delta > 0$, we obtain

$$0 \leqslant V_t(t,x) + \langle V_x(t,x), f(t,x,u) \rangle + g(t,x,u), \qquad \forall u \in U.$$

Then

$$0 \leqslant V_t + H(t,x,V_x(t,x)).$$

On the other hand, for any $\varepsilon > 0$, there exists a $u^{\delta,\varepsilon}(\cdot) \in \mathcal{U}[t, t+\delta]$ such that (with $(X^{\delta,\varepsilon}(\cdot) = X(\cdot\,; t, x, u^{\delta,\varepsilon}(\cdot)))$

$$\varepsilon > \frac{1}{\delta}\Big\{ \int_t^{t+\delta} g(s, X^{\delta,\varepsilon}(s), u^{\delta,\varepsilon}(s))ds + V(t+\delta, X^{\delta,\varepsilon}(t+\delta)) - V(t,x) \Big\}$$

$$= \frac{1}{\delta}\int_t^{t+\delta} \Big[g(s, X^{\delta,\varepsilon}(s), u^{\delta,\varepsilon}(s)) + V_t(s, X^{\delta,\varepsilon}(s))$$

$$+ \langle V_x(s, X^{\delta,\varepsilon}(s)), f(s, X^{\delta,\varepsilon}(s), u^{\delta,\varepsilon}(s)) \rangle \Big]ds$$

$$\geqslant V_t(t,x) + H(t, x, V_x(t,x)) + R(\delta, \varepsilon),$$

where

$$R(\delta, \varepsilon) = \frac{1}{\delta}\int_t^{t+\delta} \Big[g(s, X^{\delta,\varepsilon}(s), u^{\delta,\varepsilon}(s)) - g(t, x, u^{\delta,\varepsilon}(s)) + V_t(s, X^{\delta,\varepsilon}(s))$$

$$- V_t(t,x) + \langle V_x(s, X^{\delta,\varepsilon}(s)), f(s, X^{\delta,\varepsilon}(s), u^{\varepsilon,\delta}(s)) \rangle$$

$$- \langle V_x(t, x, u^{\delta,\varepsilon}(s)), f(t, x, u^{\delta,\varepsilon}(s) \rangle \Big]ds$$

$$\geqslant -\frac{1}{\delta}\int_t^{t+\delta} \omega\big(|X^{\delta,\varepsilon}(s)| \vee |x|, |X^{\delta,\varepsilon}(s) - x|\big)ds$$

$$\geqslant -\frac{1}{\delta}\int_t^{t+\delta} \omega\big(e^{LT}(1+|x|), (e^{L\delta} - 1)(1+|x|)\big)ds = o(1),$$

for some local modulus of continuity $\omega(\cdot, \cdot)$. Hence, letting $\delta \to 0$, we obtain

$$\varepsilon \geqslant V_t(t,x) + H(t, x, V_x(t,x)).$$

Then sending $\varepsilon \to 0$, one has

$$0 \geqslant V_t(t,x) + H(t, x, V_x(t,x)).$$

Our conclusion follows. $\square$

Further, we have the following result.

Proposition 2.4.4. (Verification Theorem). *Suppose* $V(\cdot, \cdot)$ *is continuously differentiable satisfying* (2.16). *Let* $u(t, x)$ *be a map satisfying*

$$\langle V_x(t,x), f(t, x, u(t, x)) \rangle + g(t, x, u(t, x))$$

$$= \inf_{u \in U} \Big[\langle V_x(t,x), f(t, x, u) \rangle + g(t, x, u) \Big], \qquad (t, x) \in [0, T] \times \mathbb{R}^n.$$

Moreover, let

$$\begin{cases} \dot{\bar{X}}(s) = f(s, \bar{X}(s), u(s, \bar{X}(s))), & s \in [t, T], \\ \bar{X}(t) = x, \end{cases} \qquad (2.18)$$

admit a solution $\bar{X}(\cdot)$ and let

$$\bar{u}(s) = u(s, \bar{X}(s)), \qquad s \in [t, T].$$

Then $(\bar{X}(\cdot), \bar{u}(\cdot))$ is an optimal pair of Problem $(OC)^T$ for the initial pair (t, x).

Proof. By the definition of $u(t, x)$ and $\bar{u}(\cdot)$, we have that along $(\bar{X}(\cdot), \bar{u}(\cdot))$, the following holds:

$$
\begin{aligned}
\langle V_x(s, \bar{X}(s)), f(s, \bar{X}(s), \bar{u}(s)) \rangle &+ g(s, \bar{X}(s), \bar{u}(s)) \rangle \\
&= H(s, \bar{X}(s), V_x(s, \bar{X}(s))) \\
&\leqslant \langle V_x(s, \bar{X}(s)), f(s, \bar{X}(s), u) \rangle + g(s, \bar{X}(s), u), \qquad \forall u \in U.
\end{aligned}
$$

We observe the following:

$$
\begin{aligned}
h(\bar{X}(T)) - V(t, x) &= V(T, \bar{X}(T)) - V(t, x) \\
&= \int_t^T \left[V_t(s, \bar{X}(s)) + \langle V_x(s, \bar{X}(s)), f(s, \bar{X}(s), \bar{u}(s)) \rangle \right] ds \\
&= -\int_t^T g(s, \bar{X}(s), \bar{u}(s)) ds.
\end{aligned}
$$

Hence,

$$V(t, x) = \int_t^T g(s, \bar{X}(s), \bar{u}(s)) ds + h(\bar{X}(T)) = J^T(t, x; \bar{u}(\cdot)),$$

which proves the optimality of $\bar{u}(\cdot)$. $\qquad\qquad\qquad\qquad\qquad$ □

From the above verification theorem, we see that roughly speaking, if we can solve HJB equation, then we can construct an optimal control in principle. However, we point out that to make this rigorous, there are some technical difficulties: The value function $V(\cdot, \cdot)$ is not necessary differentiable; even if $V(\cdot, \cdot)$ is differentiable, the map $u(t, x)$ might not be good enough to guarantee the equation (2.18) to have a unique solution. Before going further, let us look at an example.

Example 2.4.5. Consider the following control system:

$$
\begin{cases}
\dot{X}(s) = u(s), \qquad s \in [t, T], \\
X(t) = x,
\end{cases}
$$

with cost functional

$$J^T(t, x; u(\cdot)) = -|X(T)|.$$

The control domain is taken to be $U = [-1, 1]$. Since

$$X(s; t, u(\cdot)) = x + \int_t^s u(r)dr, \qquad s \in [t, T],$$

we have

$$J^T(t, x; u(\cdot)) = \left| x + \int_t^T u(r)dr \right|.$$

Thus,

$$V(t, x) = \inf_{u(\cdot) \in \mathcal{U}[t,T]} J^T(t, x; u(\cdot)) = \inf_{|y| \leqslant T-t} |x + y| = \left(|x| - T + t \right)^+,$$

where $a^+ = \max\{a, 0\}$. The optimal control is given by

$$\bar{u}(s) = -(\operatorname{sgn} x) I_{[t, (t+|x|) \wedge T]}(s), \qquad s \in [t, T],$$

where $a \wedge b = \min\{a, b\}$. We see that the value function is continuous, but not differentiable along the line $|x| = T - t$. In the current case,

$$H(t, x, p) = \inf_{u \in [-1,1]} pu = -|p|.$$

Hence, the HJB equation reads

$$\begin{cases} V_t(t, x) - |V_x(t, x)| = 0, & (t, x) \in [0, T] \times \mathbb{R}, \\ V(T, x) = |x|, & x \in \mathbb{R}. \end{cases}$$

2.5 Viscosity Solutions

Consider the following equation

$$\begin{cases} V_t(t, x) + H(t, x, V_x(t, x)) = 0, & (t, x) \in [0, T] \times \mathbb{R}^n, \\ V(T, x) = h(x), \end{cases} \tag{2.19}$$

with a general continuous map $H : [0, T] \times \mathbb{R}^n \times \mathbb{R}^n \to \mathbb{R}$, not necessarily given by (2.17). Also, $h : \mathbb{R}^n \to \mathbb{R}$ is assumed to be continuous. Such an equation is called a Hamilton-Jacoobi (HJ, for short) equation. We introduce the following notion.

Definition 2.5.1. (i) A continuous function $V(\cdot, \cdot)$ is called a *viscosity sub-solution* of HJ equation (2.19) if

$$V(T, x) \leqslant h(x), \qquad \forall x \in \mathbb{R}^n,$$

and for any continuously differentiable function $\varphi : [0, T] \times \mathbb{R}^n \to \mathbb{R}$, if $V(\cdot, \cdot) - \varphi(\cdot, \cdot)$ attains a local maximum at $(t_0, x_0) \in [0, T) \times \mathbb{R}^n$, the following inequality holds:

$$\varphi_t(t_0, x_0) + H(t_0, x_0, \varphi_x(t_0, x_0)) \geqslant 0.$$

(ii) A continuous function $V(\cdot,\cdot)$ is called a *viscosity super-solution* of HJ equation (2.19) if

$$V(T,x) \geqslant h(x), \qquad \forall x \in \mathbb{R}^n,$$

and for any continuously differentiable function $\varphi : [0,T] \times \mathbb{R}^n \to \mathbb{R}$, if $V(\cdot,\cdot) - \varphi(\cdot,\cdot)$ attains a local minimum at $(t_0,x_0) \in [0,T) \times \mathbb{R}^n$, the following inequality holds:

$$\varphi_t(t_0,x_0) + H(t_0,x_0,\varphi_x(t_0,x_0)) \leqslant 0.$$

(iii) A continuous function $V(\cdot,\cdot)$ is called a *viscosity solution* of HJ equation (2.19) if it is both viscosity sub-solution and viscosity super-solution of (2.19).

The following result shows that the notion of viscosity solution is an extension of classical solution to HJ equation (2.19).

Proposition 2.5.2. *Suppose $V(\cdot,\cdot)$ is a classical solution to HJ equation (2.19). Then it is a viscosity solution to the equation. Conversely, if $V(\cdot,\cdot)$ is a viscosity solution to HJ equation (2.19) and it is C^1, then it is a classical solution to the equation.*

Proof. First, let $V(\cdot,\cdot)$ be a classical solution to HJ equation (2.19). Then for any C^1 function $\varphi(\cdot,\cdot)$, if $V(\cdot,\cdot) - \varphi(\cdot,\cdot)$ attains a local maximum at $(t_0,x_0) \in (0,T) \times \mathbb{R}^n$, then

$$\varphi_t(t_0,x_0) = V_t(t_0,x_0), \qquad \varphi_x(t_0,x_0) = V_x(t_0,x_0).$$

Thus, we have

$$\varphi_t(t_0,x_0) + H(t_0,x_0,\varphi_x(t_0,x_0)) = V_t(t_0,x_0) + H(t_0,x_0,V_x(t_0,x_0)) = 0.$$

This means $V(\cdot,\cdot)$ is a viscosity sub-solution of HJ equation (2.19). Similarly, we can show that $V(\cdot,\cdot)$ is a viscosity super-solution to HJ equation (2.19).

Conversely, let $V(\cdot,\cdot)$ be a viscosity solution to HJ equation (2.19), which is C^1. Then by taking $\varphi(\cdot,\cdot) = V(\cdot,\cdot)$, we see that $V(\cdot,\cdot) - \varphi(\cdot,\cdot)$ attains a local maximum and minimum at every point $(t,x) \in [0,T] \times \mathbb{R}^n$. Thus, by the definition of viscosity solution, we see that $V(\cdot,\cdot)$ is a classical solution of HJ equation (2.19). $\qquad\square$

To study the uniqueness of viscosity solutions to HJ equation (2.19), we introduce the following hypothesis.

(H) The map $H : [0, T] \times \mathbb{R}^n \times \mathbb{R}^n \to \mathbb{R}$ is continuous. There exist a constant $L > 0$ and a local modulus of continuity $\omega : \mathbb{R}_+ \times \mathbb{R}_+ \to \mathbb{R}_+$ such that

$$\begin{cases} |H(t,x,p) - H(t,x,q)| \leqslant L(1 + |x|)|p - q|, \\ \qquad\qquad \forall t \in [0,T], \ x,p,q \in \mathbb{R}^n, \\ |H(t,x,p) - H(t,y,p)| \leqslant \omega\big(|x| \vee |y|, |x - y|(1 + |p|)\big), \\ \qquad\qquad \forall t \in [0,T], \ x,y,p \in \mathbb{R}^n. \end{cases} \tag{2.20}$$

We now present the following uniqueness theorem.

Theorem 2.5.3. *Suppose $H : [0, T] \times \mathbb{R}^n \times \mathbb{R}^n \to \mathbb{R}$ and $h : \mathbb{R}^n \to \mathbb{R}$ are continuous, and (H) holds. Then HJ equation (2.19) admits at most one viscosity solution.*

Proof. Suppose $V(\cdot, \cdot)$ and $\widehat{V}(\cdot, \cdot)$ are two viscosity solutions of (2.19). We want to show that

$$V(t,x) \leqslant \widehat{V}(t,x), \qquad \forall (t,x) \in [0,T] \times \mathbb{R}^n. \tag{2.21}$$

Since the positions of $V(\cdot, \cdot)$ and $\widehat{V}(\cdot, \cdot)$ are the same, by the symmetry, we can get the other direction of inequality and then the uniqueness follows.

To prove (2.21), we show that for any $x_0 \in \mathbb{R}^n$, and $T_0 = (T - \frac{1}{2L})^+$,

$$\sup_{(t,x) \in \Delta(x_0)} \big[V(t,x) - \widehat{V}(t,x) \big] \leqslant 0, \tag{2.22}$$

where

$$\Delta(x_0) = \Big\{ (t,x) \in [T_0, T] \times \mathbb{R}^n \ \big| \ |x - x_0| < L_0(t - T_0) \Big\},$$

with

$$L_0 = 2L(1 + |x_0|).$$

Once (2.22) is proved for any $x_0 \in \mathbb{R}^n$, we obtain (2.21) on

$$[T_0, T] \times \mathbb{R}^n = \bigcup_{x_0 \in \mathbb{R}^n} \overline{\Delta(x_0)}.$$

Then, replacing T by T_0 and continue the procedure. Repeating the procedure at most $[2LT] + 1$ times, (2.21) will be proved.

Now, we prove (2.22) by contradiction. Suppose

$$\sup_{(t,x) \in \Delta(x_0)} \big[V(t,x) - \widehat{V}(t,x) \big] = \bar{\sigma} > 0. \tag{2.23}$$

Note that

$$2L(t - T_0) \leqslant 2L\left[T - (T - \frac{1}{2L})^+\right] = 2L(T \wedge \frac{1}{2L}) \leqslant 1.$$

Hence, by (H), we see that for any $(t, x) \in \Delta(x_0)$ and $p, q \in \mathbb{R}^n$,

$$
\begin{aligned}
|H(t, x, p) - H(t, x, q)| &\leqslant L(1 + |x|)|p - q| \\
&\leqslant L(1 + |x_0| + |x - x_0|)|p - q| \\
&< L\left[1 + |x_0| + L_0(T - T_0)\right]|p - q| \qquad (2.24) \\
&= L\left[1 + |x_0| + 2L(1 + |x_0|)(T - T_0)\right]|p - q| \\
&\leqslant 2L(1 + |x_0|)|p - q| \equiv L_0|p - q|.
\end{aligned}
$$

Take small $\varepsilon, \delta > 0$ satisfying

$$\varepsilon + 2\delta < L_0(T - T_0).$$

Define

$$\Delta_{\varepsilon,\delta}(x_0) = \left\{(t, x) \in \Delta(x_0) \mid \langle x \rangle_\varepsilon < L_0(t - T_0) - \delta\right\},$$

where $\langle x \rangle_\varepsilon = \sqrt{|x - x_0|^2 + \varepsilon^2}$. Then, for $\varepsilon, \delta > 0$ small enough, $\Delta_{\varepsilon,\delta}(x_0)$ is nonempty since

$$\bigcup_{\varepsilon,\delta > 0} \Delta_{\varepsilon,\delta}(x_0) = \Delta(x_0) \neq \emptyset,$$

and $\Delta_{\varepsilon,\delta}(x_0)$ increases as ε or δ decreases. Thus, making use of (2.23), by shrinking ε, δ, we may assume that

$$\sup_{(t,x) \in \Delta_{\varepsilon,2\delta}(x_0)} \left[V(t, x) - \widehat{V}(t, x)\right] \geqslant \frac{\bar{\sigma}}{2} > 0.$$

Next, we take $K > 0$ to be large enough so that

$$K > \sup_{(t,x,s,y) \in \Delta(x_0)^2} \left[V(t, x) - \widehat{V}(s, y)\right],$$

and introduce $\zeta_\delta(\cdot) \in C^\infty(\mathbb{R})$ satisfying

$$\zeta_\delta(r) = \begin{cases} 0, & r \leqslant -2\delta, \\ -K, & r \geqslant -\delta, \end{cases} \qquad \zeta_\delta'(r) \leqslant 0, \qquad \forall r \in \mathbb{R}. \qquad (2.25)$$

For any $\alpha, \beta, \sigma > 0$, we define

$$
\begin{aligned}
\Phi(t, x, s, y) = {}&V(t, x) - \widehat{V}(s, y) - \frac{1}{\alpha}|x - y|^2 - \frac{1}{\beta}|t - s|^2 \\
&+ \zeta_\delta(\langle x \rangle_\varepsilon - L_0(t - T_0)) + \zeta_\delta(\langle y \rangle_\varepsilon - L_0(s - T_0)) \\
&+ \sigma(t + s) - 2\sigma T, \qquad (t, x, s, y) \in \overline{\Delta(x_0)^2}.
\end{aligned}
$$

Let $(\bar{t}, \bar{x}, \bar{s}, \bar{y}) \in \overline{\Delta_{\varepsilon,\delta}(x_0)^2}$ be a maximum of $\Phi(\cdot, \cdot, \cdot, \cdot)$ over $\overline{\Delta_{\varepsilon,\delta}(x_0)^2}$. Note that $(\bar{t}, \bar{x}, \bar{s}, \bar{y})$ depends on $\alpha, \beta, \sigma, \varepsilon, \delta > 0$.

From

$$\Phi(\bar{t}, \bar{x}, \bar{t}, \bar{x}) + \Phi(\bar{s}, \bar{y}, \bar{s}, \bar{y}) \leqslant 2\Phi(\bar{t}, \bar{x}, \bar{s}, \bar{y}),$$

one has

$$V(\bar{t}, \bar{x}) - \widehat{V}(\bar{t}, \bar{x}) + 2\zeta_\delta\big(\langle \bar{x} \rangle_\varepsilon - L_0(\bar{t} - T_0)\big) + 2\sigma\bar{t} - 2\sigma T$$
$$+ V(\bar{s}, \bar{y}) - \widehat{V}(\bar{s}, \bar{y}) + 2\zeta_\delta\big(\langle \bar{y} \rangle_\varepsilon - L_0(\bar{s} - T_0)\big) + 2\sigma\bar{s} - 2\sigma T$$
$$\leqslant 2V(\bar{t}, \bar{x}) - \widehat{2V}(\bar{s}, \bar{y}) - \frac{2}{\alpha}|\bar{x} - \bar{y}|^2 - \frac{2}{\beta}|\bar{t} - \bar{s}|^2$$
$$+ 2\zeta\big(\langle \bar{x} \rangle_\varepsilon - L_0(\bar{t} - T_0)\big) + 2\zeta_\delta\big(\langle \bar{y} \rangle_\varepsilon - L_0(\bar{s} - T_0)\big) + 2\sigma(\bar{t} + \bar{s}) - 4\sigma T,$$

which results in

$$\frac{2}{\alpha}|\bar{x} - \bar{y}|^2 + \frac{2}{\beta}|\bar{t} - \bar{s}|^2 \leqslant V(\bar{t}, \bar{x}) - V(\bar{s}, \bar{y}) + \widehat{V}(\bar{t}, \bar{x}) - \widehat{V}(\bar{s}, \bar{y})$$
$$\leqslant 2\omega_0\big(|\bar{x} - \bar{y}| + |\bar{t} - \bar{s}|\big), \tag{2.26}$$

where

$$\omega_0(r) = \frac{1}{2} \sup_{\substack{|t-s|+|x-y| \leqslant r \\ (t,x,s,y) \in \Delta(x_0)^2}} \big(|V(t,x) - V(s,y)| + |\widehat{V}(t,x) - \widehat{V}(s,y)|\big).$$

Clearly,

$$\lim_{r \to 0} \omega_0(r) = 0, \qquad \bar{\omega}_0 \equiv \sup_{r \geqslant 0} \omega_0(r) < \infty.$$

Then (2.26) implies

$$|\bar{x} - \bar{y}| \leqslant \sqrt{\alpha\bar{\omega}_0}, \qquad |\bar{t} - \bar{s}| \leqslant \sqrt{\beta\bar{\omega}_0}, \tag{2.27}$$

and thus,

$$\frac{1}{\alpha}|\bar{x} - \bar{y}|^2 + \frac{1}{\beta}|\bar{t} - \bar{s}|^2 \leqslant \omega_0\big(\sqrt{\alpha\bar{\omega}_0} + \sqrt{\beta\bar{\omega}_0}\big). \tag{2.28}$$

Next, we claim that

$$\langle \bar{x} \rangle_\varepsilon < L_0(\bar{t} - T_0) - \delta, \qquad \langle \bar{y} \rangle_\varepsilon < L_0(\bar{s} - T_0) - \delta. \tag{2.29}$$

If not, then

$$\zeta_\delta\big(\langle \bar{x} \rangle_\varepsilon - L_0(\bar{t} - T_0)\big) + \zeta_\delta\big(\langle \bar{y} \rangle_\varepsilon - L_0(\bar{s} - T_0)\big) \leqslant -K.$$

Hence, from

$$
\begin{aligned}
0 &= V(T, x_0) - \widehat{V}(T, x_0) + 2\zeta_\delta\big(\varepsilon - L_0(T - T_0)\big) \\
&= \Phi(T, x_0, T, x_0) \leqslant \Phi(\bar{t}, \bar{x}, \bar{s}, \bar{y}) \\
&\leqslant V(\bar{t}, \bar{x}) - \widehat{V}(\bar{s}, \bar{y}) - \frac{1}{\alpha}|\bar{x} - \bar{y}|^2 - \frac{1}{\beta}|\bar{t} - \bar{s}|^2 \\
&\quad + \zeta_\delta\big(\langle \bar{x} \rangle_\varepsilon - L_0(\bar{t} - T_0)\big) + \zeta_\delta\big(\langle \bar{y} \rangle_\varepsilon - L_0(\bar{s} - T_0)\big) + \sigma(\bar{t} + \bar{s}) - 2\sigma T \\
&< K - K + \sigma(\bar{t} + \bar{s}) - 2\sigma T \leqslant 0,
\end{aligned}
$$

a contradiction. Thus, (2.29) holds. Consequently, noting (2.25), for sufficiently small $\sigma > 0$, one has

$$
\begin{aligned}
V(\bar{t}, \bar{x}) - \widehat{V}(\bar{s}, \bar{y}) &= \Phi(\bar{t}, \bar{x}, \bar{s}, \bar{y}) + \frac{1}{\alpha}|\bar{x} - \bar{y}|^2 + \frac{1}{\beta}|\bar{t} - \bar{s}|^2 \\
&\quad - \zeta_\delta\big(\langle \bar{x} \rangle_\varepsilon - L_0(\bar{t} - T_0)\big) - \zeta_\delta\big(\langle \bar{y} \rangle_\varepsilon - L_0(\bar{s} - T_0)\big) \\
&\geqslant \sup_{(t,x)\in\Delta_{\varepsilon,2\delta}(x_0)} \Phi(t, x, t, x) \\
&= \sup_{(t,x)\in\Delta_{\varepsilon,2\delta}(x_0)} \big[V(t, x) - \widehat{V}(t, x) + 2\sigma(t - T)\big] \geqslant \frac{\bar{\sigma}}{4} > 0.
\end{aligned}
\tag{2.30}
$$

In the above, we have used the fact that

$$
\zeta_\delta\big(\langle x \rangle_\varepsilon - L_0(t - T_0)\big) = 0, \qquad \forall(t, x) \in \Delta_{\varepsilon,2\delta}(x_0).
$$

We now further claim that

$$
\bar{t}, \bar{s} < T.
\tag{2.31}
$$

In fact, if this is not the case, we can find $(\alpha_m, \beta_m) \to (0, 0)$ such that the corresponding maximum $(\bar{t}_m, \bar{x}_m, \bar{s}_m, \bar{y}_m) \in \overline{\Delta_{\varepsilon,\delta}(x_0)^2}$ of $\Phi(\cdot, \cdot, \cdot, \cdot)$ satisfies

$$
\bar{t}_m = T, \quad \text{or} \quad \bar{s}_m = T, \qquad m \geqslant 1.
$$

By (2.27), we have

$$
|\bar{x}_m - \bar{y}_m| \to 0, \qquad \bar{t}_m, \bar{s}_m \to T, \qquad (m \to \infty).
$$

By the compactness of $\overline{\Delta_{\varepsilon,\delta}(x_0)^2}$, we may assume that

$$
\bar{x}_m, \bar{y}_m \to \bar{x}.
$$

Consequently, by (2.30), we get

$$
\frac{\bar{\sigma}}{4} \leqslant \lim_{m\to\infty} \big[V(\bar{t}_m, \bar{x}_m) - \widehat{V}(\bar{s}_m, \bar{y}_m)\big] = V(T, \bar{x}) - \widehat{V}(T, \bar{x}) = 0,
$$

a contradiction. Hence, claim (2.31) holds. Combining (2.29) with (2.31), we obtain that

$$(\bar{t}, \bar{x}, \bar{s}, \bar{y}) \in \Delta_{\varepsilon,\delta}(x_0)^2.$$

That is, $(\bar{t}, \bar{x}, \bar{s}, \bar{y})$ is in $\Delta_{\varepsilon,\delta}(x_0)^2$, the interior of $\overline{\Delta_{\varepsilon,\delta}(x_0)^2}$. Hence, the map

$$(t, x) \mapsto V(t, x) - \left\{ \widehat{V}(\bar{s}, \bar{y}) + \frac{1}{\alpha}|x - \bar{y}|^2 + \frac{1}{\beta}|t - \bar{s}|^2 - \zeta_\delta(\langle x \rangle_\varepsilon - L_0(t - T_0)) \right.$$

$$\left. - \zeta_\delta(\langle \bar{y} \rangle_\varepsilon - L_0(\bar{s} - T_0)) - \sigma(t + \bar{s}) + 2\sigma T \right\}$$

attains a local maximum at $(\bar{t}, \bar{x}) \in \Delta_{\varepsilon,\delta}(x_0)$. It follows from Definition 2.5.1 that

$$\frac{2}{\beta}(\bar{t} - \bar{s}) + \zeta_\delta'(X_\varepsilon))L_0 - \sigma$$
$$+ H\left(\bar{t}, \bar{x}, \frac{2}{\alpha}(\bar{x} - \bar{y}) - \zeta_\delta'(X_\varepsilon)\frac{\bar{x} - x_0}{\langle \bar{x} \rangle_\varepsilon}\right) \geqslant 0, \tag{2.32}$$

with

$$X_\varepsilon = \langle \bar{x} \rangle_\varepsilon - L_0(\bar{t} - T_0).$$

Likewise, the map

$$(s, y) \mapsto \widehat{V}(s, y) - \left\{ V(\bar{t}, \bar{x}) - \frac{1}{\alpha}|y - \bar{x}|^2 - \frac{1}{\beta}|s - \bar{t}|^2 + \zeta_\delta(\langle \bar{x} \rangle_\varepsilon - L_0(\bar{t} - T_0)) \right.$$

$$\left. + \zeta_\delta(\langle y \rangle_\varepsilon - L_0(s - T_0)) + \sigma(\bar{t} + s) - 2\sigma T \right\}$$

attains a local minimum at $(\bar{s}, \bar{y}) \in \Delta_{\varepsilon,\delta}(x_0)$. Hence, by Definition 2.5.1,

$$\frac{2}{\beta}(\bar{t} - \bar{s}) - \zeta_\delta'(Y_\varepsilon)L_0 + \sigma$$
$$+ H\left(\bar{s}, \bar{y}, \frac{2}{\alpha}(\bar{x} - \bar{y}) + \zeta_\delta'(Y_\varepsilon)\frac{\bar{y} - x_0}{\langle \bar{y} \rangle_\varepsilon}\right) \leqslant 0, \tag{2.33}$$

with

$$Y_\varepsilon = \langle \bar{y} \rangle_\varepsilon - L_0(\bar{s} - T_0).$$

Combining (2.32) and (2.33), we obtain

$$2\sigma \leqslant L_0\left[\zeta_\delta'(X_\varepsilon) + \zeta_\delta'(Y_\varepsilon)\right] + H\left(\bar{t}, \bar{x}, \frac{2}{\alpha}(\bar{x} - \bar{y}) - \zeta_\delta'(X_\varepsilon)\frac{\bar{x} - x_0}{\langle \bar{x} \rangle_\varepsilon}\right)$$

$$- H\left(\bar{s}, \bar{y}, \frac{2}{\alpha}(\bar{x} - \bar{y}) + \zeta_\delta'(Y_\varepsilon)\frac{\bar{y} - x_0}{\langle \bar{y} \rangle_\varepsilon}\right).$$

Along a suitable sequence of $\beta \downarrow 0$, we will have a limit for the corresponding sequence $(\bar{t}, \bar{x}, \bar{s}, \bar{y})$. For convenience, let us denote the limit still by

$(\bar{t}, \bar{x}, \bar{s}, \bar{y})$. Then by (2.27), it is necessary that $\bar{t} = \bar{s}$. Hence, pass to the limit along this sequence, the above becomes (making use of (2.24))

$$
\begin{aligned}
2\sigma \leqslant {} & L_0\big[\zeta_\delta'(X_\varepsilon) + \zeta_\delta'(Y_\varepsilon)\big] + H\Big(\bar{t}, \bar{x}, \frac{2}{\alpha}(\bar{x} - \bar{y}) - \zeta_\delta'(X_\varepsilon)\frac{\bar{x} - x_0}{\langle \bar{x} \rangle_\varepsilon}\Big) \\
& - H\Big(\bar{t}, \bar{y}, \frac{2}{\alpha}(\bar{x} - \bar{y}) + \zeta_\delta'(Y_\varepsilon)\frac{\bar{y} - x_0}{\langle \bar{y} \rangle_\varepsilon}\Big) \\
\leqslant {} & L_0\big[\zeta_\delta'(X_\varepsilon) + \zeta_\delta'(Y_\varepsilon)\big] + L_0\big(|\zeta_\delta'(X_\varepsilon)| + |\zeta_\delta'(Y_\varepsilon)|\big) \\
& + \Big|H\Big(\bar{t}, \bar{x}, \frac{2}{\alpha}(\bar{x}-\bar{y}) - \zeta_\delta'(X_\varepsilon)\frac{\bar{x}-x_0}{\langle \bar{x} \rangle_\varepsilon}\Big) - H\Big(\bar{t}, \bar{y}, \frac{2}{\alpha}(\bar{x}-\bar{y}) - \zeta_\delta'(X_\varepsilon)\frac{\bar{x}-x_0}{\langle \bar{x} \rangle_\varepsilon}\Big)\Big| \\
\leqslant {} & \omega\Big(|\bar{x}| \vee |\bar{y}|, |\bar{x} - \bar{y}|\big(1 + \big|\frac{2}{\alpha}(\bar{x} - \bar{y}) - \zeta_\delta'(X_\varepsilon)\frac{\bar{x}-x_0}{\langle \bar{x} \rangle_\varepsilon}\big|\big)\Big) \\
\leqslant {} & \omega\Big(|\bar{x}| \vee |\bar{y}|, |\bar{x} - \bar{y}|\big(1 + |\zeta_\delta'(X_\varepsilon)|\big) + 2\frac{|\bar{x} - \bar{y}|^2}{\alpha}\Big).
\end{aligned}
$$

In the above, we have used the fact that

$$
\zeta_\delta'(r) \leqslant 0, \qquad \forall r \in \mathbb{R},
$$

and the second condition in (2.20). Now, let $\alpha \to 0$, by (2.28), we obtain

$$
2\sigma \leqslant 0,
$$

a contradiction. This completes the proof. $\qquad\qquad\qquad\qquad\qquad\qquad$ $\square$

Note that for optimal control problems, one has

$$
H(t, x, p) = \inf_{u \in U}\Big[\langle p, f(t, x, u)\rangle + g(t, x, u)\Big], \quad (t, x, p) \in [0, T] \times \mathbb{R}^n \times \mathbb{R}^n.
$$

Thus, under (C5), we have

$$
|H(t, x, p) - H(t, x, q)| \leqslant L(1 + |x|)|p - q|, \quad t \in [0, T], \ x, p, q \in \mathbb{R}^n,
$$

and

$$
\begin{aligned}
|H(t, x, p) - H(t, y, p)| &\leqslant L|x - y||p| + \omega(|x| \vee |y|, |x - y|) \\
&\leqslant L(1 + |p|)|x - y| + \omega\big(|x| \vee |y|, (1 + |p|)|x - y|\big) \\
&\equiv \bar{\omega}\big(|x| \vee |y|, (1 + |p|)|x - y|\big),
\end{aligned}
$$

with

$$
\bar{\omega}(r, \rho) = L\rho + \omega(r, \rho), \qquad r, \rho \in \mathbb{R}_+.
$$

Now, we present the following characterization of the value function of Problem $(OC)^T$.

Theorem 2.5.4. *Let* (C5) *hold. Then the value function* $V(\cdot, \cdot)$ *of Problem* $(OC)^T$ *is the unique viscosity of the corresponding HJB equation.*

Proof. It suffices to show that the value function $V(\cdot,\cdot)$ is a viscosity solution to HJB equation.

For any continuously differentiable function $\varphi(\cdot,\cdot)$, let $V(\cdot,\cdot) - \varphi(\cdot,\cdot)$ attains a local maximum at $(t_0, x_0) \in [0, T) \times \mathbb{R}^n$. Pick any $u \in U$ and let $u(\cdot) = u$ be the constant control. Let $X^u(\cdot) \equiv X(\cdot; t_0, x_0, u)$. For any $\tau > t_0$ with $\tau - t_0 > 0$ small,

$$V(\tau, X^u(\tau)) - \varphi(\tau, X^u(\tau)) \leqslant V(t_0, x_0) - \varphi(t_0, x_0).$$

Then by Theorem 2.4.2,

$$
\begin{aligned}
0 &\leqslant \varphi(\tau, X^u(\tau)) - \varphi(t_0, x_0) - \Big[V(\tau, X^u(\tau)) - V(t_0, x_0)\Big] \\
&= \varphi(\tau, X^u(\tau)) - \varphi(t_0, x_0) - V(\tau, X^u(\tau)) \\
&\quad + \inf_{u(\cdot) \in \mathcal{U}[t_0,\tau]} \left[\int_{t_0}^\tau g(s, X^u(s), u(s))ds + V(\tau, X^u(\tau))\right] \\
&\leqslant \varphi(\tau, X^u(\tau)) - \varphi(t_0, x_0) + \int_{t_0}^\tau g(s, X^u(s), u)ds.
\end{aligned}
$$

Dividing by $\tau - t_0$ and sending $\tau \to t_0$, we obtain

$$0 \leqslant \varphi_t(t_0, x_0) + \langle \varphi_x(t_0, x_0), f(t_0, x_0, u) \rangle + g(t_0, x_0, u), \qquad \forall u \in U.$$

Consequently,

$$\varphi_t(t_0, x_0) + H(t_0, x_0, \varphi_x(t_0, x_0)) \geqslant 0.$$

Thus, $V(\cdot,\cdot)$ is a viscosity sub-solution of HJB equation.

Next, let $V(\cdot,\cdot) - \varphi(\cdot,\cdot)$ attain a local minimum at $(t_0, x_0) \in [0, T) \times \mathbb{R}^n$. By Theorem 2.4.2, for any small $\varepsilon > 0$ and $\tau > t_0$ with $\tau - t_0 > 0$ small, one can find a $u^{\varepsilon,\tau}(\cdot) \in \mathcal{U}[t_0, \tau]$ such that

$$V(t_0, x_0) + \varepsilon(\tau - t_0) > V(\tau, X^{\varepsilon,\tau}(\tau)) + \int_{t_0}^\tau g(s, X^{\varepsilon,\tau}(s), u^{\varepsilon,\tau}(s))ds,$$

where $X^{\varepsilon,\tau}(\cdot) = X(\cdot; t_0, x_0, u^{\varepsilon,\tau}(\cdot))$. By Proposition 2.1.1, we know that

$$|X^{\varepsilon,\tau}(s) - x_0| \leqslant \left[e^{L(s-t_0)} - 1\right](1 + |x_0|). \tag{2.34}$$

This implies that when $\tau - t_0 > 0$ is small enough, $(\tau, X^{\tau,\varepsilon}(\tau))$ will be close enough to (t_0, x_0), uniformly in (ε, τ). Hence,

$$
\begin{aligned}
0 &\geqslant \varphi(\tau, X^{\varepsilon,\tau}(\tau)) - \varphi(t_0, x_0) - \Big[V(\tau, X^{\varepsilon,\tau}(\tau)) - V(t_0, x_0)\Big] \\
&\geqslant \int_{t_0}^\tau \Big[\varphi_t(s, X^{\varepsilon,\tau}(s)) + \langle \varphi_x(s, X^{\varepsilon,\tau}(s)) \rangle \\
&\qquad + g(s, X^{\varepsilon,\tau}(s), u^{\varepsilon,\tau}(s))\Big]ds - \varepsilon(\tau - t_0) \\
&\geqslant \int_{t_0}^\tau \Big[\varphi_t(s, X^{\varepsilon,\tau}(s)) + H\big(s, X^{\varepsilon,\tau}(s), \varphi_x(s, X^{\varepsilon,\tau}(s))\big)\Big]ds - \varepsilon(\tau - t_0).
\end{aligned}
$$

Note that

$$|H(s, x, p) - H(s, x_0, p_0)|$$
$$\leqslant |H(s, x, p) - H(s, x, p_0)| + |H(s, x, p_0) - H(s, x_0, p_0)|$$
$$\leqslant L(1 + |x|)|p - p_0| + L(1 + |p_0|)|x - x_0| + \omega(|x_0| \vee |x|, |x - x_0|).$$

Hence, dividing $(\tau - t_0) > 0$ and then sending $\tau \to t_0$, making use of (2.34), we obtain

$$0 \geqslant \varphi_t(t_0, x_0) + H(t_0, x_0, \varphi_x(t_0, x_0)) - \varepsilon.$$

Then letting $\varepsilon \to 0$, we see that $V(\cdot, \cdot)$ is a viscosity super-solution of HJB equation. $\qquad\qquad\square$

A careful observation on the proof of Theorem 2.5.3 tells us that a very small modification will lead to an interesting comparison result. More precisely, let us consider another HJ equation:

$$\begin{cases} \widehat{V}_t(t, x) + \widehat{H}(t, x, V_x(t, x)) = 0, & (t, x) \in [0, T] \times \mathbb{R}^n, \\ \widehat{V}(T, x) = \widehat{h}(x). \end{cases} \qquad (2.35)$$

We have the following result whose detailed proof is left to the readers.

Theorem 2.5.5. *Suppose* $H, \widehat{H} : [0, T] \times \mathbb{R}^n \times \mathbb{R}^n \to \mathbb{R}$ *and* $h, \widehat{h} : \mathbb{R}^n \to \mathbb{R}$ *are continuous, and* (H) *holds for* $H(\cdot, \cdot, \cdot)$ *and* $\widehat{H}(\cdot, \cdot, \cdot)$. *Suppose*

$$\begin{cases} H(t, x, p) \leqslant \widehat{H}(t, x, p), & \forall (t, x, p) \in [0, T] \times \mathbb{R}^n \times \mathbb{R}^n, \\ h(x) \leqslant \widehat{h}(x), & \forall x \in \mathbb{R}^n. \end{cases}$$

Let $V(\cdot, \cdot)$ *be a viscosity sub-solution of* (2.19) *and* $\widehat{V}(\cdot, \cdot)$ *be a viscosity super-solution of* (2.35). *Then the following comparison holds:*

$$V(t, x) \leqslant \widehat{V}(t, x), \qquad \forall (t, x) \in [0, T] \times \mathbb{R}^n.$$

2.6 Autonomous Systems — Controllability Problems

We now consider the following equation:

$$\dot{X}(s) = f(X(s), u(s)), \qquad s \geq 0, \qquad (2.36)$$

which is called an *autonomous control system*. Note that since the right-hand side does not explicitly depend on s, the initial time moment is essentially irrelevant, by which we mean the following: if $X(\cdot; t, x, u(\cdot))$ is the solution of (2.36) for some $(t, x) \in \mathbb{R}_+ \times \mathbb{R}^n$ and $u(\cdot) \in \mathcal{U}[t, \infty)$ with

$$X(t; t, x, u(\cdot)) = x,$$

then

$$X(s; t, x, u(\cdot)) = X(s - t; 0, x, u(\cdot - t)), \qquad \forall s \geqslant t.$$

Hence, for autonomous system (2.36), we need only consider the initial time $t = 0$, namely, we need only consider

$$\begin{cases} \dot{X}(s) = f(X(s), u(s)), & s \geqslant 0, \\ X(0) = x. \end{cases} \tag{2.37}$$

Hereafter, we denote $X(\cdot; x, u(\cdot))$ the solution to (2.37) corresponding to $(x, u(\cdot)) \in \mathbb{R}^n \times \mathcal{U}[0, \infty)$.

For convenience, we modify (C1) as follows.

(C1)′ The map $f : \mathbb{R}^n \times U \to \mathbb{R}^n$ is measurable and there exists a constant $L > 0$ such that

$$\begin{cases} |f(x, u) - f(\bar{x}, u)| \leqslant L|x - \bar{x}|, & \forall x, \bar{x} \in \mathbb{R}^n, \ u \in U, \\ |f(0, u)| \leqslant L, & \forall u \in U. \end{cases}$$

By Proposition 2.1.1, under (C1)′, for any $(x, u(\cdot)) \in \mathbb{R}^n \times \mathcal{U}[0, \infty)$, state equation (2.37) admits a unique solution $X(\cdot) \equiv X(\cdot; x, u(\cdot))$, and for any $u(\cdot) \in \mathcal{U}[0, \infty)$, $x, \bar{x} \in \mathbb{R}^n$ and $s \geq 0$, the following hold:

$$\begin{cases} |X(s; x, u(\cdot))| \leqslant e^{Ls}(1 + |x|) - 1, \\ |X(s; x, u(\cdot)) - x| \leqslant (e^{Ls} - 1)(1 + |x|), \\ |X(s; x, u(\cdot)) - X(s; \bar{x}, u(\cdot))| \leqslant e^{Ls}|x - \bar{x}|. \end{cases} \tag{2.38}$$

2.6.1 *Controllability*

We now let $M \subseteq \mathbb{R}^n$ be a fixed target set which is closed and let

$$\Omega = M^c,$$

which is open. It is clear that

$$\partial M = \partial \Omega,$$

which will be assumed to be C^1 below. Note that when $M = \emptyset$, $\Omega = \mathbb{R}^n$, and $\partial \Omega = \emptyset$. Let us introduce some concepts for system (2.36).

Definition 2.6.1. Let $M \subseteq \mathbb{R}^n$ be a fixed non-empty closed target set.

(i) System (2.36) is said to be *controllable* from the initial state $x \in \mathbb{R}^n$ to M if there exist a $\tau \geq 0$ and a control $u(\cdot) \in \mathcal{U}[0, \tau]$ such that

$$X(\tau; x, u(\cdot)) \in M.$$

(ii) System (2.36) is said to be *locally controllable* to M if there exists an open neighborhood $\mathcal{O}(M)$ of M such that (2.36) is controllable from any initial state $x \in \mathcal{O}(M)$ to M.

(iii) System (2.36) is said to be *small time locally controllable* (STLC, for short) to M if for any $x \in \partial M$ and $\varepsilon > 0$, there exists a $\delta > 0$ such that for any $\bar{x} \in B_\delta(x)$, there exists a $u(\cdot) \in \mathcal{U}[0, \varepsilon]$ satisfying

$$X(\bar{\tau}; \bar{x}, u(\cdot)) \in M,$$

for some $\tau \in [0, \varepsilon]$.

(iv) System (2.36) is said to be *globally controllable to* M if the system is controllable to M from any initial state $x \in \mathbb{R}^n$.

(v) System (2.36) is said to be *completely controllable* if for any $\bar{x} \in \mathbb{R}^n$, the system is controllable from any initial state $x \in \mathbb{R}^n$ to the singleton target set $\{\bar{x}\}$.

(vi) System (2.36) is said to be *completely non-controllable* to M if the system is not controllable to M from any $x \notin M$.

Note that if $M = \emptyset$, then any system is completely non-controllable to M. We will see that this is not just a trivial case, and it will be an important case. To study various controllabilities defined above, let us introduce the following.

Definition 2.6.2. For any non-empty set $\Omega_0 \subseteq \mathbb{R}^n$, the set

$$\mathcal{R}(s; \Omega_0) = \Big\{ X(s; x, u(\cdot)) \mid x \in \Omega_0, \ u(\cdot) \in \mathcal{U}[0, s] \Big\}$$

is called the *reachable set* of system (2.36) at s from Ω_0, and the set

$$\mathcal{R}([0, s]; \Omega_0) = \bigcup_{r \in [0, s]} \mathcal{R}(r; \Omega_0)$$

is called the *reachable set* of system (2.36) on $[0, s]$ from Ω_0. We denote

$$\mathcal{R}(\Omega_0) \equiv \mathcal{R}(\mathbb{R}_+; \Omega_0) = \bigcup_{r \in \mathbb{R}_+} \mathcal{R}(r; \Omega_0) = \bigcup_{s > 0} \mathcal{R}([0, s]; \Omega_0),$$

which is the set of all points that the state of the system can reach from Ω_0 on $\mathbb{R}_+$. When $\Omega_0 = \{x\}$ is a singleton, we simply denote $\mathcal{R}(s; \{x\})$, $\mathcal{R}([0, s]; \{x\})$, and $\mathcal{R}(\{x\})$ by $\mathcal{R}(s; x)$, $\mathcal{R}([0, s]; x)$, and $\mathcal{R}(x)$, respectively.

We point out that the reachable sets are completely determined by the system (2.36) and they have nothing to do with the target set. From the above definition, we see that the following proposition is true.

Proposition 2.6.3. *Let* (C1)$'$ *hold and* $M \subseteq \mathbb{R}^n$ *be a non-empty closed set.*

(i) *For any non-empty set* $\Omega_0 \subseteq \mathbb{R}^n$, *the map* $s \mapsto \mathcal{R}([0,s]; \Omega_0)$ *is non-decreasing, i.e.,*

$$\mathcal{R}([0,s_1]; \Omega_0) \subseteq \mathcal{R}([0,s_2]; \Omega_0), \qquad \forall 0 \leqslant s_1 < s_2.$$

(ii) *Suppose there exists a* $u_0 \in U$ *such that*

$$f(x, u_0) = 0, \qquad \forall x \in \mathbb{R}^n.$$

Then for any non-empty set $\Omega_0 \subseteq \mathbb{R}^n$,

$$\mathcal{R}(s; \Omega_0) = \mathcal{R}([0,s]; \Omega_0), \qquad \forall s \geqslant 0.$$

(iii) *System* (2.36) *is controllable from* $x \in \mathbb{R}^n$ *to* M *if and only if*

$$\mathcal{R}(x) \cap M \neq \emptyset.$$

(iv) *System* (2.36) *is locally controllable to* $M \subseteq \mathbb{R}^n$ *if and only if there exists an open neighborhood* $\mathcal{O}(M)$ *of* M *such that*

$$\mathcal{R}(x) \cap M \neq \emptyset, \qquad \forall x \in \mathcal{O}(M).$$

(v) *System* (2.36) *is STLC to* M *if and only if for any* $\varepsilon > 0$, *there exists an open neighborhood* $\mathcal{O}(M)$ *of* M *such that*

$$\mathcal{R}([0,\varepsilon]; x) \cap M \neq \emptyset, \qquad \forall x \in \mathcal{O}(M).$$

(vi) *System* (2.36) *is globally controllable to* M *if and only if*

$$\mathcal{R}(x) \cap M \neq \emptyset, \qquad \forall x \in \mathbb{R}^n.$$

(vii) *System* (2.36) *is completely controllable if and only if*

$$\mathcal{R}(x) = \mathbb{R}^n, \qquad \forall x \in \mathbb{R}^n.$$

(viii) *System* (2.36) *is completely non-controllable to* M *if and only if*

$$\mathcal{R}(x) \cap M = \emptyset, \qquad \forall x \notin M.$$

Next, we call

$$\mathcal{C}(s; M) = \Big\{ x \in \mathbb{R}^n \mid \mathcal{R}([0,s]; x) \cap M \neq \emptyset \Big\}$$

the *controllable set* of system (2.36) to the target set M within $[0,s]$, and call

$$\mathcal{C}(M) \equiv \mathcal{C}(\infty; M) = \bigcup_{s \geq 0} \mathcal{C}(s; M) = \Big\{ x \in \mathbb{R}^n \mid \mathcal{R}(x) \cap M \neq \emptyset \Big\}$$

the *controllable set* of system (2.36) to the target set M. Then we have the following result.

Proposition 2.6.4. *Let* $(C1)'$ *hold and* $M \subseteq \mathbb{R}^n$ *be a non-empty closed set.*

(i) *System* (2.36) *is controllable from* $x \in \mathbb{R}^n$ *to* M *if and only if*

$$x \in \mathcal{C}(M).$$

(ii) *System* (2.36) *is locally controllable to* $M \subseteq \mathbb{R}^n$ *if and only if there exists an open neighborhood* $\mathcal{O}(M)$ *of* M *such that*

$$\mathcal{O}(M) \subseteq \mathcal{C}(M).$$

(iii) *System* (2.36) *is STLC to* M *if and only if for any* $\varepsilon > 0$, *there exists an open neighborhood* $\mathcal{O}(M)$ *of* M *such that*

$$\mathcal{O}(M) \subseteq \mathcal{C}(\varepsilon; M).$$

In the case that ∂M *is compact, system* (2.36) *is STLC to* M *if and only if for any* $\varepsilon > 0$, *there exists a* $\delta > 0$ *such that*

$$B_\delta(M) \subseteq \mathcal{C}(\varepsilon; M).$$

(iv) *System* (2.36) *is globally controllable to* M *if and only if*

$$\mathcal{C}(M) = \mathbb{R}^n.$$

(v) *System* (2.36) *is completely controllable if and only if*

$$\mathcal{C}(\{x\}) = \mathbb{R}^n, \qquad \forall x \in \mathbb{R}^n.$$

(vi) *System* (2.36) *is completely non-controllable if and only if*

$$\mathcal{C}(M) = M.$$

The above result transforms the various controllability problems into the characterization of various controllable sets. From the above, we see that for given system (2.36), the following implications hold:

On the other hand, it is clear that

$$M \subseteq \mathcal{C}(M).$$

More interestingly, we could ask how big the set $\mathcal{C}(M) \setminus M$ is. We will explore this by means of optimal control in a later subsection. Now, let us present the following result.

Proposition 2.6.5. *Let* (C1)$'$ *hold and* $M \subseteq \mathbb{R}^n$ *be a non-empty closed set. Suppose system* (2.36) *is locally controllable to* M. *Then* $\mathcal{C}(M)$ *is open.*

Proof. Suppose $x \in \mathcal{C}(M)$. Then there exists a $u(\cdot) \in \mathcal{U}[0, \infty)$ and a $\tau > 0$ such that

$$X(\tau; x, u(\cdot)) \in M.$$

By (C1)$'$, we know that for any $\bar{x} \in \mathbb{R}^n$,

$$|X(\tau; x, u(\cdot)) - X(\tau; \bar{x}, u(\cdot))| \leqslant Le^{L\tau}|x - \bar{x}|.$$

On the other hand, by the local controllability, there exists an open neighborhood $\mathcal{O}(M)$ such that any points in $\mathcal{O}(M)$ are controllable to M. Hence, by choosing $\delta > 0$ small enough, we will have

$$X(\tau; \bar{x}, u(\cdot)) \in \mathcal{O}(M) \subseteq \mathcal{C}(M), \qquad \forall \bar{x} \in B_\delta(x).$$

Then

$$B_\delta(x) \subseteq \mathcal{C}(M),$$

proving our claim. $\qquad\qquad\qquad\qquad\qquad\qquad\qquad\qquad\qquad\qquad\qquad$ $\square$

To further explore the controllability of the system (2.36) with the target set M, we need to impose some conditions on M.

• Let $M \subseteq \mathbb{R}^n$ be the closure of a *domain* (a connected open set) such that for any $x \in \partial M$, there exist a $\delta > 0$ and a C^1 function $\varphi : B_\delta(x) \to \mathbb{R}$ with $|\varphi_x(y)| > 0$ for all $y \in B_\delta(x)$ such that

$$\begin{cases} M \cap B_\delta(x) = \{z \in B_\delta(x) \mid \varphi(z) \leqslant 0\}, \\ \partial M \cap B_\delta(x) = \{z \in B_\delta(x) \mid \varphi(z) = 0\}. \end{cases}$$

In this case, we say that ∂M is C^1. We define

$$\nu(y) = \frac{\varphi_x(y)}{|\varphi_x(y)|}, \qquad y \in \partial M \cap B_\delta(x)$$

which is the *unit outward normal* of M at $y \in \partial M$, and it is independent of the choice of $\varphi(\cdot)$. One can show that when ∂M is C^1, $\nu : \partial M \to \partial B_1(0)$ is continuous. We call $\nu(\cdot)$ the *unit outward normal map* of M on ∂M.

• Let $M \subseteq \mathbb{R}^n$ be a convex and closed set. For any $x \in \partial M$, we define the *outward normal cone* of M at x by the following:

$$N(x) = \big\{ \xi \in \mathbb{R}^n \mid \langle \xi, z - x \rangle \leqslant 0, \quad \forall z \in M \big\}.$$

Let us look at two important special cases.

(i) Suppose M is the closure of an open convex set with ∂M being C^1. If $\nu : \partial M \to \partial B_1(0)$ is the outward normal map, then

$$N(x) = \{\lambda\nu(x) \mid \lambda \geqslant 0\}.$$

(ii) Suppose M is a *linear manifold*, by which, we mean that

$$M = x_0 + M_0,$$

where M_0 is a subspace and $\bar{x} \in \mathbb{R}^n$. We may also call such an M a *translated subspace*. Then

$$N(x) = M_0^\perp, \qquad \forall x \in M,$$

where $M_0^\perp$ is the orthogonal complementary of M_0. In fact, $x \in M$ implies $x_0 - x \in M_0$. Thus, $\nu \in N(x)$ if and only if

$$\langle \nu, (x_0 + z) - x \rangle = \langle \nu, z + (x_0 - x) \rangle \leqslant 0, \qquad \forall z \in M_0,$$

which is equivalent to $\nu \in M_0^\perp$. In particular, if $M = \{x_0\}$ is a singleton, then $M_0 = \{0\}$ and

$$N(\{x_0\}) = \mathbb{R}^n.$$

We now present the following result.

Theorem 2.6.6. *Let (C1)' hold. Let $M \subseteq \mathbb{R}^n$ be the closure of a domain with a C^1 boundary ∂M. Suppose the following holds:*

$$\inf_{u \in U} \langle \nu(x), f(x, u) \rangle \leqslant -\mu, \qquad \forall x \in \partial M, \tag{2.39}$$

for some $\mu > 0$, where $\nu : \partial M \to \partial B_1(0)$ is the outward normal map of M. Then system (2.36) is STLC to M. Further, if in addition, ∂M is compact and there exists a differentiable function $\psi : M^c \to (0, \infty)$ such that

$$\begin{cases} \psi(x) \geqslant d(x, M) \equiv \inf_{z \in M} |x - z|, & \forall x \in M^c, \\ \inf_{u \in U} \langle \psi_x(x), f(x, u) \rangle \leqslant -\beta d(x, M), & x \in M^c, \end{cases} \tag{2.40}$$

for some $\beta > 0$. Then system (2.36) is globally controllable to M.

Proof. Since ∂M is C^1, for any $\bar{x} \in \partial M$, there exist a $\delta = \delta(\bar{x}) \in (0, 1)$, and a C^1 map $\varphi : B_\delta(\bar{x}) \to \mathbb{R}$ such that

$$\begin{cases} M \cap B_\delta(\bar{x}) = \{x \in B_\delta(\bar{x}) \mid \varphi(x) \leqslant 0\}, \\ \partial M \cap B_\delta(\bar{x}) = \{x \in B_\delta(\bar{x}) \mid \varphi(x) = 0\}, \end{cases}$$

and
$$0 < \sigma \leqslant |\varphi_x(x)| \leqslant \bar{\sigma}, \qquad \forall x \in B_\delta(\bar{x}),$$
for some constants $\bar{\sigma} > \sigma > 0$. Now, by (2.39), one can find a $u_0 \in U$ such that
$$\langle \varphi_x(\bar{x}), f(\bar{x}, u_0) \rangle = |\varphi_x(\bar{x})| \langle \nu(\bar{x}), f(\bar{x}, u_0) \rangle \leqslant -\mu\sigma.$$
Then for any $x \in B_\delta(\bar{x}) \setminus M$, under control $u(s) = u_0$ $(s \geqslant 0)$, the state trajectory $X(\cdot\,; x, u(\cdot))$ satisfies

$$
\begin{aligned}
\varphi(X(s)) &= \varphi(x) + \int_0^s \langle \varphi_x(X(r)), f(X(r), u(r)) \rangle \, dr \\
&\leqslant \bar{\sigma}|x - \bar{x}| + \int_0^s \langle \varphi_x(\bar{x}), f(\bar{x}, u_0) \rangle \, dr \\
&\quad + \int_0^s |\langle \varphi_x(X(r)), f(X(r), u_0) \rangle - \langle \varphi_x(\bar{x}), f(\bar{x}, u_0) \rangle| \, dr \\
&\leqslant \bar{\sigma}|x - \bar{x}| - \mu\sigma s + \int_0^s \omega(|X(r) - \bar{x}|) \, dr \\
&\leqslant \bar{\sigma}|x - \bar{x}| - \Big[\mu\sigma - \omega\big((e^{Ls} - 1)(2 + |\bar{x}|) + |x - \bar{x}|\big)\Big] s,
\end{aligned}
$$

where $\omega(\cdot)$ is a modulus of continuity for the map $x \mapsto \langle \varphi_x(x), f(x, u_0) \rangle$ on $\bar{B}_\delta(\bar{x})$. Thus, for any $\varepsilon > 0$, we first let $0 < \bar{\varepsilon} \leqslant \varepsilon$ such that
$$\omega\Big((e^{L\bar{\varepsilon}} - 1)(2 + |\bar{x}|) + \frac{\bar{\varepsilon}\mu\sigma}{4\bar{\sigma}}\Big) \leqslant \frac{\mu\sigma}{2}.$$
Then let $0 < \bar{\delta} < \frac{\bar{\varepsilon}\mu\sigma}{4\bar{\sigma}}$. With such a $\bar{\delta} = \bar{\delta}(\varepsilon) > 0$, for any $B_{\bar{\delta}}(\bar{x}) \setminus M$, and $s \in [0, \bar{\varepsilon}]$, one has

$$
\begin{aligned}
\varphi(X(s)) &\leqslant \bar{\sigma}|x - \bar{x}| - \Big[\mu\sigma - \omega\big((e^{Ls} - 1)(2 + |\bar{x}|) + |x - \bar{x}|\big)\Big] s \\
&\leqslant \bar{\sigma}\bar{\delta} - \Big[\mu\sigma - \omega\big((e^{L\bar{\varepsilon}} - 1)(2 + |\bar{x}|) + \bar{\delta}\big)\Big]\bar{\varepsilon} \\
&\leqslant \frac{\bar{\varepsilon}\mu\sigma}{4} - \frac{\mu\sigma}{2}\bar{\varepsilon} = -\frac{\bar{\varepsilon}\mu\sigma}{4} < 0.
\end{aligned}
$$

Thus, there must be some $\tau \in [0, \bar{\varepsilon})$ such that
$$\varphi(X(\tau)) = 0,$$
which is equivalent to $X(\tau; x, u_0) \in M$. This proves that the system (2.36) is STLC.

Now, if (2.40) holds, then by Filippov's Lemma, we may find a control $u(\cdot) \in \mathcal{U}[0, \infty)$ such that

$$
\begin{aligned}
d(X(s), M) &\leqslant \psi(X(s)) = \psi(x) + \int_0^s \langle \psi_x(X(r), f(X(r), u(r)) dr \rangle \\
&\leqslant \psi(x) - \int_0^s \frac{\beta}{2} d(X(r), M) dr.
\end{aligned}
$$

Hence, by Gronwall's inequality,

$$d(X(s), M) \leqslant \psi(x) e^{-\frac{\beta}{2} s}, \qquad \forall s \geq 0.$$

Then by the STLC of the system to the target set M, we see that system (2.36) is globally controllable to M. $\qquad\qquad\qquad\qquad\qquad\qquad\qquad$ □

Using a similar idea, we can prove the following result.

Theorem 2.6.7. *Let* (C1)$'$ *hold. Let* $M \subseteq \mathbb{R}^n$ *be the closure of a domain with a* C^1 *boundary* ∂M. *Suppose there exists an* $x \in \partial M$ *such that the following holds:*

$$\inf_{u \in U} \langle \nu(x), f(x, u) \rangle > 0.$$

Then system (2.36) *is not STLC to* M *in a neighborhood of* x. *Further, if*

$$\inf_{u \in U} \langle \nu(x), f(x, u) \rangle > 0, \qquad \forall x \in \partial M, \tag{2.41}$$

then system (2.36) *is completely non-controllable.*

Intuitively, condition (2.39) implies that M can be reached by the state from any direction, whereas, condition (2.41) implies that the target cannot be reached from any direction. These are two extremal cases. More interesting cases are those in between, namely, the target set M can be reached from some directions but not from all directions. Here is an example.

Example 2.6.8. Consider the following 2-dimensional system:

$$\begin{cases} \dot{X}_1(s) = 2X_2(s), \\ \dot{X}_2(s) = u(s). \end{cases}$$

We may also write the above as

$$\begin{pmatrix} \dot{X}_1(s) \\ \dot{X}_2(s) \end{pmatrix} = \begin{pmatrix} 2X_2(s) \\ u(s) \end{pmatrix}.$$

Let $M = \bar{B}_1(0)$ and let $U = [-1, 1]$. Then for any $x \in \partial M$, $|x| = 1$ and

$$\nu(x) = x.$$

Hence,

$$\begin{aligned} \inf_{u \in U} \langle \nu(x), f(x, u) \rangle &= \inf_{|u| \leqslant 1} \langle \begin{pmatrix} x_1 \\ x_2 \end{pmatrix}, \begin{pmatrix} 2x_2 \\ u \end{pmatrix} \rangle \\ &= \inf_{|u| \leqslant 1} (2x_1 x_2 + x_2 u) = 2x_1 x_2 - |x_2|. \end{aligned}$$

From the above, we see that

$$\partial M = \Gamma_- \cup \Gamma_+,$$

with

$$\begin{cases} \Gamma_- = \{(x_1, x_2) \in \partial M \mid 2x_1 x_2 - |x_2| < 0\}, \\ \Gamma_+ = \{x_1, x_2) \in \partial M \mid 2x_1 x_2 - |x_2| \geqslant 0\} = \partial M \setminus \Gamma_-. \end{cases}$$

The target set M cannot be reached through Γ_+ which can be more explicitly described by the following:

$$\Gamma_+ = \left\{(x_1, x_2) \in \partial M \mid x_1 \geqslant \frac{1}{2}, x_2 \geqslant 0\right\} \cup \left\{(x_1, x_2) \in \partial M \mid x_1 \leqslant -\frac{1}{2}, x_2 \leqslant 0\right\}.$$

We note that Theorems 2.6.6 and 2.6.7 above require that target sets have non-empty interior. Now, let us look at a case that the target set has an empty interior. Let the control system take the following form:

$$\dot{X}(t) = AX(t) + Bu(t), \qquad t \geqslant 0 \tag{2.42}$$

with $A \in \mathbb{R}^{n \times n}$ and $B \in \mathbb{R}^{n \times m}$. The set $\mathcal{U}[0, \infty)$ of admissible controls is defined by the following:

$$\mathcal{U}[0, \infty) = \left\{u : [0, \infty) \to \mathbb{R}^m \mid \int_0^T |u(t)| dt < \infty, \quad \forall T > 0\right\}.$$

Then for any initial state $x \in \mathbb{R}^n$ and control $u(\cdot) \in \mathcal{U}[0, \infty)$, the corresponding state trajectory $X(\cdot)$ is given by the following:

$$X(t) = e^{At} x + \int_0^t e^{A(t-s)} Bu(s) ds, \qquad t \geqslant 0,$$

where e^{at} is the fundamental matrix of A. Let us first present the following result which is a consequence of Caley–Hamilton Theorem.

Lemma 2.6.9. *For any given $t > 0$,*

$$\mathcal{R}\big((B, AB, \cdots, A^{n-1} B)\big) = \text{span}\left\{e^{-As} Bu \mid u \in \mathbb{R}^m, \ s \in [0, t]\right\}. \tag{2.43}$$

Proof. By Caley–Hamilton Theorem, we have

$$e^{-As} B = \sum_{k=0}^{n-1} \mu_k(s) A^k B, \qquad s \in [0, t],$$

for some differentiable functions $\mu_k(\cdot)$. Thus,

$$y^T A^k B = 0, \qquad k \geqslant 0 \quad \Rightarrow \quad y^T e^{-As} B, \qquad s \in [0, t].$$

On the other hand, by differentiating and then evaluating at $s = 0$, we have

$$y^T e^{-As} B = 0, \qquad s \in [0, t] \quad \Rightarrow \quad y^T A^k B = 0, \qquad \forall k \geqslant 0.$$

Hence,

$$\left[\mathcal{R}\big((B, AB, \cdots, A^{n-1}B)\big)\right]^{\perp} = \left[\text{span}\left\{e^{-As}Bu \mid u \in \mathbb{R}^m,\ s \in [0, t]\right\}\right]^{\perp}.$$

Then (2.43) follows since both sides of it are closed. $\qquad\square$

Let the target set $M \subseteq \mathbb{R}^n$ be a *linear manifold*, i.e.,

$$M = M_0 + x_0, \tag{2.44}$$

where M_0 is a linear subspace of $\mathbb{R}^n$ with dim $M_0 \leq n - 1$. Thus, M has an empty interior. In what follows, we let $\Pi_{M_0^{\perp}} : \mathbb{R}^n \to M_0^{\perp}$ be the orthogonal projection onto $M_0^{\perp}$. For state equation (2.42) and target set (2.44), we have the following result.

Theorem 2.6.10. *System (2.42) is STLC to the target M defined by (2.44) if and only if*

$$\Pi_{M_0^{\perp}} \mathcal{R}\big((B, AB, \cdots, A^{n-1}B)\big) = M_0^{\perp}. \tag{2.45}$$

In this case, system (2.42) must also be globally controllable to the target set M, within any time duration $[0, T]$, which is referred to as the global small time controllability to M.

Proof. For any $t > 0$, define

$$\mathbf{W}(t) \equiv \int_0^t \Pi_{M_0^{\perp}} e^{A(t-s)} BB^T e^{A^T(t-s)} \Pi_{M_0^{\perp}}\, ds.$$

We claim that under (2.45), $\mathbf{W}(t)$ is positive definite on $M_0^{\perp}$. In fact, we first have

$$\langle \mathbf{W}(t)y, y \rangle = \int_0^t |B^T e^{A^T(t-s)} y|^2 ds \geqslant 0, \qquad \forall y \in M_0^{\perp}.$$

On the other hand, if the above is zero for some $y \in M_0^{\perp}$, one has

$$B^T e^{A^T(t-s)} y = 0, \qquad s \in [0, t].$$

Then

$$y \in \mathcal{N}\left(\begin{pmatrix} B^T \\ (AB)^T \\ \vdots \\ (A^{n-1}B)^T \end{pmatrix} \Pi_{M_0^{\perp}} \right) = \mathcal{R}\left(\Pi_{M_0^{\perp}}(B, AB, \cdots, A^{n-1}) \right)^{\perp}$$

$$= (M_0^{\perp})^{\perp} = M_0.$$

Thus,

$$y \in M_0^{\perp} \cap M_0 = \{0\}.$$

Hence, $\mathbf{W}(t)$ is positive definite on $M_0^\perp$. Consequently, $\mathbf{W}(t)$ is invertible as an operator on $M_0^\perp$, which is denoted by $\mathbf{W}(t)^\dagger$, as an element in $\mathbb{R}^{n\times n}$. Therefore, for any $x \in \mathbb{R}^n$, by choosing

$$u(s) = -B^T e^{A^T(t-s)} \Pi_{M_0^\perp} \mathbf{W}(t)^\dagger \Pi_{M_0^\perp} \left[e^{At}x - x_0 \right], \qquad s \in [0,t],$$

one has

$$\Pi_{M_0^\perp}\left[X(t;x,u(\cdot)) - x_0 \right] = \Pi_{M_0^\perp}\left[e^{At}x - x_0 + \int_0^t e^{A(t-s)}Bu(s)ds \right]$$

$$= \Pi_{M_0^\perp}\left[e^{At}x - x_0 \right]$$

$$-\left(\int_0^t \Pi_{M_0^\perp} e^{A(t-s)} BB^T e^{A^T(t-s)} \Pi_{M_0^\perp} ds \right) \mathbf{W}(t)^\dagger \Pi_{M_0^\perp}\left[e^{At}x - x_0 \right] = 0.$$

Hence,

$$X(t;x,u(\cdot)) \in x_0 + M_0 = M,$$

which means that the system is globally small time controllable to M.

Conversely, suppose (2.45) fails, then there exists an $\eta \in M_0^\perp$, $|\eta| = 1$ such that

$$0 = \eta^T \Pi_{M_0^\perp} A^k B = \eta^T A^k B, \qquad k \geq 0.$$

This implies that

$$\eta^T \int_0^t e^{A(t-s)}Bu(s)ds = 0, \qquad \forall u(\cdot) \in \mathcal{U}[0,\infty).$$

Now, by the STLC, for $x_0 \in M = \partial M$, for any $\varepsilon > 0$, there exists a $\delta > 0$ such that for any $x \in B_\delta(x_0)$, there exists a $u(\cdot) \in \mathcal{U}[0,\varepsilon]$ such that

$$\Pi_{M_0^\perp}\left[X(\tau;,x,u(\cdot)) - x_0 \right] = 0.$$

We take $x = x_0 \pm \frac{\delta}{2}\eta \in B_\delta(x_0)$. Then for some $u_\pm(\cdot) \in \mathcal{U}[0,\varepsilon]$, and $0 < \tau_\pm < \varepsilon$, one has

$$\Pi_{M_0^\perp}\left[e^{A\tau_\pm}\left(x_0 \pm \frac{\delta}{2}\eta \right) - x_0 \right] + \Pi_{M_0^\perp} \int_0^{\tau_\pm} e^{A(\tau_\pm - s)} Bu_\pm(s)ds = 0.$$

Multiplying the above by η^T from left, we have

$$0 = \eta^T \Pi_{M_0^\perp}\left[e^{A\tau_\pm}\left(x_0 \pm \frac{\delta}{2}\eta \right) - x_0 \right] = \pm\frac{\delta}{2}\eta^T e^{A\tau_\pm}\eta + \eta^T(e^{A\tau_\pm} - I)x_0.$$

Thus,

$$\eta^T e^{A\tau_\pm}\eta = \mp\frac{2}{\delta}\eta^T(e^{A\tau_\pm} - I)x_0 = \mp\frac{2}{\delta}\sum_{k=1}^\infty \frac{\tau_\pm^k}{k!}\eta^T A^k x_0.$$

Since the left-hand side is positive for $\varepsilon > 0$ small, the right-hand side must not be zero. Let $\ell \geqslant 1$ such that

$$\eta^T A^\ell x_0 \neq 0, \quad \eta^T A^k x_0 = 0, \quad 1 \leqslant k \leqslant \ell - 1.$$

Then

$$0 < \eta^T e^{A\tau_\pm} \eta = \mp \frac{2}{\delta} \frac{\tau_\pm^\ell}{\ell!} \eta^T A^\ell x_0 \big(1 + o(1)\big),$$

which implies that $\eta^T A^\ell x_0 > 0$ and $\eta^T A^\ell x_0 < 0$. This is impossible. $\quad\square$

Note that in the case $M_0 = \{0\}$, condition (2.45) is equivalent to the following:

$$\mathcal{R}\Big((B, AB, \cdots, A^{n-1}B)\Big) = \mathbb{R}^n,$$

which is equivalent to the following:

$$\text{rank}\,(B, AB, \cdots, A^{n-1}B) = n. \tag{2.46}$$

This is called the *Kalman rank condition*. From the above, we have the following corollary.

Corollary 2.6.11. *System (2.42) is completely controllable if and only if (2.46) holds.*

It is natural to ask if (2.45) is necessary for global controllability (not STLC) of system (2.42) to the linear manifold target set M, in general. The answer is negative when $\dim M_0 > 0$. Here is a counterexample.

Example 2.6.12. Consider

$$\dot{X}(t) = AX(t) + Bu(t), \qquad t \geqslant 0,$$

with

$$A = \begin{pmatrix} 0 & 1 \\ -1 & 0 \end{pmatrix}, \qquad B = \begin{pmatrix} 0 \\ 0 \end{pmatrix},$$

and let

$$M = M_0 = \{(\lambda, 0) \mid \lambda \in \mathbb{R}\}.$$

Then

$$M_0^\perp = \{(0, \mu) \mid \mu \in \mathbb{R}\},$$

and

$$\Pi_{M_0^\perp} \mathcal{R}\Big((B, AB)\Big) = \{0\} \neq M_0^\perp.$$

Thus, (2.45) fails. However, for any $x = (x_1, x_2)^T$, the state trajectory is given by the following:

$$X(t) = \begin{pmatrix} \cos t & \sin t \\ -\sin t & \cos t \end{pmatrix} \begin{pmatrix} x_1 \\ x_2 \end{pmatrix}.$$

Clearly, there is a $\tau \in [0, 2\pi]$ such that

$$X(\tau; x, u(\cdot)) \in M.$$

Hence, the system is globally controllable to M. But we see that the system is not STLC to M.

2.6.2   *Time optimal control problem*

For the fixed target set M which is assumed to be nonempty and closed, we recall the first hitting time:

$$\mathcal{T}_M(x; u(\cdot)) = \min\left\{s \in \mathbb{R}_+ \mid X(s; x, u(\cdot)) \in M\right\} \equiv \mathcal{T}(x; u(\cdot)), \qquad (2.47)$$

with the convention that $\min \phi \overset{\Delta}{=} +\infty$. Note that in the case

$$\left\{s \in \mathbb{R}_+ \mid X(s; x, u(\cdot)) \in M\right\} \neq \emptyset,$$

the above minimum is achieved since M is closed. Also, it is clear that

$$\mathcal{T}(x; u(\cdot)) = 0, \qquad \forall(x, u(\cdot)) \in M \times \mathcal{U}[0, \infty); \qquad (2.48)$$

and the controllable set of the system (to the target set M) can be characterized by

$$\mathcal{C}(M) \equiv \left\{x \in \mathbb{R}^n \mid X(s; x, u(\cdot)) \in M,\right.$$

$$\left. \text{for some } s \in \mathbb{R}_+, \, u(\cdot) \in \mathcal{U}[0, \infty)\right\}$$

$$= \left\{x \in \mathbb{R}^n \mid \inf_{u(\cdot) \in \mathcal{U}[0, \infty)} \mathcal{T}(x; u(\cdot)) < \infty\right\}.$$

In the current case, for any $x \in \mathcal{C}(M)$, we let

$$\mathcal{U}_x = \left\{u(\cdot) \in \mathcal{U}[0, \infty) \mid X(\tau; x, u(\cdot)) \in M, \text{ for some } \tau \in [0, \infty)\right\}.$$

We pose the following problem.

   Problem (T)′. For any $x \in \mathcal{C}(M)$, find a $\bar{u}(\cdot) \in \mathcal{U}_x$ such that

$$\mathcal{T}(x; \bar{u}(\cdot)) = \inf_{u(\cdot) \in \mathcal{U}_x} \mathcal{T}(x; u(\cdot)) \equiv \mathcal{T}(x). \qquad (2.49)$$

This is a special case of Problem (T) stated in Subsection 1.3.1. Any $\bar{u}(\cdot) \in \mathcal{U}_x$ satisfying (2.49) is called a *time optimal control*, and $\mathcal{T}(x)$ is called the *minimum time* reaching M from x. Naturally, we define

$$\mathcal{T}(x) = \infty, \qquad \forall x \in \mathbb{R}^n \setminus \mathcal{C}(M).$$

With the above defined $\mathcal{T}(x)$, we see that $\mathcal{C}(M)$ can be characterized by the following:

$$\mathcal{C}(M) = \left\{x \in \mathbb{R}^n \mid \mathcal{T}(x) < \infty\right\} \equiv \mathcal{D}(\mathcal{T}). \qquad (2.50)$$

The right-hand side $\mathcal{D}(\mathcal{T})$ of (2.50) is called the *domain* of the function $\mathcal{T} : \mathbb{R}^n \to [0, \infty]$. We note that since $\mathcal{T}(\cdot)$ can take value $+\infty$ somewhere, the characterization (2.50) of $\mathcal{C}(M)$ is not very ideal since the characterization of $\mathcal{T}(\cdot)$ is not very convenient. To get a better representation of $\mathcal{C}(M)$, we introduce the following functional

$$J(x;u(\cdot)) = \int_0^{\mathcal{T}(x;u(\cdot))} e^{-s}ds = 1 - e^{-\mathcal{T}(x;u(\cdot))}, \qquad \forall u(\cdot) \in \mathcal{U}[0,\infty),$$

with the convention that $e^{-\infty} = 0$. Then we impose the following problem which is equivalent to Problem (T′).

Problem (T)″. Find a $\bar{u}(\cdot) \in \mathcal{U}[0,\infty)$ such that

$$J(x;\bar{u}(\cdot)) = \inf_{u(\cdot) \in \mathcal{U}[0,\infty)} J(x;u(\cdot)) \equiv V(x).$$

For convenience, the above is also referred to as a *time optimal control problem*. We have the following result.

Theorem 2.6.13. *Let* (C1)′ *hold. Let* $M \subseteq \mathbb{R}^n$ *be closed and nonempty. Let* (2.36) *be STLC to* M. *Then* $V(\cdot)$ *is continuous on* $\mathbb{R}^n$ *and*

$$\begin{cases} V(x) = 0, & \forall x \in M, \\ 0 < V(x) < 1, & \forall x \in \mathcal{C}(M) \setminus M, \\ V(x) = 1, & \forall x \in \mathbb{R}^n \setminus \mathcal{C}(M). \end{cases} \tag{2.51}$$

Proof. First of all, for any $x \in \mathcal{C}(M)$, there exists a $u(\cdot) \in \mathcal{U}[0,\infty)$ such that

$$\mathcal{T}(x;u(\cdot)) < \infty,$$

which yields

$$0 \leqslant V(x) \leqslant J(x;u(\cdot)) = 1 - e^{-\mathcal{T}(x;u(\cdot))} < 1.$$

Further, since $M \subseteq \mathcal{C}(M)$, (2.48) trivially holds, which leads to

$$V(x) = 0, \qquad \forall x \in M.$$

Also, by our convention, we see that

$$V(x) = 1, \qquad \forall x \in \mathbb{R}^n \setminus \mathcal{C}(M).$$

Hence, (2.51) holds.

We now show that $V(\cdot)$ is continuous on $\mathbb{R}^n$. For any $\bar{x} \in \mathcal{C}(M) \setminus M$, one has $\mathcal{T}(\bar{x}) \in (0,\infty)$. Thus, for any $\varepsilon \in (0,1)$, we can find a $u_\varepsilon(\cdot) \in \mathcal{U}[0,\infty)$ and an $s_\varepsilon \in (0, \mathcal{T}(\bar{x}) + \varepsilon]$ such that

$$\bar{\xi}^\varepsilon \equiv X(s_\varepsilon;\bar{x},u_\varepsilon(\cdot)) \in \partial M.$$

Next, for any $x \in \mathbb{R}^n$, by (2.38),

$$|X(s_\varepsilon;x,u_\varepsilon(\cdot)) - \bar{\xi}_\varepsilon| \leqslant e^{L(\mathcal{T}(\bar{x})+1)}|x - \bar{x}|.$$

Thus, by STLC of the system, we can find a $\delta > 0$ such that for any $x \in B_\delta(\bar{x})$,

$$X\big(\tau_\varepsilon; X(s_\varepsilon; x, u_\varepsilon(\cdot)), v_\varepsilon(\cdot)\big) = X(s_\varepsilon + \tau_\varepsilon; x, u_\varepsilon(\cdot) \oplus v_\varepsilon(\cdot)) \in M,$$

for some $\tau_\varepsilon \in [0, \varepsilon]$ and $v_\varepsilon(\cdot) \in \mathcal{U}[0, \infty)$, where

$$u_\varepsilon(\cdot) \oplus v_\varepsilon(\cdot) = u_\varepsilon(\cdot) I_{[0, s_\varepsilon)}(\cdot) + v_\varepsilon(\cdot - s_\varepsilon) I_{[s_\varepsilon, \infty)}(\cdot).$$

Therefore,

$$\mathcal{T}(x) \leqslant s_\varepsilon + \tau_\varepsilon \leqslant \mathcal{T}(\bar{x}) + 2\varepsilon, \qquad \forall x \in B_\delta(\bar{x}).$$

This implies

$$V(x) = 1 - e^{-\mathcal{T}(x)} \leqslant 1 - e^{-[\mathcal{T}(\bar{x}) + 2\varepsilon]} = V(\bar{x}) + e^{-\mathcal{T}(\bar{x})}(1 - e^{-2\varepsilon}).$$

Hence,

$$\varlimsup_{x \to \bar{x}} V(x) \leqslant V(\bar{x}),$$

i.e., $x \mapsto V(x)$ is *upper semi-continuous* on $\mathcal{C}(M)$. Now, suppose $V(\cdot)$ is not continuous at some $\bar{x} \in \mathcal{C}(M)$. Then there exists a sequence $x_k \in \mathcal{C}(M)$ with $x_k \to \bar{x}$ such that for some $\varepsilon_0 > 0$,

$$\mathcal{T}(x_k) \leqslant \mathcal{T}(\bar{x}) - \varepsilon_0, \qquad \forall k \geqslant 1.$$

This implies that for each $k \geqslant 1$, there exists an $s_k < \mathcal{T}(\bar{x}) - \frac{\varepsilon_0}{2}$, and $u_k(\cdot) \in \mathcal{U}[0, \infty)$ such that

$$X(s_k; x_k, u_k(\cdot)) \in M.$$

On the other hand, by (2.38) again,

$$|X(s_k; \bar{x}, u_k(\cdot)) - X(s_k; x_k, u_k(\cdot))| \leqslant e^{L\mathcal{T}(\bar{x})} |\bar{x} - x_k|,$$

and

$$|X(s_k; \bar{x}, u_k(\cdot))| \leqslant e^{L\mathcal{T}(\bar{x})}(1 + |\bar{x}|) - 1.$$

Hence, we may assume that

$$X(s_k; \bar{x}; u_k(\cdot)) \to \bar{\xi}, \qquad k \to \infty,$$

and

$$X(s_k; x_k, u_k(\cdot)) \to \bar{\xi} \in \partial M, \qquad k \to \infty.$$

Now, by STLC, for the point $\bar{\xi}$ and the given $\varepsilon_0 > 0$, we can find a $\delta > 0$ such that

$$B_\delta(\bar{\xi}) \subseteq \mathcal{C}\Big(\frac{\varepsilon_0}{2}; M\Big).$$

For $k \geqslant 1$ large enough, we have
$$X(s_k; \bar{x}, u_k(\cdot)) \in B_\delta(\bar{\xi}).$$
Hence, there exists a $v_k(\cdot) \in \mathcal{U}[0, \infty)$ such that for some $\tau_k \in [0, \frac{\varepsilon_0}{2}]$,
$$X(\tau_k; X(s_k; \bar{x}, u_k(\cdot)), v_k(\cdot)) = X(s_k + \tau_k; \bar{x}, u_k(\cdot) \oplus v_k(\cdot)) \in M,$$
where
$$u_k(\cdot) \oplus v_k(\cdot) = u_k(\cdot) I_{[0, s_k)}(\cdot) + v_k(\cdot - s_k) I_{[\tau_k, \infty)}(\cdot).$$
Hence,
$$\mathcal{T}(\bar{x}) \leqslant s_k + \tau_k \leqslant \mathcal{T}(\bar{x}) - \varepsilon_0 + \frac{\varepsilon_0}{2} = \mathcal{T}(\bar{x}) - \frac{\varepsilon_0}{2},$$
which is a contradiction. Hence, $V(\cdot)$ is continuous on $\mathcal{C}(M)$.

To complete the proof of continuity of $V(\cdot)$ on $\mathbb{R}^n$, we need to show that for any $\bar{x} \in \partial \mathcal{C}(M)$,
$$\lim_{x \to \bar{x}} V(x) = 1. \tag{2.52}$$
We prove it by contradiction. Suppose the above is not the case, i.e., there exists a sequence $x_k \in \mathcal{C}(M)$ and an $\varepsilon_0 \in (0, 1)$ such that
$$|x_k - \bar{x}| \leqslant \frac{1}{k}, \quad V(x_k) \leqslant 1 - \varepsilon_0, \quad \forall k \geqslant 1.$$
Then there are $u_k(\cdot) \in \mathcal{U}[0, \infty)$ and $t_k \in [0, T_0]$ with $T_0 < \infty$ such that
$$X(t_k; x_k, u_k(\cdot)) \in M, \quad k \geqslant 1.$$
Consequently,
$$d(X(t_k; \bar{x}, u_k(\cdot)), M) \leqslant |X(t_k; \bar{x}, u_k(\cdot)) - X(t_k; x_k, u_k(\cdot))|$$
$$\leqslant e^{Lt_k} |\bar{x} - x_k| \leqslant \frac{e^{LT_0}}{k}, \quad k \geqslant 1.$$
Also,
$$|X(t_k; \bar{x}, u_k(\cdot))| \leqslant e^{Lt_k}(1 + |\bar{x}|) \leqslant e^{LT_0}(1 + |\bar{x}|), \quad k \geqslant 1.$$
Thus, when k is large enough, the point $X(t_k; \bar{x}, u_k(\cdot))$ will be in a neighborhood of M in which every point will be in $\mathcal{C}(M)$. Namely, $\bar{x} \in \mathcal{C}(M)$. Since $\mathcal{C}(M)$ is open, we conclude that $\bar{x} \notin \partial \mathcal{C}(M)$, a contradiction. This completes the proof. □

From the above result, we have the following simple corollary whose proof is straightforward.

Corollary 2.6.14. *Let* (C1)$'$ *hold and* $M \subseteq \mathbb{R}^n$ *be closed and nonempty. Suppose the system is STLC to* M. *Then* $\mathcal{T}(\cdot)$ *is continuous in* $\mathcal{C}(M)$ *and*
$$\lim_{x \to \partial \mathcal{C}(M)} \mathcal{T}(x) = \infty.$$
From Theorem 2.6.13, we also see that $\mathcal{C}(M)$ is characterized by the value function $V(\cdot)$. Thus, we need the characterization of $V(\cdot)$, which will be carried out in the following subsection.

2.6.3   *Optimal control problem with first exit time*

Again, we let $M \subseteq \mathbb{R}^n$ be a closed set and $\Omega = M^c$ be non-empty (and open, therefore, $M \neq \mathbb{R}^n$). Note that Ω is allowed to be unbounded, and it is even allowed that $\Omega = \mathbb{R}^n$ (which amounts to having $M = \phi$). For any $(x, u(\cdot)) \in \mathbb{R}^n \times \mathcal{U}[0, \infty)$, we recall

$$\mathcal{T}(x; u(\cdot)) = \inf\{t \in [0, \infty) \mid X(t; x, u(\cdot)) \in M\}$$
$$\equiv \inf\{t \in [0, \infty) \mid X(t; x, u(\cdot)) \notin \Omega\},$$

with the convention that $\inf \phi = \infty$. Therefore,

$$\mathcal{T}(x, u(\cdot)) = \infty, \qquad \forall (x, u(\cdot)) \in \mathbb{R}^n \times \mathcal{U}[0, \infty), \qquad (2.53)$$

provided either $\Omega = \mathbb{R}^n$, which is equivalent to $M = \phi$, or

$$X(t; x, u(\cdot)) \in \Omega, \qquad \forall t \geq 0.$$

We now call $\mathcal{T}(x, u(\cdot))$ the *first exit time* from Ω. Let us introduce the following cost functional

$$J^\tau(x; u(\cdot)) = \int_0^{\mathcal{T}(x, u(\cdot))} e^{-\lambda s} g(X(s; x, u(\cdot)), u(s)) ds$$
$$+ e^{-\lambda \mathcal{T}(x, u(\cdot))} h\big(X(\mathcal{T}(x; u(\cdot)))\big),$$

with the convention that

$$J^\tau(x; u(\cdot)) = \int_0^\infty e^{-\lambda s} g(X(s; x, u(\cdot)), u(s)) ds, \qquad (2.54)$$

whenever $\mathcal{T}(x, u(\cdot)) = \infty$. In the above, $g : \mathbb{R}^n \times U \to \mathbb{R}$, $h : \mathbb{R}^n \to \mathbb{R}$, and $\lambda > 0$ is some constant called the *discount rate*. The following is concerned with the functions $g(\cdot)$ and $h(\cdot)$.

(C2)$'$ The map $g : \mathbb{R}^n \times U \to \mathbb{R}$ is measurable, with $x \mapsto g(x, u)$ being continuous uniformly in $u \in U$, and the map $h : \mathbb{R}^n \to \mathbb{R}$ is continuous. There exist some constants $\lambda, \mu, L_0, L_1 \geq 0$ with the property that

$$\lambda > \begin{cases} 0, & L_1 = 0, \text{ or } \Omega \text{ is bounded,} \\ \mu L, & L_1 > 0 \text{ and } \Omega \text{ is unbounded,} \end{cases}$$

with $L > 0$ being the constant appear in (C1)$'$, such that

$$\begin{cases} |g(x, u)| \leq L_0 + L_1 |x|^\mu, & (x, u) \in \mathbb{R}^n \times U, \\ |h(x)| \leq L_0 + L_1 |x|^\mu, & x \in \mathbb{R}^n. \end{cases}$$

By the continuity of $x \mapsto (g(x, u), h(x))$, we may assume the following:

$$|g(x, u) - g(\bar{x}, u)| + |h(x) - h(\bar{x})| \leq \omega(|x| \vee |\bar{x}|, |x - \bar{x}|),$$
$$(x, u) \in \mathbb{R}^n \times U,$$

for some local modulus of continuity.

Let (C1)′–(C2)′ hold. In the case that Ω is bounded, for any $(x, u(\cdot)) \in \Omega \times \mathcal{U}[0, \infty)$, and $s \in [0, \tau(x, u(\cdot)))$, we have

$$e^{-\lambda s}|g(X(s), u(s))| \leqslant e^{-\lambda s}\Big(L_0 + L_1|X(s)|^\mu\Big)$$
$$\leqslant e^{-\lambda s}\Big(L_0 + L_1 \sup_{y \in \Omega} |y|^\mu\Big),$$

and

$$e^{-\lambda T}|h(X(T))| \leqslant e^{-\lambda T}\Big(L_0 + L_1 \sup_{y \in \Omega} |y|^\mu\Big).$$

In the case that $L_1 = 0$, regardless of Ω bounded or not,

$$e^{-\lambda s}|g(X(s), u(s))| \leqslant L_0 e^{-\lambda s},$$

and

$$e^{-\lambda T}|h(X(T))| \leqslant L_0 e^{-\lambda T}.$$

In the case that Ω is unbounded, for any $(x, u(\cdot)) \in \Omega \times \mathcal{U}[0, \infty)$ and $s \in \mathcal{U}[0, \mathcal{T}(x, u(\cdot)))$, we have

$$e^{-\lambda s}|g(X(s), u(s))| \leqslant e^{-\lambda s}\Big(L_0 + L_1|X(s)|^\mu\Big)$$
$$\leqslant L_0 e^{-\lambda s} + L_1 e^{-\lambda s}\Big[e^{Ls}(1 + |x|) - 1\Big]^\mu \qquad (2.55)$$
$$\leqslant L_0 e^{-\lambda s} + L_1 e^{-(\lambda - \mu L)s}(1 + |x|)^\mu,$$

and

$$e^{-\lambda T}|h(X(T))| \leqslant L_0 e^{-\lambda T} + L_1 e^{-(\lambda - \mu L)T}(1 + |x|)^\mu.$$

Thus, the map $(x, u(\cdot)) \mapsto J(x; u(\cdot))$ is well-defined. Then we can pose the following optimal control problem.

Problem (OC)$^\tau$. For given $x \in \Omega$, find $\bar{u}(\cdot) \in \mathcal{U}[0, \infty)$ such that

$$J^\tau(x; \bar{u}(\cdot)) = \inf_{u(\cdot) \in \mathcal{U}[0, \infty)} J^\tau(x; u(\cdot)) \equiv V(x). \qquad (2.56)$$

Similar to Problem (OC) stated in Subsection 1.3.1, we call $\bar{u}(\cdot)$ satisfying (2.56) an *optimal control*, the corresponding $\bar{X}(\cdot) \equiv X(\cdot\,; x, \bar{u}(\cdot))$ and $(\bar{X}(\cdot), \bar{u}(\cdot))$ are called the *optimal state trajectory* and *optimal pair*, respectively. The function $V(\cdot)$ is called the *value function*. We note that in the current case, $V(\cdot)$ is independent of t. Also, we define

$$V(\bar{x}) = h(\bar{x}), \qquad \forall \bar{x} \in \partial\Omega.$$

The following collects some basic properties of the value function.

Theorem 2.6.15. *Let* (C1)′–(C2)′ *hold. Then the following holds:*

$$|V(x)| \leqslant K(1 + |x|^\mu), \qquad \forall x \in \Omega. \tag{2.57}$$

Further, if the system is completely non-controllable to Ω^c, *then* $V(\cdot)$ *is continuous in* Ω. *If the system is STLC to* Ω^c *with the following additional compatibility condition: There exists an open neighborhood* $\mathcal{O}(\partial\Omega)$ *of* $\partial\Omega$ *such that*

$$\inf_{u \in U} \left[g(x, u) + \langle h_x(x), f(x, u) \rangle - \lambda h(x) \right] \geqslant 0, \qquad \forall x \in \mathcal{O}(\partial\Omega). \tag{2.58}$$

Then $V(\cdot)$ *is continuous on* $\bar{\Omega}$.

Proof. We only prove the case that Ω is unbounded and $L_1 > 0$ (thus, $\lambda > \mu L_1$ from (C2)′). The other two cases are easier and left to the readers.

We first prove (2.57). For any $(x, u(\cdot)) \in \Omega \times \mathcal{U}[0, \infty)$, one has

$$\begin{aligned}
\left| J^\tau(x; u(\cdot)) \right| &\leqslant \int_0^{\mathcal{T}(x, u(\cdot))} e^{-\lambda s} |g(s, X(s), u(s))| ds \\
&\quad + e^{-\lambda \mathcal{T}(x, u(\cdot))} h\big(X(\mathcal{T}(x, u(\cdot)))\big) \\
&\leqslant \int_0^\infty \left[L_0 e^{-\lambda s} + L_1 e^{-(\lambda - \mu L)s}(1 + |x|)^\mu \right] ds \\
&\quad + e^{-\lambda \mathcal{T}(x, u(\cdot))} \left(L_0 + L_1 |X(\mathcal{T}(x, u(\cdot)))|^\mu \right) \\
&\leqslant \frac{L_0}{\lambda} + \frac{L_1}{\lambda - \mu L}(1 + |x|)^\mu + L_0 + L_1 e^{-(\lambda - \mu L)\mathcal{T}(x, u(\cdot))}(1 + |x|)^\mu \\
&\leqslant K(1 + |x|^\mu).
\end{aligned}$$

This implies (2.57).

Now, we prove the continuity of $V(\cdot)$. First, let the system be completely non-controllable. Then (2.53) and (2.54) hold. Hence, for any $x, \bar{x} \in \Omega$, we have (pick $T > 0$ large and denote $X(\cdot) = X(\cdot\,; x, u(\cdot))$, $\bar{X}(\cdot) = X(\cdot\,; \bar{x}, u(\cdot))$)

$$\begin{aligned}
\left| J^\tau(x; u(\cdot)) - J^\tau(\bar{x}; u(\cdot)) \right| &\leqslant \int_0^\infty e^{-\lambda s} |g(X(s), u(s)) - g(\bar{X}(s), u(s))| ds \\
&\leqslant \int_0^T e^{-\lambda s} \omega\big(|X(s)| \vee |\bar{X}(s)|, |X(s) - \bar{X}(s)|\big) ds \\
&\quad + 2 \int_T^\infty \left(L_0 e^{-\lambda s} + L_1 e^{-(\lambda - \mu L)s}\big[(1 + |x| \vee |\bar{x}|)^\mu\big] \right) ds \\
&\leqslant \frac{1 - e^{-\lambda T}}{\lambda} \omega\big(e^{LT}(1 + |x| \vee |\bar{x}|), e^{LT}|x - \bar{x}|\big) \\
&\quad + \frac{2L_0}{\lambda} e^{-\lambda T} + \frac{2L_1 \big[(1 + |x| \vee |\bar{x}|)^\mu\big]}{\lambda - \mu L} e^{-(\lambda - \mu L)T}.
\end{aligned}$$

Thus,

$$|V(x) - V(\bar{x})| \leqslant \frac{1 - e^{-\lambda T}}{\lambda} \, \omega\big(e^{LT}(1 + |x| \vee |\bar{x}|), e^{LT}|x - \bar{x}|\big)$$
$$+ \frac{2L_0}{\lambda} e^{-\lambda T} + \frac{2L_1\big[(1 + |x| \vee |\bar{x}|)^{\mu}\big]}{\lambda - \mu L} e^{-(\lambda - \mu L)T}.$$

The above will lead to the continuity of $V(\cdot)$ in $\bar{B}_R(0)$, for any $R > 0$. In fact, for any $\varepsilon > 0$, we first take $T > 0$ large enough so that

$$\frac{2L_0}{\lambda} e^{-\lambda T} + \frac{2L_1(1 + R)^{\mu}}{\lambda - \mu L} e^{-(\lambda - \mu L)T} < \frac{\varepsilon}{2}.$$

Then find $\delta > 0$ such that

$$\frac{1 - e^{-\lambda T}}{\lambda} \, \omega\big(e^{LT}(1 + R), e^{LT}\delta\big) < \frac{\varepsilon}{2}.$$

Consequently,

$$|V(x) - V(\bar{x})| < \varepsilon, \qquad \forall x, \bar{x} \in \bar{B}_R(0), \; |x - \bar{x}| < \delta,$$

proving the continuity of $V(\cdot)$ in Ω.

Next, let the system be STLC and (2.58) hold.

We first show that $V(\cdot)$ is continuous at each point on $\partial\Omega$, i.e., for any $\bar{x} \in \partial\Omega$,

$$\lim_{x \to \bar{x}} V(x) = V(\bar{x}) \equiv h(\bar{x}).$$

To this end, we fix any $\bar{x} \in \partial\Omega$. By STLC, we know that for any $\varepsilon > 0$, there exists a $\delta > 0$, such that for any $x \in B_{\delta}(\bar{x}) \cap \Omega$, one can find a $u_{\varepsilon}(\cdot) \in \mathcal{U}[0, T]$ and a $\tau_{\varepsilon} \in [0, \varepsilon]$,

$$X(\tau_{\varepsilon}; x, u_{\varepsilon}(\cdot)) \in \partial\Omega.$$

Then, by denoting $X_{\varepsilon}(\cdot) = X(\cdot; x, u_{\varepsilon}(\cdot))$, we have

$$V(x) - h(\bar{x}) \leqslant J^{\tau}(x; u_{\varepsilon}(\cdot)) - h(\bar{x})$$
$$= \int_0^{\tau_{\varepsilon}} e^{-\lambda s} g(X_{\varepsilon}(s), u_{\varepsilon}(s)) ds + e^{-\lambda \tau_{\varepsilon}} h(X(\tau_{\varepsilon})) - h(\bar{x})$$
$$\leqslant \varepsilon \Big[L_0 + L_1 e^{L\varepsilon\mu}(1 + |x|)^{\mu}\Big] + \omega\big(|X_{\varepsilon}(\tau_{\varepsilon})| \vee |x|, |X_{\varepsilon}(\tau_{\varepsilon}) - x|\big)$$
$$+ e^{-\tau_{\varepsilon}} h(x) - h(\bar{x})$$
$$\leqslant \varepsilon \Big[L_0 + L_1 e^{L\varepsilon\mu}(1 + |x|)^{\mu}\Big] + \omega\big(e^{L\varepsilon\mu}(1 + |x|)^{\mu}, (e^{L\varepsilon\mu} - 1)(1 + |x|)^{\mu}\big)$$
$$+ e^{-\tau_{\varepsilon}} h(x) - h(\bar{x}).$$

Hence, we obtain

$$\overline{\lim_{x \to \bar{x}}} V(x) \leqslant h(\bar{x}),$$

which means that $V(\cdot)$ is upper semi-continuous at each point of $\partial\Omega$.

Now, suppose $V(\cdot)$ is not continuous at $\bar{x} \in \partial\Omega$. Then we may assume that there exists a sequence $x_k \to \bar{x}$, $x_k \in \Omega$ such that

$$h(\bar{x}) - \varepsilon_0 \geqslant V(x_k), \qquad \forall k \geqslant 1,$$

for some $\varepsilon_0 > 0$. By STLC, there exists a $u_k(\cdot) \in \mathcal{U}[0,\infty)$ and $\tau_k \to 0$ such that with $X_k(\cdot) = X(\cdot; x_k, u_k(\cdot))$, one has

$$X_k(\tau_k) \in \partial\Omega,$$

and (making use of (2.58))

$$\begin{aligned}
h(\bar{x}) - \varepsilon_0 \geqslant V(x_k) &\geqslant J(x_k; u_k(\cdot)) - \frac{\varepsilon_0}{2} \\
&= \int_0^{\tau_k} e^{-\lambda s} g(X_k(s), u_k(s)) ds + e^{-\lambda \tau_k} h(X_k(\tau_k)) - \frac{\varepsilon_0}{2} \\
&= h(x_k) - \frac{\varepsilon_0}{2} + \int_0^{\tau_k} e^{-\lambda s} \Big[g(X_k(s), u_k(s)) \\
&\quad + h_x(X_k(s)) f(X_k(s), u_k(s)) - \lambda h(X_k(s)) \Big] ds \\
&\geqslant h(x_k) - \frac{\varepsilon_0}{2}.
\end{aligned}$$

Sending $k \to \infty$ in the above will lead to a contradiction. Hence, we have the continuity of $V(\cdot)$ at each point of $\partial\Omega$.

We now prove the continuity of $V(\cdot)$ on $\bar{\Omega}$. Let $x, \bar{x} \in \bar{\Omega}$ and $|x - \bar{x}| < \delta$ for $\delta > 0$ small. Without loss of generality, let

$$V(x) \geqslant V(\bar{x}).$$

For any $\varepsilon > 0$, there exists a $\bar{u}_\varepsilon(\cdot) \in \mathcal{U}[0, \infty)$ such that

$$V(\bar{x}) \geqslant J^\tau(\bar{x}; \bar{u}_\varepsilon(\cdot)) - \varepsilon.$$

Denoting $\bar{X}_\varepsilon(\cdot) = X(\cdot; \bar{x}, \bar{u}_\varepsilon(\cdot))$ and $\bar{\mathcal{T}}_\varepsilon = \mathcal{T}(\bar{x}, \bar{u}_\varepsilon(\cdot)) \in (0, \infty]$. For this given $\varepsilon > 0$, we let $T > 0$ large enough such that

$$= e^{-\lambda T} L_0 \frac{1 + \lambda}{\lambda} + e^{-(\lambda - \mu L)T} \frac{1 + \lambda - \mu L}{\lambda - \mu L} (1 + |x| \vee |\bar{x}|)^\mu < \frac{\varepsilon}{2}.$$

We consider two cases.

Case 1. $\bar{\mathcal{T}}_\varepsilon > T$, allowing $\bar{\mathcal{T}}_\varepsilon = \infty$. Since

$$|X(T; x, \bar{u}_\varepsilon(\cdot)) - \bar{X}_\varepsilon(T)| \leqslant e^{LT} |x - \bar{x}|,$$

by STLC, we know that for $x \in B_\delta(\bar{x})$ with $\delta > 0$ small,
$$\mathcal{T}_\varepsilon \equiv \mathcal{T}(x, \bar{u}_\varepsilon(\cdot)) > T.$$
Then
$$0 \leqslant V(x) - V(\bar{x}) \leqslant J^\tau(x; \bar{u}_\varepsilon(\cdot)) - J^\tau(x; \bar{u}_\varepsilon(\cdot)) + \varepsilon$$
$$\leqslant \int_0^T e^{-\lambda s}\Big[g(X_\varepsilon(s), \bar{u}_\varepsilon(s)) - g(\bar{X}_\varepsilon(s), \bar{u}_\varepsilon(s))\Big]ds$$
$$+\int_T^{\mathcal{T}_\varepsilon} e^{-\lambda s}g(X_\varepsilon(s), u_\varepsilon(s))ds - \int_T^{\overline{\mathcal{T}}_\varepsilon} e^{-\lambda s}g(\bar{X}_\varepsilon(s), \bar{u}_\varepsilon(s))ds$$
$$+e^{-\lambda \mathcal{T}_\varepsilon}h(X_\varepsilon(\mathcal{T}_\varepsilon)) - e^{-\lambda \overline{\mathcal{T}}_\varepsilon}h(\bar{X}_\varepsilon(\overline{\mathcal{T}}_\varepsilon)) + \varepsilon$$
$$\leqslant \int_0^T e^{-\lambda s}\omega\big(|X_\varepsilon(s)| \vee |\bar{X}_\varepsilon(s)|, |X_\varepsilon(s) - \bar{X}_\varepsilon(s)|\big)ds$$
$$+2\int_T^\infty e^{-\lambda s}\Big(L_0 + L_1 e^{\mu L s}(1 + |x| \vee |\bar{x}|)^\mu\Big)ds$$
$$+2e^{-\lambda T}\Big(L_0 + L_1 e^{\mu L T}(1 + |x| \vee |\bar{x}|)^\mu\Big) + \varepsilon$$
$$\leqslant \int_0^T e^{-\lambda s}\omega\big(e^{LT}(1 + |x| \vee |\bar{x}|), e^{LT}|x - \bar{x}|\big)ds$$
$$+2\Big[e^{-\lambda T}L_0\frac{1+\lambda}{\lambda} + e^{-(\lambda-\mu L)T}\frac{1+\lambda-\mu L}{\lambda-\mu L}(1 + |x| \vee |\bar{x}|)^\mu\Big] + \varepsilon$$
$$\leqslant \omega\big(e^{LT}(1 + |x| \vee |\bar{x}|), e^{LT}|x - \bar{x}|\big)T + 2\varepsilon.$$
Hence, we may find a further shrunk $\delta > 0$ such that
$$0 \leqslant V(x) - V(\bar{x}) \leqslant 3\varepsilon, \qquad \forall |x - \bar{x}| < \delta.$$

Case 2. $\overline{\mathcal{T}}_\varepsilon \leqslant T$. Then
$$X(\overline{\mathcal{T}}_\varepsilon; \bar{x}, \bar{u}_\varepsilon(\cdot)) \in \partial\Omega.$$
There are two subcases:

(i) $\mathcal{T}_\varepsilon \leqslant \overline{\mathcal{T}}_\varepsilon$. In this case, we denote
$$\xi_\varepsilon = X_\varepsilon(\mathcal{T}_\varepsilon) \in \partial\Omega, \quad \bar{\xi}_\varepsilon = \bar{X}_\varepsilon(\mathcal{T}_\varepsilon), \quad \bar{v}_\varepsilon(\cdot) = \bar{u}_\varepsilon(\cdot + \mathcal{T}_\varepsilon).$$
Then
$$V(\bar{x}) \geqslant J^\tau(\bar{x}; \bar{u}_\varepsilon(\cdot)) - \varepsilon$$
$$= \int_0^{\mathcal{T}_\varepsilon} e^{-\lambda s}g(\bar{X}_\varepsilon(s), \bar{u}_\varepsilon(s))ds + \int_{\mathcal{T}_\varepsilon}^{\overline{\mathcal{T}}_\varepsilon} e^{-\lambda s}g(\bar{X}_\varepsilon(s), \bar{u}_\varepsilon(s))ds$$
$$+e^{-\lambda \overline{\mathcal{T}}_\varepsilon}h(\bar{X}_\varepsilon(\overline{\mathcal{T}}_\varepsilon)) - \varepsilon$$
$$= \int_0^{\mathcal{T}_\varepsilon} e^{-\lambda s}g(\bar{X}_\varepsilon(s), \bar{u}_\varepsilon(s))ds + e^{-\lambda \mathcal{T}_\varepsilon}J^\tau(\bar{\xi}_\varepsilon; \bar{v}_\varepsilon(\cdot)) - \varepsilon.$$

Hence,

$$0 \leqslant V(x) - V(\bar{x}) \leqslant J^\tau(x; \bar{u}_\varepsilon(\cdot)) - J^\tau(\bar{x}; \bar{u}_\varepsilon(\cdot)) + \varepsilon$$

$$= \int_0^{\mathcal{T}_\varepsilon} e^{-\lambda s}\Big[g(X_\varepsilon(s), \bar{u}_\varepsilon(s)) - g(\bar{X}_\varepsilon(s), \bar{u}_\varepsilon(s))\Big]ds$$

$$-\int_{\overline{\mathcal{T}}_\varepsilon}^{\overline{\mathcal{T}}_\varepsilon} e^{-\lambda s}g(\bar{X}_\varepsilon(s), \bar{u}_\varepsilon(s))ds + e^{-\lambda \mathcal{T}_\varepsilon}h(X_\varepsilon(\mathcal{T}_\varepsilon)) - e^{-\lambda \overline{\mathcal{T}}_\varepsilon}h(\bar{X}_\varepsilon(\overline{\mathcal{T}}_\varepsilon)) + \varepsilon$$

$$\leqslant \int_0^{\mathcal{T}_\varepsilon} e^{-\lambda s}\omega\big(|X_\varepsilon(s)| \vee |\bar{X}_\varepsilon(s)|, |X_\varepsilon(s) - \bar{X}_\varepsilon(s)|\big)ds$$

$$+e^{-\lambda \mathcal{T}_\varepsilon}\Big[h(\xi_\varepsilon) - V(\bar{\xi}_\varepsilon)\Big] + \varepsilon$$

$$\leqslant \omega\big(e^{LT}(1 + |x| \vee |\bar{x}|), e^{LT}|x - \bar{x}|\big)T + V(\xi_\varepsilon) - V(\bar{\xi}_\varepsilon) + \varepsilon.$$

(ii) $\overline{\mathcal{T}}_\varepsilon < \mathcal{T}_\varepsilon$. In this case, we have

$$V(x) \leqslant \int_0^{\overline{\mathcal{T}}_\varepsilon} e^{-\lambda s}g(X_\varepsilon(s), \bar{u}_\varepsilon(s))ds + e^{-\lambda \overline{\mathcal{T}}_\varepsilon}J^\tau(\xi_\varepsilon; \bar{u}^\varepsilon(\cdot + \overline{\mathcal{T}}_\varepsilon)),$$

which leads to

$$V(x) \leqslant \int_0^{\overline{\mathcal{T}}_\varepsilon} e^{-\lambda s}g(X_\varepsilon(s), \bar{u}_\varepsilon(s))ds + e^{-\lambda \overline{\mathcal{T}}_\varepsilon}V(\xi_\varepsilon).$$

Then

$$0 \leqslant V(x) - V(\bar{x}) \leqslant \int_0^{\overline{\mathcal{T}}_\varepsilon} e^{-\lambda s}g(X_\varepsilon(s), \bar{u}_\varepsilon(s))ds + e^{\lambda \overline{\mathcal{T}}_\varepsilon}V(\xi_\varepsilon)$$

$$-J^\tau(\bar{x}; \bar{u}_\varepsilon(\cdot)) + \varepsilon$$

$$\leqslant \int_0^{\overline{\mathcal{T}}_\varepsilon} e^{-\lambda s}\Big[g(X_\varepsilon(s), \bar{u}_\varepsilon(s)) - g(\bar{X}_\varepsilon(s), \bar{u}_\varepsilon(s))\Big]ds$$

$$+e^{-\lambda \overline{\mathcal{T}}_\varepsilon}\Big[V(\xi_\varepsilon) - h(\bar{\xi}_\varepsilon)\Big] + \varepsilon$$

$$\leqslant \omega\big(e^{LT}(1 + |x| \vee |\bar{x}|), e^{LT}|x - \bar{x}|\big)T + |V(\xi_\varepsilon) - V(\bar{\xi}_\varepsilon)| + \varepsilon.$$

Therefore, in both subcases, one can find a $\delta > 0$, small enough, such that (using the continuity of $V(\cdot)$ on $\partial\Omega$),

$$0 \leqslant V(x) - V(\bar{x}) \leqslant 2\varepsilon,$$

proving the continuity of $V(\cdot)$ on $\bar{\Omega}$.   □

An important special case of the above is the following:

$$g(x, u) = 1, \quad h(x) = 0, \quad \forall(x, u) \in \mathbb{R}^n \times U.$$

This case corresponds to the time optimal control problem. Note that for this case, (2.58) is automatically true. Therefore, Theorem 2.6.15 recovers Theorem 2.6.13 in some sense.

We have the following optimality principle and HJB equation.

Theorem 2.6.16. *Let* (C1)'–(C2)' *hold. Suppose* $\Omega \subseteq \mathbb{R}^n$ *is an open set such that the system is either completely non-controllable or STLC to* Ω^c. *Then for any* $x \in \Omega$ *and* $t \in (0, \mathcal{T}(x))$, *the following holds:*

$$V(x) = \inf_{u(\cdot) \in \mathcal{U}[0,\infty)} \left[\int_0^t e^{-\lambda s} g(X(s; x, u(\cdot)), u(s)) ds \right.$$
$$\left. + e^{-\lambda t} V\big(X(t; x, u(\cdot))\big) \right].$$

In the case that $V(\cdot) \in C^1(\mathbb{R}^n)$, *the following HJB equation is satisfied:*

$$\begin{cases} \lambda V(x) - H(x, V_x(x)) = 0, & x \in \Omega, \\ V(x) = h(x), & x \in \partial\Omega, \end{cases} \tag{2.59}$$

where

$$H(x, p) = \inf_{u \in U} \left[\langle p, f(x, u) \rangle + g(x, u) \right], \qquad x, p \in \mathbb{R}^n.$$

Sketch of the Proof. For any $x \in \Omega$ with $\mathcal{T}(x) = \infty$, we have

$$J^\tau(x; u(\cdot)) = \int_0^\infty e^{-\lambda s} g(X(s; x, u(\cdot)), u(s)) ds$$
$$= \int_0^t e^{-\lambda s} g(X(s; x, u(\cdot)), u(s)) ds + e^{-\lambda t} J^\tau\big(X(t; x, u(\cdot)); u(\cdot + t)\big).$$

In the case that $\mathcal{T}(x) < \infty$, for any $t \in (0, \mathcal{T}(x))$, we have

$$J^\tau(x; u(\cdot)) = \int_0^{\mathcal{T}(x)} e^{-\lambda s} g(X(s; x, u(\cdot)), u(s)) ds$$
$$+ e^{-\lambda \mathcal{T}(x)} h\big(X(\mathcal{T}(x); x, u(\cdot))\big)$$
$$= \int_0^t e^{-\lambda s} g(X(s; x, u(\cdot)), u(s)) ds + e^{-\lambda t} J^\tau\big(X(t; x, u(\cdot)), u(\cdot + t)\big).$$

Then following the same arguments as the proof of Theorem 2.4.2, we can prove the first part of the above theorem.

Now, in the case that $V(\cdot) \in C^1(\mathbb{R}^n)$, similar to the proof of Proposition 2.4.3, we can obtain the HJB equation (2.59). $\qquad\square$

Readers are encouraged to give a complete proof of the above result.

2.6.4   *Stationary HJB equations*

In this subsection, we consider the following equation:

$$\begin{cases} \lambda V(x) - H(x, V_x(x)) = 0, & x \in \Omega, \\ V(x) = h(x), & x \in \partial\Omega, \end{cases} \tag{2.60}$$

where $\lambda > 0$, $\Omega \subseteq \mathbb{R}^n$ is a domain with a C^1 boundary $\partial\Omega$, and $H : \Omega \times \mathbb{R}^n \to \mathbb{R}$, $h : \partial\Omega \to \mathbb{R}$ are given maps. We call the above a *stationary HJB equation*. Note that Ω is allowed to be unbounded and it could be $\Omega = \mathbb{R}^n$. We introduce the following definition which is comparable with Definition 2.5.1.

Definition 2.6.17. (i) A continuous function $V(\cdot)$ is called a *viscosity sub-solution* of HJB equation (2.60) if

$$V(x) \leqslant h(x), \qquad \forall x \in \partial\Omega,$$

and for any C^1 function $\varphi : \Omega \to \mathbb{R}$, if $V(\cdot) - \varphi(\cdot)$ attains a local maximum at $x_0 \in \Omega$, the following holds:

$$\lambda V(x_0) - H(x_0, \varphi_x(x_0)) \leqslant 0.$$

(ii) A continuous function $V(\cdot)$ is called a *viscosity super-solution* of HJB equation (2.60) if

$$V(x) \geqslant h(x), \qquad \forall x \in \partial\Omega,$$

and for any C^1 function $\varphi : \Omega \to \mathbb{R}$, if $V(\cdot) - \varphi(\cdot)$ attains a local minimum at $x_0 \in \Omega$, the following holds:

$$\lambda V(x_0) - H(x_0, \varphi_x(x_0)) \geqslant 0.$$

(iii) A continuous function $V(\cdot)$ is called a *viscosity solution* of HJB equation (2.60) if it is both a viscosity sub-solution and a viscosity super-solution of (2.60).

Next, for any $\mu \geqslant 0$, we let

$$Q_\mu(\bar\Omega; \mathbb{R}) = \Big\{ \varphi(\cdot) \in \bigcap_{K>0} C\big(\overline{\Omega \cap B_K(0)}; \mathbb{R}\big) \ \Big| \ \sup_{x \in \Omega} \frac{|\varphi(x)|}{\langle x \rangle^\mu} < \infty \Big\},$$

where $\langle x \rangle = \sqrt{1 + |x|^2}$. Note that elements in $Q_\mu(\bar\Omega; \mathbb{R})$ have a specific growth rate as $|x| \to \infty$. This will play a crucial role below for the case Ω is unbounded. Whereas, in the case Ω is bounded,

$$Q_\mu(\bar\Omega; \mathbb{R}) = C(\bar\Omega; \mathbb{R}), \qquad \forall \mu \geqslant 0.$$

Thus, for such a case, it is not necessary to introduce $Q_\mu(\bar\Omega; \mathbb{R})$.

The following gives the uniqueness of viscosity solution to HJB equation (2.60).

Theorem 2.6.18. *Let* $H : \Omega \times \mathbb{R}^n \to \mathbb{R}$ *and* $h : \partial\Omega \to \mathbb{R}$ *be continuous. There exist constants* $L, \bar{L} > 0$ *and a local modulus of continuity* $\omega(\cdot, \cdot)$ *such that*

$$\begin{cases} |H(x,p) - H(y,p)| \leqslant \omega\big(|x| \vee |y|, |x-y|(1+|p|)\big), \\ |H(x,p) - H(x,q)| \leqslant \big(\bar{L} + L|x|\big)|p - q|, \\ \qquad\qquad\qquad \forall x, y, p, q \in \mathbb{R}^n. \end{cases} \tag{2.61}$$

Suppose $V(\cdot), \widehat{V}(\cdot) \in \bigcup_{\mu < \frac{\lambda}{L}} Q_\mu(\bar{\Omega}; \mathbb{R})$ *are viscosity solutions to* (2.60). *Then* $V(\cdot, \cdot) = \widehat{V}(\cdot, \cdot)$.

Proof. We prove the case that $\Omega \neq \mathbb{R}^n$, and it is unbounded. The cases that Ω is bounded or $\Omega = \mathbb{R}^n$ can be proved similarly. Suppose there exists an $x_0 \in \Omega$ such that

$$V(x_0) - \widehat{V}(x_0) > 0.$$

By our assumption, we can find some $0 < \mu < \frac{\lambda}{L}$ such that

$$\lim_{|x| \to \infty} \frac{|V(x)| + |\widehat{V}(x)|}{\langle x \rangle^\mu} = 0.$$

Then for any $\alpha > 0$,

$$\lim_{|x| + |y| \to \infty} \left[V(x) - \widehat{V}(y) - \alpha\big(\langle x \rangle^\mu + \langle y \rangle^\mu\big) \right] = -\infty.$$

Thus, there exists a $K(\alpha) > 0$, such that

$$V(x) - \widehat{V}(y) - \alpha\big(\langle x \rangle^\mu + \langle y \rangle^\mu\big)$$
$$\leqslant V(x_0) - \widehat{V}(x_0) - 2\alpha \langle x_0 \rangle^\mu - 2, \qquad \forall (x,y) \in \bar{\Omega}^2 \setminus \Omega_\alpha^2,$$

where

$$\Omega_\alpha = \Omega \cap B_{K(\alpha)}(0).$$

Since $\partial\Omega$ is C^1, $\partial\Omega_\alpha$ is Lipschitz. Thus, there exists a $\sigma \in (0, 1)$ depending on $K(\alpha)$, such that one can find a Lipschitz continuous function $\theta^\alpha : \mathbb{R}^n \to \mathbb{R}$ which is C^1 on $\mathbb{R}^n \setminus \partial\Omega_\alpha$, satisfying

$$\theta^\alpha(x) = \begin{cases} \sigma, & x \in \Omega \cap B_{K(\alpha)}(0), \ d_{\partial\Omega}(x) \geqslant 2\sigma, \\ -\sigma, & x \in \Omega^c \cap B_{K(\alpha)}(0), \ d_\Omega(x) \geqslant 2\sigma, \end{cases} \tag{2.62}$$

and

$$\begin{cases} |\theta^{\alpha}(x)| \leqslant \sigma, & x \in \mathbb{R}^{n}, \\ |\theta_{x}^{\alpha}(x)| \leqslant 1, & x \in \mathbb{R}^{n} \setminus \partial\Omega_{\alpha}. \end{cases}$$

Note that we only require (2.62) on $B_{K(\alpha)}(0)$. In general when Ω is unbounded, even if $\partial\Omega$ is C^{1}, one does not necessarily have (2.62) for a uniform $\sigma > 0$ independent of $B_{K(\alpha)}(0)$ since the shape of $\partial\Omega$ could be "sharper and sharper" as $|x| \to \infty$. Now, for any $\varepsilon, \alpha, \beta \in (0,1)$, we define

$$\Phi(x,y) = V(x) - \widehat{V}(y) - \frac{1}{\varepsilon}|x-y|^{2} - \alpha\big(\langle x \rangle^{\mu} + \langle y \rangle^{\mu}\big)$$
$$- \beta\big(\theta^{\alpha}(x) + \theta^{\alpha}(y)\big), \qquad \forall(x,y) \in \bar{\Omega} \times \bar{\Omega}.$$

Since

$$\lim_{|x|+|y| \to \infty, x,y \in \Omega} \Phi(x,y) = -\infty,$$

there exists an $(\bar{x}, \bar{y}) \in \bar{\Omega} \times \bar{\Omega}$, depending on $\varepsilon, \alpha, \beta$, at which $\Phi(\cdot, \cdot)$ attains a global maximum. Note that for any $(x,y) \in \bar{\Omega}^{2} \setminus \Omega_{\alpha}^{2}$,

$$\Phi(x,y) = V(x) - \widehat{V}(y) - \tfrac{1}{\varepsilon}|x-y|^{2} - \alpha\big(\langle x \rangle^{\mu} + \langle y \rangle^{\mu}\big) - \beta\big(\theta^{\alpha}(x) + \theta^{\alpha}(y)\big)$$
$$\leqslant V(x_{0}) - \widehat{V}(x_{0}) - 2\alpha\langle x_{0} \rangle^{\mu} - 2$$
$$= \Phi(x_{0}, x_{0}) - 2\big(1 - \beta\theta^{\alpha}(x_{0})\big) < \Phi(x_{0}, x_{0}).$$

Note here that

$$\beta\theta^{\alpha}(x_{0}) \leqslant \beta\sigma < 1.$$

Hence, one has

$$(\bar{x}, \bar{y}) \in \Omega_{\alpha}^{2}.$$

This also implies that the bound of $(\bar{x}, \bar{y})$ only depends on $\alpha \in (0,1)$, independent of $\varepsilon, \beta \in (0,1)$. Next, from

$$\Phi(\bar{x}, \bar{x}) + \Phi(\bar{y}, \bar{y}) \leqslant 2\Phi(\bar{x}, \bar{y}),$$

one gets

$$V(\bar{x}) - \widehat{V}(\bar{x}) + V(\bar{y}) - \widehat{V}(\bar{y}) - 2\alpha\big(\langle \bar{x} \rangle^{\mu} + \langle \bar{y} \rangle^{\mu}\big) - 2\beta\big(\theta^{\alpha}(\bar{x}) + \theta^{\alpha}(\bar{y})\big)$$
$$\leqslant 2V(\bar{x}) - 2\widehat{V}(\bar{y}) - 2\alpha\big(\langle \bar{x} \rangle^{\mu} + \langle \bar{y} \rangle^{\mu}\big) - \frac{2}{\varepsilon}|\bar{x}-\bar{y}|^{2} - 2\beta\big(\theta^{\alpha}(\bar{x}) + \theta^{\alpha}(\bar{y})\big),$$

which implies

$$\frac{2}{\varepsilon}|\bar{x}-\bar{y}|^{2} \leqslant V(\bar{x}) - V(\bar{y}) + \widehat{V}(\bar{x}) - \widehat{V}(\bar{y}) \leqslant \omega_{\alpha}\big(|\bar{x}-\bar{y}|\big),$$

where $\omega_\alpha(\cdot)$ is a modulus of continuity of the map $x \mapsto V(\cdot) + \widehat{V}(\cdot)$ on $\bar{\Omega}_\alpha$. Thus,

$$\frac{|\bar{x} - \bar{y}|^2}{\varepsilon} = o(1), \qquad \text{as } \varepsilon \to 0. \tag{2.63}$$

We claim that $\bar{x}, \bar{y} \notin \partial\Omega$. If not, say $\bar{x} \in \partial\Omega$, then

$$h(\bar{x}) - \widehat{V}(\bar{y}) - \frac{1}{\varepsilon}|\bar{x} - \bar{y}|^2 - \alpha\big(\langle \bar{x} \rangle^\mu + \langle \bar{y} \rangle^\mu\big) - \beta\theta^\alpha(\bar{y})$$

$$= \Phi(\bar{x}, \bar{y}) = \sup_{x, y \in \Omega_\alpha} \Phi(x, y) \geq \sup_{x \in \Omega_\alpha} \Phi(x, x)$$

$$= \sup_{x \in \Omega_\alpha} \Big[V(x) - \widehat{V}(x) - 2\alpha \langle x \rangle^\mu - 2\beta\theta^\alpha(x) \Big]$$

$$\geqslant V(x_0) - \widehat{V}(x_0) - 2\alpha \langle x_0 \rangle^\mu - 2\beta\theta^\alpha(x_0) > 0,$$

provided $\alpha, \beta > 0$ small enough. This cannot be true if $\varepsilon > 0$ is small enough, by (2.63). Thus, $(\bar{x}, \bar{y}) \in \Omega \times \Omega$. Then the map

$$x \mapsto V(x) - \Big[\widehat{V}(\bar{y}) + \frac{1}{\varepsilon}|x - \bar{y}|^2 + \alpha\big(\langle x \rangle^\mu + \langle \bar{y} \rangle^\mu\big) + \beta\big(\theta^\alpha(x) + \theta^\alpha(\bar{y})\big) \Big]$$

attains a local maximum at $\bar{x}$. By Definition 2.6.17,

$$\lambda V(\bar{x}) - H\Big(\bar{x}, \frac{2}{\varepsilon}(\bar{x} - \bar{y}) + \alpha\mu \langle \bar{x} \rangle^{\mu-2}\bar{x} + \beta\theta_x^\alpha(\bar{x})\Big) \leqslant 0.$$

Likewise, the map

$$y \mapsto \widehat{V}(y) - \Big[V(\bar{x}) - \frac{1}{\varepsilon}|y - \bar{x}|^2 - \alpha\big(\langle \bar{x} \rangle^\mu + \langle y \rangle^\mu\big) - \beta\big(\theta^\alpha(\bar{x}) + \theta^\alpha(y)\big) \Big]$$

attain a local minimum at $\bar{y}$. Thus,

$$\lambda \widehat{V}(\bar{y}) - H\Big(\bar{y}, \frac{2}{\varepsilon}(\bar{x} - \bar{y}) - \alpha\mu \langle \bar{y} \rangle^{\mu-2}\bar{y} - \beta\theta_x^\alpha(\bar{y})\Big) \geqslant 0.$$

Consequently, making use of (2.61), one obtains

$$\lambda\big(V(\bar{x}) - \widehat{V}(\bar{y})\big) \leqslant H\Big(\bar{x}, \frac{2}{\varepsilon}(\bar{x} - \bar{y}) + \alpha\mu \langle \bar{x} \rangle^{\mu-2}\bar{x} + \beta\theta_x^\alpha(\bar{x})\Big)$$

$$-H\Big(\bar{y}, \frac{2}{\varepsilon}(\bar{x} - \bar{y}) - \alpha\mu \langle \bar{y} \rangle^{\mu-2}\bar{y} - \beta\theta_x(\bar{y})\Big)$$

$$\leqslant H\Big(\bar{x}, \frac{2}{\varepsilon}(\bar{x} - \bar{y})\Big) - H\Big(\bar{y}, \frac{2}{\varepsilon}(\bar{x} - \bar{y})\Big)$$

$$+H\Big(\bar{x}, \frac{2}{\varepsilon}(\bar{x} - \bar{y}) + \alpha\mu \langle \bar{x} \rangle^{\mu-2}\bar{x} + \beta\theta_x^\alpha(\bar{x})\Big) - H\Big(\bar{x}, \frac{2}{\varepsilon}(\bar{x} - \bar{y})\Big)$$

$$+H\Big(\bar{y}, \frac{2}{\varepsilon}(\bar{x} - \bar{y})\Big) - H\Big(\bar{y}, \frac{2}{\varepsilon}(\bar{x} - \bar{y}) - \alpha\mu \langle \bar{y} \rangle^{\mu-2}\bar{y} - \beta\theta_x^\alpha(\bar{y})\Big)$$

$$\leqslant \omega\big(|\bar{x}| \vee |\bar{y}|, |\bar{x} - \bar{y}|(1 + \frac{2}{\varepsilon}|\bar{x} - \bar{y}|)\big)$$

$$+\big(\bar{L} + L|\bar{x}|\big)\big(\alpha\mu \langle \bar{x} \rangle^{\mu-2}|\bar{x}| + \beta\big) + \big(\bar{L} + L|\bar{y}|\big)\big(\alpha\mu \langle \bar{y} \rangle^{\mu-2}|\bar{y}| + \beta\big)$$

$$\leqslant \omega\big(K(\alpha), |\bar{x} - \bar{y}| + \frac{2|\bar{x} - \bar{y}|^2}{\varepsilon}\big) + \beta(L + \bar{L})\big(\langle \bar{x} \rangle + \langle \bar{y} \rangle\big)$$

$$+\alpha\mu\big(L\big(\langle \bar{x} \rangle^\mu + \langle \bar{y} \rangle^\mu\big) + \bar{L}\big(\langle \bar{x} \rangle^{\mu-1} + \langle \bar{y} \rangle^{\mu-1}\big)\big).$$

Hence, it follows that

$$0 < V(x_0) - \widehat{V}(x_0) = \Phi(x_0, x_0) + 2\alpha \langle x_0 \rangle^\mu + 2\beta\theta^\alpha(x_0)$$
$$\leqslant \Phi(\bar{x}, \bar{y}) + 2\alpha \langle x_0 \rangle^\mu + 2\beta\theta^\alpha(x_0)$$
$$= V(\bar{x}) - \widehat{V}(\bar{y}) - \frac{1}{\varepsilon}|\bar{x} - \bar{y}|^2 - \alpha(\langle \bar{x} \rangle^\mu + \langle \bar{y} \rangle^\mu) - \beta(\theta^\alpha(\bar{x}) + \theta^\alpha(\bar{y}))$$
$$+ 2\alpha \langle x_0 \rangle^\mu + 2\beta\theta^\alpha(x_0)$$
$$\leqslant \frac{1}{\lambda}\omega\big(K(\alpha), |\bar{x} - \bar{y}| + \frac{2|\bar{x} - \bar{y}|^2}{\varepsilon}\big) + \frac{\beta(L + \bar{L})}{\lambda}(\langle \bar{x} \rangle + \langle \bar{y} \rangle)$$
$$+ \frac{\alpha\mu L}{\lambda}(\langle \bar{x} \rangle^\mu + \langle \bar{y} \rangle^\mu) + \frac{\alpha\mu\bar{L}}{\lambda}(\langle \bar{x} \rangle^{\mu-1} + \langle \bar{y} \rangle^{\mu-1})$$
$$- \alpha(\langle \bar{x} \rangle^\mu + \langle \bar{y} \rangle^\mu) - \beta(\theta^\alpha(\bar{x}) + \theta^\alpha(\bar{y})) + 2\alpha \langle x_0 \rangle^\mu + 2\beta\theta^\alpha(x_0).$$

Then fix $\alpha \in (0,1)$ and send $\varepsilon, \beta \to 0$. By (2.63), we see that $(\bar{x}, \bar{y}) \equiv (\bar{x}_{\varepsilon,\beta}, \bar{y}_{\varepsilon,\beta}) \to (\widetilde{x}, \widetilde{x})$. Thus,

$$0 < V(x_0) - \widehat{V}(x_0) \leqslant 2\alpha\Big[\Big(1 - \frac{\mu L}{\lambda} + \frac{\mu\bar{L}}{\lambda\langle\widetilde{x}\rangle}\Big)\langle\widetilde{x}\rangle^\mu + \langle x_0 \rangle^\mu\Big].$$

Since $0 < \mu < \frac{\lambda}{L}$, the term $[\cdots]$ in the above is bounded from above. Hence, sending $\alpha \to 0$, we end up with a contradiction. Therefore, we must have

$$V(x) \leqslant \widehat{V}(x), \qquad \forall x \in \Omega.$$

By symmetry, we have the uniqueness. $\qquad\qquad\qquad\qquad\qquad\qquad\square$

Note that under (C1)′ and (C2)′, the Hamiltonian $H(\cdot, \cdot)$ for Problem (OC)$^\tau$ satisfies (2.61). Thus, we have the following result.

Corollary 2.6.19. *Let* (C1)′ *and* (C2)′ *hold. Then the value function* $V(\cdot)$ *of Problem* (OC)$^\tau$ *is the unique viscosity solution to the corresponding HJB equation* (2.59). *In particular, the value function* $V(\cdot)$ *of Problem* (T)″ *is the unique viscosity solution to the following HJB equation:*

$$\begin{cases} V(x) - \inf\limits_{u \in U} V_x(x)f(x, u) - 1 = 0, & x \in \Omega, \\ V(x) = 0, & x \in \Omega. \end{cases}$$

Combining the above with Theorem 2.6.13, we obtain a characterization of the controllable set $\mathcal{C}(M)$ in terms of dynamic programming method.

Further, similar to Theorem 2.5.5, we have the following comparison theorem for stationary HJ equations.

Theorem 2.6.20. *Let* $H, \widehat{H} : \mathbb{R}^n \times \mathbb{R}^n \to \mathbb{R}$ *and* $h, \widehat{h} : \mathbb{R}^n \to \mathbb{R}$ *be continuous such that* (2.61) *is satisfied by* $H(\cdot, \cdot)$ *and* $\widehat{H}(\cdot, \cdot)$. *Let*

$$h(x) \leqslant \widehat{h}(x), \qquad x \in \mathbb{R}^n,$$

and

$$H(x,p) \leqslant \widehat{H}(x,p), \qquad \forall x,p \in \mathbb{R}^n.$$

Let $V(\cdot)$ be a viscosity sub-solution of (2.60) and $\widehat{V}(\cdot)$ is a viscosity super-solution of the following:

$$\begin{cases} \lambda \widehat{V}(x) - H(x,\widehat{V}_x(x)) = 0, & x \in \Omega, \\ \widehat{V}(x) = \widehat{h}(x), & x \in \partial\Omega, \end{cases}$$

Then

$$V(x) \leqslant \widehat{V}(x), \qquad x \in \Omega.$$

2.7   Viability Problems

We still consider the state equation (2.36). Let us introduce the following definition.

Definition 2.7.1. Let $\Omega \subseteq \mathbb{R}^n$ be a non-empty set. System (2.36) is said to be *viable* with respect to Ω if for any $x \in \Omega$, there exists a $u(\cdot) \in \mathcal{U}[0,\infty)$ such that

$$X(t; x, u(\cdot)) \in \Omega, \qquad \forall t \geqslant 0.$$

In this case, the set $\Omega \subseteq \mathbb{R}^n$ is also said to enjoy the *viability property* with respect to the state equation (2.36).

Theorem 2.7.2. *Let $(C1)'$ hold. Let $\Omega \subseteq \mathbb{R}^n$ be a C^1 domain, with $\nu : \partial\Omega \to \partial B_1(0)$ being its outward normal map.*

(i) Suppose system (2.36) is viable with respect to Ω. Then

$$\sup_{u \in U} \langle \nu(x), f(x,u) \rangle \leqslant 0, \qquad \forall x \in \partial\Omega. \tag{2.64}$$

(ii) Suppose

$$\sup_{u \in U} \langle \nu(x), f(x,u) \rangle \leqslant -\mu, \qquad \forall x \in \partial\Omega, \tag{2.65}$$

for some $\mu > 0$. Then system (2.36) is viable with respect to $\bar{\Omega}$.

Proof. (i) Suppose (2.64) fails. Then there exists an $x_0 \in \partial\Omega$ such that

$$\sup_{u \in U} \langle \nu(y), f(y,u) \rangle \geqslant \varepsilon > 0, \qquad \forall y \in B_\delta(x_0) \cap \partial\Omega,$$

for some $\varepsilon, \delta > 0$. Then mimicking the proof of Theorem 2.6.6, taking $M = \Omega^c$, we can show that there exists a $\delta' \in (0, \delta)$ such that for any $x \in B_{\delta'}(x_0) \cap \Omega$, and any $u(\cdot) \in \mathcal{U}[0,\infty)$, there exists a $\tau > 0$ such that

$$X(\tau; x, u(\cdot)) \in M \equiv \Omega^c,$$

which is equivalent to

$$X(\tau; x, u(\cdot)) \notin \Omega. \qquad (2.66)$$

This contradicts the viability of (2.36) with respect to Ω. Hence, (2.64) must be true.

(ii) Suppose (2.65) holds. But system (2.36) is not viable with respect to Ω. Then, there exists some $x \in \Omega$ such that for any $u(\cdot) \in \mathcal{U}[0, \infty)$, one must have (2.66). Let

$$\bar{\tau} = \inf\{t > 0 \mid X(t; x, u(\cdot)) \in \partial\Omega\} > 0.$$

Then for any $\varepsilon \in (0, \bar{\tau})$, one has (with $X(\cdot) = X(\cdot\,; x, u(\cdot))$)

$$0 > -d\big(X(\bar{\tau} - \varepsilon), \Omega^c\big)^2 = d\big(X(\bar{\tau}), \Omega^c\big)^2 - d\big(X(\bar{\tau} - \varepsilon), \Omega^c\big)^2$$
$$= 2\int_{\bar{\tau}-\varepsilon}^{\bar{\tau}} d(X(s), \Omega^c)\,\langle\,\partial d(X(s), \Omega^c), f(X(s), u(s))\,\rangle\,ds.$$

Note that $-\nu(X(\bar{\tau}))$ is the outward normal of Ω^c. Thus,

$$|\partial d(X(s), \Omega^c) + \nu(X(\bar{\tau}))| = o(1), \qquad s \to \bar{\tau}.$$

Hence, the above leads to

$$0 > 2\int_{\bar{\tau}-\varepsilon}^{\bar{\tau}} d(X(s), \Omega^c)\,\langle\,\partial d(X(s), \Omega^c), f(X(s), u(s))\,\rangle\,ds$$
$$> -2\int_{\bar{\tau}-\varepsilon}^{\bar{\tau}} d(X(s), \Omega^c)\Big[\,\langle\,\nu(X(\bar{\tau})), f(X(\bar{\tau}), u(s))\,\rangle + o(1)\Big]ds$$
$$\geqslant 2\int_{\bar{\tau}-\varepsilon}^{\bar{\tau}} d(X(s), \Omega^c)\big[\mu + o(1)\big]ds > 0,$$

provided $\varepsilon > 0$ is small enough, which is a contradiction. Thus, (2.36) is viable with respect to Ω. $\qquad\square$

We now give a characterization of viability.

Theorem 2.7.3. *Let* (C1)$'$ *hold, and* $\Omega \subseteq \mathbb{R}^n$ *be non-empty. Suppose for each* $x \in \bar{\Omega}$, $f(x, U)$ *is convex and compact. Then* $\bar{\Omega}$ *enjoys the viability property with respect to system* (2.36) *if and only if the map* $x \mapsto d_\Omega(x)^2$ *satisfies*

$$2L d_\Omega(x)^2 - \inf_{u \in U}[d_\Omega(x)^2]_x f(x, u) \geqslant 0, \quad x \in \mathbb{R}^n, \qquad (2.67)$$

where L *is the Lipschitz constant appears in* (C1)$'$.

Proof. Note that for any fixed $x_0 \in \Omega$,

$$0 \leqslant d_\Omega(x)^2 \equiv \inf_{y \in \Omega} |x - y|^2 \leqslant |x - x_0|^2$$

$$\leqslant \frac{1 + \varepsilon_0}{\varepsilon_0} |x_0|^2 + (1 + \varepsilon_0)|x|^2, \qquad \forall x \in \mathbb{R}^n.$$

Thus, under (C1)$'$, by (2.38), for any $\beta > 0$, the following is well-defined:

$$J^\infty(x; u(\cdot)) = \beta \int_0^\infty e^{-(2L+\beta)s} d_\Omega\big(X(s; x, u(\cdot))\big)^2 ds,$$

and we may formulate an optimal control problem with the above cost functional $J^\infty(x; u(\cdot))$ (and with the empty target set so that $\mathcal{T}(x; u(\cdot)) = \infty$ for all $(x, u(\cdot)) \in \mathbb{R}^n \times \mathcal{U}[0, \infty)$). By defining

$$V(x) = \inf_{u(\cdot) \in \mathcal{U}[0,\infty)} J^\infty(x; u(\cdot)),$$

we know that $V(\cdot)$ is the unique viscosity solution to the following HJB equation:

$$(2L + \beta)V(x) - \inf_{u \in U} V_x(x) f(x, u) - \beta d_\Omega(x)^2 = 0, \qquad x \in \mathbb{R}^n.$$

Now, if $d_\Omega(\cdot)^2$ satisfies (2.67), it is a viscosity super-solution of (2.67), then by comparison (Theorem 2.6.20), one has

$$V(x) \leqslant d_\Omega(x)^2, \qquad \forall x \in \mathbb{R}^n.$$

Thus,

$$V(x) = 0, \qquad \forall x \in \bar{\Omega}.$$

On the other hand, for $x \in \bar{\Omega}$, since $f(x, U)$ is convex and compact, there exists an optimal control $\bar{u}(\cdot)$ such that

$$0 = V(x) = \int_0^\infty e^{-(2L+\beta)s} d_\Omega(X(s; x, \bar{u}(\cdot)))^2 ds.$$

This implies that for any $x \in \bar{\Omega}$,

$$X(s; x, \bar{u}(\cdot)) \in \bar{\Omega}, \qquad \forall s \geqslant 0.$$

Thus, system (2.36) is viable with respect to $\bar{\Omega}$.

Conversely, for any $x \in \bar{\Omega}$, $d_\Omega(x)^2 = 0$ and $\partial_x[d_\Omega(x)^2] = 0$. Thus, $d_\Omega(\cdot)^2$ satisfies (2.67) in the classical sense on Ω. Next, let $x \in \mathbb{R}^n \setminus \bar{\Omega}$. Then there exists an $\bar{x} \in \bar{\Omega}$ such that

$$|x - \bar{x}| = d_\Omega(x).$$

Since $\bar{\Omega}$ has viability property, there exists a $\bar{u}(\cdot)$ such that

$$\bar{X}(t) \equiv X(t; \bar{x}, \bar{u}(\cdot)) \in \Omega, \qquad t \geqslant 0.$$

For any $\varepsilon > 0$ small enough, define

$$\tau_\varepsilon = \varepsilon \wedge \inf\{s \geqslant 0 \mid |X(s) - x| > \varepsilon\} \wedge \inf\{s \geqslant 0 \mid |\bar{X}(s) - \bar{x}| > \varepsilon\},$$

where $X(\cdot) \equiv X(\cdot; x, \bar{u}(\cdot))$. We observe

$$
\begin{aligned}
d_\Omega(X(\tau_\varepsilon))^2 - d_\Omega(x)^2 &= 2 \int_0^{\tau_\varepsilon} d_\Omega(X(s)) \left\langle \partial_x d_\Omega(X(s)), f(X(s), \bar{u}(s)) \right\rangle ds \\
&= 2 \int_0^{\tau_\varepsilon} \left[d_\Omega(x) \left\langle \partial_x d_\Omega(x), f(x, \bar{u}(s)) \right\rangle + o(1) \right] ds \\
&\geqslant 2\tau_\varepsilon \inf_{u \in U} \left[d_\Omega(x) \left\langle \partial_x d_\Omega(x), f(x, u) \right\rangle \right] + o(\varepsilon).
\end{aligned}
$$

On the other hand, for any $x \in \Omega$, since $\bar{X}(\tau_\varepsilon) \in \Omega$, we have

$$d_\Omega(X(\tau_\varepsilon))^2 \leqslant |X(\tau_\varepsilon) - \bar{X}(\tau_\varepsilon)|^2,$$

and

$$
\begin{aligned}
&|X(\tau_\varepsilon) - \bar{X}(\tau_\varepsilon)|^2 \\
&= |x - \bar{x}|^2 + 2 \int_0^{\tau_\varepsilon} \left\langle X(s) - \bar{X}(s), f(X(s), \bar{u}(s)) - f(\bar{X}(s), \bar{u}(s)) \right\rangle ds \\
&\leqslant |x - \bar{x}|^2 + 2L \int_0^{\tau_\varepsilon} |X(s) - \bar{X}(s)|^2 ds.
\end{aligned}
$$

By Gronwall's inequality, one has

$$|X(\tau_\varepsilon) - \bar{X}(\tau_\varepsilon)|^2 \leqslant e^{2L\tau_\varepsilon} |x - \bar{x}|^2.$$

Hence,

$$
\begin{aligned}
&2\tau_\varepsilon \inf_{u \in U} d_\Omega(x) \left\langle \partial_x d_\Omega(x), f(x, u) \right\rangle + o(\varepsilon) \\
&\leqslant d_\Omega(X(\tau_\varepsilon))^2 - d_\Omega(x)^2 \leqslant |X(\tau_\varepsilon) - \bar{X}(\tau_\varepsilon)|^2 - d_\Omega(x)^2 \\
&\leqslant e^{2L\tau_\varepsilon} |x - \bar{x}|^2 - d_\Omega(x)^2 = \left[e^{2L\tau_\varepsilon} - 1 \right] d_\Omega(x)^2.
\end{aligned}
$$

Dividing τ_ε and sending $\varepsilon \to 0$, we obtain

$$2d_\Omega(x) \inf_{u \in U} \left\langle \partial_x d_\Omega(x), f(x, u) \right\rangle \leqslant 2L d_\Omega(x)^2,$$

which implies (2.67). $\qquad\qquad\qquad\qquad\qquad\qquad\qquad\qquad\quad \square$

2.8   Non-Uniqueness of Solutions to HJ Equations*

From previous sections, it seems to give people an impression that viscosity solutions to HJB equation are always unique. Such an impression should be corrected. In this section, we will look at the non-uniqueness issue of HJB equations. First, let us look at HJB equation of form

$$V(x) + H(x, V_x(x)) = 0, \qquad x \in \mathbb{R}^n.$$

This kind of equation appears when the optimal control is posed on $[0, \infty)$ with time-invariant coefficients (see Section 2.6). Further, in the case that the control domain is unbounded, the map $p \mapsto H(x, p)$ could be super-linear.

Example 2.8.1. Consider HJB equation: $(q > 1)$

$$V(x) - \frac{q-1}{q} |V_x(x)|^q = 0, \qquad x \in \mathbb{R}^n. \tag{2.68}$$

Clearly, $V(x) \equiv 0$ is a solution. Also, if we let

$$V(x) = \frac{q-1}{q} |x|^{\frac{q}{q-1}}, \qquad x \in \mathbb{R}^n, \tag{2.69}$$

then

$$|V_x(x)| = |x|^{\frac{1}{q-1}}, \qquad x \in \mathbb{R}^n,$$

and $V(\cdot)$ is another (viscosity) solution to (2.68). One can show that the viscosity solution is unique within the class of continuous functions $V(\cdot)$ satisfying

$$\sup_{x,y \in \mathbb{R}^n, x \neq y} \frac{|V(x) - V(y)|}{(1 + |x|^{\mu-1} + |y|^{\mu-1})} < \infty,$$

with $1 < \mu < q$, for which the function (2.69) is excluded. The proof is a modification of that for Theorem 2.6.18.

Next, we look at a time-varying case.

Example 2.8.2. Consider the following HJ equation:

$$\begin{cases} V_t(t, x) + b(x) V_x(t, x) = 0, & (t, x) \in [0, T] \times \mathbb{R}, \\ V(T, x) = h(x), & x \in \mathbb{R}. \end{cases} \tag{2.70}$$

Suppose $X(\cdot\,; x)$ is a solution to the following:

$$\begin{cases} \dot{X}(t) = b(X(t)), & t \in [0, T], \\ X(0) = x, \end{cases} \tag{2.71}$$

and let

$$V(t, x) = h(X(T - t; x)), \qquad (t, x) \in [0, T] \times \mathbb{R}.$$

We claim that $V(\cdot, \cdot)$ is a viscosity solution to HJ equation (2.70). In fact, it is clear that the terminal condition is satisfied. Also, let $V(\cdot, \cdot) - \varphi(\cdot, \cdot)$ attain a local maximum at (t_0, x_0). Then

$$V(\bar{t}, \bar{x}) - \varphi(\bar{t}, \bar{x}) \leqslant V(t_0, x_0) - \varphi(t_0, x_0), \qquad \forall (\bar{t}, \bar{x}) \text{ near } (t_0, x_0).$$

In particular, the above holds for

$$\bar{t} = t, \quad \bar{x} = X(t - t_0; x_0),$$

with t near t_0. Observe that

$$\begin{aligned}
V(\bar{t}, \bar{x}) = V(t, X(t - t_0; x_0)) &= h\big(X(T - t; X(t - t_0; x_0))\big) \\
&= h(X(T - t_0; x_0)) = V(t_0, x_0).
\end{aligned}$$

Hence,

$$\varphi(t, X(t - t_0; x_0)) \geqslant \varphi(t_0, x_0), \qquad \forall t \text{ near } t_0.$$

Consequently,

$$\begin{aligned}
0 = \frac{d}{dt} \varphi(t, X(t - t_0; x_0))\Big|_{t=t_0} \\
= \varphi_t(t_0, X(0; x_0)) + \varphi_x(t_0, X(0; x_0)) b(X(0; x_0)) \\
= \varphi_t(t_0, x_0) + b(x_0) \varphi_x(t_0, x_0).
\end{aligned}$$

This means that $V(\cdot, \cdot)$ is a viscosity sub-solution to (2.70). In the same way, we can show that $V(\cdot, \cdot)$ is also a viscosity super-solution to (2.70).

Next, suppose $f : \mathbb{R} \to \mathbb{R}$ is continuously differentiable, strictly increasing, and $f(\mathbb{R}) = \mathbb{R}$. which implies that $f^{-1} : \mathbb{R} \to \mathbb{R}$ exists. Let

$$b(x) = f'(f^{-1}(x)), \qquad \forall x \in \mathbb{R}.$$

Then we claim that

$$X(t; x) = f(t + f^{-1}(x)), \qquad (t, x) \in \mathbb{R} \times \mathbb{R}$$

is a solution to ODE (2.71). In fact,

$$\begin{aligned}
\frac{d}{dt} X(t; x) = f'(t + f^{-1}(x)) &= f'(f^{-1}(f(t + f^{-1}(x)))) \\
&= (f' \circ f^{-1})(X(t; x)) = b(X(t; x)),
\end{aligned}$$

and

$$X(0; x) = f(f^{-1}(x)) = x.$$

This proves our claim.

Suppose we can construct two different continuously differentiable and strictly increasing functions $f, \bar{f} : \mathbb{R} \to \mathbb{R}$ such that

$$b(y) \equiv f'(f^{-1}(y)) = \bar{f}'(\bar{f}^{-1}(y)), \qquad y \in \mathbb{R}.$$

Then we will have two different viscosity solutions to the HJ equation (2.70):

$$V(t, x) = h\big(f(T - t + f^{-1}(x))\big), \qquad (t, x) \in [0, T] \times \mathbb{R},$$

and

$$\bar{V}(t, x) = h\big(\bar{f}(T - t + \bar{f}^{-1}(x))\big), \qquad (t, x) \in [0, T] \times \mathbb{R}.$$

Therefore, viscosity solutions to HJ equation (2.70) are not unique in the class of continuous functions.

Now, we construct two different functions $f, \bar{f} : \mathbb{R} \to \mathbb{R}$ satisfying the above-mention conditions.

Let $K \subseteq [0, 1]$ be a Cantor set with a strictly positive Lebesgue measure. Let

$$g(x) = d(x, K)^4, \qquad x \in [0, 1].$$

Then

$$g(x) = 0, \quad \text{iff} \quad x \in K,$$

and

$$g'(x) = 4d(x, K)^3, \qquad g''(x) = 12d(x, K)^2, \qquad x \in [0, 1],$$

which are bounded. We extend $g(\cdot)$ to $\mathbb{R}$ so that

$$\begin{cases} 0 \leqslant g(x) \leqslant 1, & x \in \mathbb{R}, \\ 0 < g(x) \leqslant 1, & x \in (-1, 0) \cup (1, 2), \\ g(x) \geqslant \dfrac{1}{2}, & x \in (\infty, -1] \cup [2, \infty), \\ |g'(x)| + |g''(x)| \leqslant C, & x \in \mathbb{R}. \end{cases}$$

Next, we define

$$f(x) = \int_0^x g(s)ds, \qquad x \in \mathbb{R}.$$

Then

$$f'(x) = g(x) \geqslant 0, \qquad \forall x \in \mathbb{R},$$

and

$$|f(x_1) - f(x_2)| \leqslant |x_1 - x_2|, \qquad x_1, x_2 \in \mathbb{R}.$$

Thus, $f(\cdot)$ is uniformly Lipschitz. Further, since K does not contain any interval, for any $x_1 < x_2$, with $|[x_1, x_2] \cap [0, 1]| > 0$,

$$\left| \left([x_1, x_2] \cap [0, 1] \right) \setminus K \right| > 0,$$

where $|A|$ is the Lebesgue measure of the set A. Hence, for any $x_1 < x_2$,

$$f(x_2) - f(x_1) = \int_{x_1}^{x_2} g(s)ds$$

$$= \int_{[x_1, x_2] \setminus [0,1]} g(s)ds + \int_{[x_1, x_2] \cap [0,1] \setminus K} d(s, K)^4 ds > 0,$$

which means that $f(\cdot)$ is strictly increasing on $\mathbb{R}$. Further, since $g(x) \geqslant \frac{1}{2}$ for $x \in \mathbb{R} \setminus [-1, 2]$, we see that

$$\lim_{x \to \infty} f(x) = \infty, \qquad \lim_{x \to -\infty} f(x) = -\infty.$$

Therefore, $f^{-1} : \mathbb{R} \to \mathbb{R}$ is well-defined and continuous. Moreover, for any $x_1, x_2 \in (-\infty, -1]$, with $x_1 < x_2$,

$$f(x_2) - f(x_1) = \int_{x_1}^{x_2} g(s)ds \geqslant \frac{1}{2}(x_2 - x_1),$$

which implies

$$|f^{-1}(y_1) - f^{-1}(y_2)| \leqslant 2|y_1 - y_2|, \qquad \forall y_1, y_2 \in (-\infty, -1].$$

Likewise,

$$|f^{-1}(y_1) - f^{-1}(y_2)| \leqslant 2|y_1 - y_2|, \qquad \forall y_1, y_2 \in [2, \infty).$$

Hence, $f^{-1}(\cdot)$ is uniformly continuous on $\mathbb{R}$.

Next, let

$$\alpha(x) = x + \int_0^x I_K(s)ds, \qquad x \in \mathbb{R}.$$

Then $\alpha(\cdot)$ is continuous and strictly increasing on $\mathbb{R}$ with $\alpha(\mathbb{R}) = \mathbb{R}$. Thus, $\alpha^{-1} : \mathbb{R} \to \mathbb{R}$ is well-defined, strictly increasing and continuous. Further, for any $x_1 < x_2$,

$$\alpha(x_2) - \alpha(x_1) = x_2 - x_1 + \int_{x_1}^{x_2} I_K(s)ds \geqslant x_2 - x_1.$$

This implies that

$$|\alpha^{-1}(z_1) - \alpha^{-1}(z_2)| \leqslant |z_1 - z_2|, \qquad \forall z_1, z_2 \in \mathbb{R}.$$

Namely, $\alpha^{-1}(\cdot)$ is uniformly Lipschitz. Now, we define

$$\bar{f}(z) = f(\alpha^{-1}(z)), \qquad \forall z \in \mathbb{R}. \tag{2.72}$$

Then $\bar{f} : \mathbb{R} \to \mathbb{R}$ is uniformly Lipschitz and strictly increasing with $\bar{f}(\mathbb{R}) = \mathbb{R}$. Apparently, $\bar{f}(\cdot)$ and $f(\cdot)$ are different. Further,

$$\bar{f}^{-1}(y) = \alpha(f^{-1}(y)), \qquad y \in \mathbb{R}$$

is uniformly continuous, and

$$\alpha^{-1}(\bar{f}^{-1}(y)) = f^{-1}(y), \qquad \forall y \in \mathbb{R}.$$

We now claim that

$$f'(f^{-1}(y)) = \bar{f}'(\bar{f}^{-1}(y)), \qquad \forall y \in \mathbb{R}. \tag{2.73}$$

To show this, we split it into two cases.

Case 1. Let $\bar{y} \in \mathbb{R}$ such that

$$\bar{x} \equiv f^{-1}(\bar{y}) \notin K.$$

Since K is closed, the above implies that there exists a $\delta > 0$ such that

$$\mathcal{O}_\delta(\bar{x}) \cap K = \emptyset,$$

where $\mathcal{O}_\delta(\bar{x}) = (\bar{x} - \delta, \bar{x} + \delta)$. Hence,

$$\alpha(x) = x + \int_0^x I_K(s)ds = x + \int_0^{\bar{x}} I_K(s)ds, \qquad \forall x \in \mathcal{O}_\delta(\bar{x}).$$

This implies that $\alpha(\cdot)$ is differentiable at $\bar{x} \equiv f^{-1}(\bar{y})$ and

$$\alpha'(\bar{x}) \equiv \alpha'(f^{-1}(\bar{y})) = 1.$$

Consequently, by implicit function theorem, $\alpha^{-1}(\cdot)$ is differentiable at $\alpha(\bar{x}) \equiv \alpha(f^{-1}(\bar{y})) = \bar{f}^{-1}(\bar{y}) \equiv \bar{z}$ with

$$(\alpha^{-1})'(\bar{z}) = (\alpha^{-1})'(\alpha(\bar{x})) = \frac{1}{\alpha'(\bar{x})} = 1.$$

Hence, by (2.72), $\bar{f}(\cdot)$ is differentiable at $\bar{z} = \bar{f}^{-1}(\bar{y})$ with

$$\bar{f}'(\bar{f}^{-1}(\bar{y})) = f'(\alpha^{-1}(\bar{f}^{-1}(\bar{y})))(\alpha^{-1})'(\bar{f}^{-1}(\bar{y}))$$
$$= f'(f^{-1}(\bar{y}))(\alpha^{-1})'(\bar{z}) = f'(f^{-1}(\bar{y})).$$

This shows that (2.73) holds for Case 1.

Case 2. Let $\bar{y} \in \mathbb{R}$ such that

$$\bar{x} \equiv f^{-1}(\bar{y}) \in K \subseteq [0, 1].$$

Observe the following:

$$|\bar{f}(z) - \bar{f}(\bar{f}^{-1}(\bar{y}))| = |f(\alpha^{-1}(z)) - f(\alpha^{-1}(\bar{f}^{-1}(\bar{y})))|$$

$$\leqslant \left| \int_0^1 f'\left(\alpha^{-1}(\bar{f}^{-1}(\bar{y})) + \lambda[\alpha^{-1}(z) - \alpha^{-1}(\bar{f}^{-1}(\bar{y}))]\right) d\lambda \right| \left| \alpha^{-1}(z) - \alpha^{-1}(\bar{f}^{-1}(\bar{y})) \right|$$

$$\leqslant \left| \int_0^1 f'\left(f^{-1}(\bar{y}) + \lambda[\alpha^{-1}(z) - \alpha^{-1}(\bar{f}^{-1}(\bar{y}))]\right) d\lambda \right| \left| z - \bar{f}^{-1}(\bar{y}) \right|.$$

Hence, by the continuity of $f'(\cdot)$ and the assumption that $\bar{x} \in K$,

$$\varlimsup_{z \to \bar{f}^{-1}(\bar{y})} \left| \frac{\bar{f}(z) - \bar{f}(\bar{f}^{-1}(\bar{y}))}{z - \bar{f}^{-1}(\bar{y})} \right|$$

$$\leqslant \varlimsup_{z \to \bar{f}^{-1}(\bar{y})} \left| \int_0^1 f'\left(f^{-1}(\bar{y}) + \lambda[\alpha^{-1}(z) - \alpha^{-1}(\bar{f}^{-1}(\bar{y}))]\right) d\lambda \right|$$

$$= |f'(f^{-1}(\bar{y}))| = |f'(\bar{x})| = d(\bar{x}, K)^4 = 0,$$

which implies

$$\bar{f}'(\bar{f}^{-1}(\bar{y})) = 0 = f'(f^{-1}(\bar{y})).$$

This implies that (2.73) holds for Case 2 as well.

Note that $b(\cdot)$ must not be globally Lipschitz, since otherwise, the solution to (2.71) with given initial condition has to be unique. Also, because $b(\cdot)$ is not Lipschitz, the second condition in (2.20) fails.

The uniqueness of viscosity solution to the HJ equation is essentially a proper compatibility problem of the Hamiltonian (or the coefficients of the state equation and cost functional, and the control domain of the control problem) with the class of the functions to which the viscosity solution belongs. In Section 2.4, under (C5), we have (2.20). In this case, the class of functions to which the viscosity solution belongs is $C([0,T] \times \mathbb{R}^n)$. Note that in Example 2.8.2,

$$H(t, x, p) = b(x)p, \qquad (t, x, p) \in [0, T] \times \mathbb{R}^n \times \mathbb{R}^n.$$

Since $b(\cdot)$ is not globally Lipschitz continuous, the second condition in (2.20) is not satisfied. In fact, for this case, we only have

$$|H(t, x, p) - H(t, y, p)| = |b(x) - b(y)||p|$$

$$\leqslant \omega(|x| \vee |y|, |x - y|)|p|,$$

which is different from the second relation in (2.20). Note that the appearance of $|x - y|(1 + |p|)$ plays a crucial role in the proof of uniqueness of viscosity solutions.

This causes the uniqueness to fail in the class $C([0,T] \times \mathbb{R}^n)$. However, one can show that (2.70) still have a unique viscosity solution in the class of all functions $v(\cdot, \cdot) \in C([0,T] \times \mathbb{R}^n)$ such that for some $L > 0$,

$$|v(t, x) - v(t, y)| \leqslant L|x - y|, \qquad \forall t \in [0, T], \ x, y \in \mathbb{R}^n.$$

2.9   Brief Historic Remarks

In the early 1950s, L. S. Pontryagin[1] of the Steklov Institute of Mathematics (in the former Soviet Union), requested by the Board of Directors of the institute, started a seminar on applied problems of mathematics [96]. Theoretical engineers, including someone working in certain military units were invited as speakers. Before long, the accumulated problems led to the formulation of two major mathematical problems: singular perturbation of ordinary differential equations, and differential pursuit and evasion games. The latter led to the time-optimal control problems, and then to the general optimal control problems. In the investigation of optimal control problems, Pontryagin realized that the classical Euler-Lagrange equation approach could not be applied to solve the formulated optimal control problems. They "*must invent a new calculus of variations*" (according to Gamkrelidze [52]). In 1956, *Pontryagin's maximum principle* was firstly announced (and derived) ([18]). The first proof was outlined in 1958 ([17]) and a detailed proof was published in 1960 ([19], see also [90], [91]). A more systematic presentation was carried out in the book *Mathematical Theory of Optimal Processes* in 1960 ([97]). For some interesting history about the birth of maximum principle, see Pesch–Bulirsch [87], Sussmann-Willems [113], Gamkrelidze [52], Pesch–Plail [88], and Pesch [89]. Also, we mention some earlier closely related works of Carathéodory [26], Hestenes [54], [55].

The original proof of the maximum principle was very technical. Thanks to the Ekeland's variational principle [38, 39], and the improved spike variation techniques due to Li–Yao [68] and Fattorini [42, 43] (see also Li–Yong [69, 70]), the proof becomes much easier now. The presentation for the proof of maximum principle (Theorem 2.3.1) here is based on that in [73] (see also [128]).

On the other hand, in the early 1950s, at the RAND Corporation (in USA), Bellman introduced *dynamic programming* method in 1950s ([5], [6], [7]). Some interesting historic stories can be found in Dreyfus [37]. For the continuous-time case, when the value function is differentiable, it satisfies the Hamilton-Jacobi-Bellman equation. However, on one hand, the value function is usually not differentiable, on the other hand, even for very smooth coefficients, the (nonlinear) first order HJB equation may have no classical solution. Hence, for a long time, dynamic programming method remained non-rigorous, for continuous-time optimal control problems. In

---

[1]By that time, Pontryagin was already a worldwide well-known topologist.

1983, Crandall and Lions ([30], [31], [32]) introduced the notion of viscosity solution for HJB equations, which can be used to characterize the value function as the unique viscosity solution of the corresponding HJB equation, under very general conditions.

The idea of the proofs for Theorems 2.5.3 and 2.6.18 are mainly based on the work of Ishii [60].

The existence of optimal control is pretty standard. Our Theorem 2.2.1 is a modification of a similar result found in the book by Berkovitz ([10]). Theorems 2.6.6 and 2.6.7 on the controllability are inspired by some relevant results for differential pursuit games found in the author's work [120]. Theorem 2.6.10 seems to be new, which is inspired by the approach found in the author's work on evadable sets ([121]). Theorem 2.6.13 on the characterization of controllable set is based on the work of Peng and Yong [85]. Theorem 2.6.15 is a modification of a result from the book by Bardi and Capuzzo-Dolcetta [1]. Theorem 2.7.2 on the viability was inspired by a similar result on differential evasion games found in the author's work [119]. Theorem 2.7.3 is a deterministic version of a result found in [24].

Although in most of interesting cases, viscosity solutions to relevant HJ equations are unique, examples of non-uniqueness of viscosity solutions exist, which was pointed out in the first work of Crandall and Lions [30] on viscosity solutions. Our Example 2.8.2 is based on the presentation of Biton [16]. As pointed out in [30] and [16], more general situations can be found in the book by Beck [4].

Chapter 3

Two-Person Zero Sum Differential Games — A General Consideration

3.1 Strategies

For convenience, let us briefly recall some relevant material presented in Section 1.3.2. Consider

$$\begin{cases} \dot{X}(s) = f(s, X(s), u_1(s), u_2(s)), & s \in [t, T], \\ X(t) = x, \end{cases} \tag{3.1}$$

for some map $f : [0, T] \times \mathbb{R}^n \times U_1 \times U_2 \to \mathbb{R}^n$, with U_1 and U_2 being metric spaces. In the above, $X(\cdot)$ is the state trajectory, and $u_1(\cdot)$ and $u_2(\cdot)$ are controls taken by two involved persons called the players. For convenience, we label them as Players 1 and 2, respectively. For $i = 1, 2$, let the set of admissible controls for Player i be given by

$$\mathcal{U}_i[t, T] = \{ u_i : [t, T] \to U_i \mid u_i(\cdot) \text{ is measurable} \}.$$

We introduce the following standing assumption.

(DG1) The map $f : [0, T] \times \mathbb{R}^n \times U_1 \times U_2 \to \mathbb{R}^n$ is continuous and there exists a constant $L > 0$ such that

$$\begin{cases} |f(t, x_1, u_1, u_2) - f(t, x_2, u_1, u_2)| \leqslant L|x_1 - x_2|, \\ \qquad\qquad (t, u_1, u_2) \in [0, T] \times U_1 \times U_2, \ x_1, x_2 \in \mathbb{R}^n, \\ |f(t, 0, u_1, u_2)| \leqslant L, \qquad (t, u_1, u_2) \in [0, T] \times U_1 \times U_2. \end{cases}$$

Similar to Proposition 2.1.1, we have the following result.

Proposition 3.1.1. Let (DG1) hold. Then, for any initial pair $(t, x) \in [0, T] \times \mathbb{R}^n$, and any control pair $(u_1(\cdot), u_2(\cdot)) \in \mathcal{U}_1[t, T] \times \mathcal{U}_2[t, T]$, there exists a unique solution $X(\cdot) \equiv X(\cdot\,; t, x, u_1(\cdot), u_2(\cdot))$ to (3.1). Moreover, the following estimates hold:

$$
\begin{cases}
|X(s;t,x,u_1(\cdot),u_2(\cdot))| \leqslant e^{L(s-t)}(1+|x|) - 1, \\
|X(s;t,x,u_1(\cdot),u_2(\cdot)) - x| \leqslant \left[e^{L(s-t)} - 1\right](1+|x|), \\
\quad\quad 0 \leqslant t \leqslant s \leqslant T, \ x \in \mathbb{R}^n, \ u_i(\cdot) \in \mathcal{U}_i[0,T],
\end{cases}
\tag{3.2}
$$

and

$$
|X(s;t,x_1,u_1(\cdot),u_2(\cdot)) - X(s;t,x_2,u_1(\cdot),u_2(\cdot))| \leqslant e^{L(s-t)}|x_1 - x_2|,
\tag{3.3}
$$
$$
0 \leqslant t \leqslant s \leqslant T, \ x_1, x_2 \in \mathbb{R}^n, \ u_i(\cdot) \in \mathcal{U}_i[0,T].
$$

Next, we introduce cost functionals for each player: For $i = 1, 2$,

$$
J_i(t,x;u_1(\cdot),u_2(\cdot)) = \int_0^T g_i(s,X(s),u_1(s),u_2(s))ds + h_i(X(T)),
\tag{3.4}
$$

where $g_i : [0,T] \times \mathbb{R}^n \times U_1 \times U_2 \to \mathbb{R}$ and $h_i : \mathbb{R}^n \to \mathbb{R}$ are some given maps, for which we introduce the following hypothesis.

(DG2) For $i = 1, 2$, the maps $g_i : [0,T] \times \mathbb{R}^n \times U_1 \times U_2 \to \mathbb{R}$ and $h_i : \mathbb{R}^n \to \mathbb{R}$ are continuous and there exists a local modulus of continuity $\omega : \mathbb{R}_+ \times \mathbb{R}_+ \to \mathbb{R}_+$ such that

$$
|g_i(s,x_1,u_1,u_2) - g_i(s,x_2,u_1,u_2)| + |h_i(x_1) - h_i(x_2)|
$$
$$
\leqslant \omega\big(|x_1| \vee |x_2|, |x_1 - x_2|\big), \quad \forall(s,u_1,u_2) \in \mathbb{R}_+ \times U_1 \times U_2, x_1, x_2 \in \mathbb{R}^n,
$$

and

$$
\sup_{(s,u_1,u_2)\in\mathbb{R}_+\times U_1 \times U_2} |g_i(s,0,u_1,u_2)| \equiv g_i^0 < \infty.
$$

It is clear that under (DG1)–(DG2), the maps $(t,x,u_1(\cdot),u_2(\cdot)) \mapsto J_i(t,x;u_1(\cdot),u_2(\cdot))$ are well-defined. Hence, we can pose the following problem.

Problem (DG). For given initial pair $(t,x) \in [0,T) \times \mathbb{R}^n$, Player i wants to choose a control $\bar{u}_i(\cdot) \in \mathcal{U}_i[t,T]$ such that $J_i(t,x,u_1(\cdot),u_2(\cdot))$ is minimized.

We refer to Problem (DG) as a *two-person differential game.*

Let us first mimic game theory (Chapter 1) and optimal control theory (Chapter 2) to approach Problem (DG). We have the following natural definitions.

Definition 3.1.2. Given initial pair $(t,x) \in [0,T) \times \mathbb{R}^n$.

(i) A pair $(\bar{u}_1(\cdot), \bar{u}_2(\cdot)) \in \mathcal{U}_1[t, T] \times \mathcal{U}_2[t, T]$ is called an *open-loop Pareto optimum* of Problem (DG) if there exists no other pair $(u_1(\cdot), u_2(\cdot)) \in \mathcal{U}_1[t, T] \times \mathcal{U}_2[t, T]$ such that

$$J_i(t, x; u_1(\cdot), u_2(\cdot)) \leqslant J_i(t, x; \bar{u}_1(\cdot), \bar{u}_2(\cdot)), \qquad i = 1, 2,$$

and at least one of the two inequalities is strict.

(ii) A pair $(\bar{u}_1(\cdot), \bar{u}_2(\cdot)) \in \mathcal{U}_1[t, T] \times \mathcal{U}_2[t, T]$ is called an *open-loop Nash equilibrium* of Problem (DG) if

$$\begin{cases} J_1(t, x; \bar{u}_1(\cdot), \bar{u}_2(\cdot)) \leqslant J_1(t, x; u_1(\cdot), \bar{u}_2(\cdot)), & \forall u_1(\cdot) \in \mathcal{U}_1[t, T], \\ J_2(t, x; \bar{u}_1(\cdot), \bar{u}_2(\cdot)) \leqslant J_2(t, x; \bar{u}_1(\cdot), u_2(\cdot)), & \forall u_2(\cdot) \in \mathcal{U}_2[t, T]. \end{cases}$$

With the above definitions, a two-person differential game can be treated as a two-person game and one might try to use the theory presented in Chapter 1 to discuss Problem (DG). However, it is not trivial in general. The reason is the following: one needs to introduce some topology on each $\mathcal{U}_i[t, T]$ so that they are compact metric spaces, and under these topologies, the maps $(u_1(\cdot), u_2(\cdot)) \mapsto J_i(t, x; u_1(\cdot), u_2(\cdot))$ are continuous. Such an idea works for some special cases, for example, in the case that the state equation is linear and the cost functionals are quadratic. We will present some relevant results in a later chapter. In general, however, the above mentioned approach seems to be a little difficult.

On the other hand, from a practical point of view, the above "open-loop" approach seems to be a little questionable. Suppose, say, Player 1 and Player 2 have opposite goals. Then at any time $s \in [t, T)$, the value of the control $u_1(\tau)$ for $\tau \in (s, T]$ by Player 1 will not be revealed to Player 2 (and actually, Player 1 might not know what the value $u_1(\tau)$ will be at time moment $s < \tau$). The same situation exists if one exchanges the positions of Players 1 and 2. From such a simple observation, we immediately see that differential game is actually much more complicated than optimal control problem (which can be regarded as a single-player differential game).

In order to take into account of the above consideration in the study of Problem (DG), we introduce the following notion.

Definition 3.1.3. For given $t \in [0, T] \times \mathbb{R}^n$, a map $\alpha_1 : \mathcal{U}_2[t, T] \to \mathcal{U}_1[t, T]$ is called an *Elliott–Kalton strategy* (E–K strategy, for short) of Player 1 on $[t, T]$ if it is *non-anticipating* in the following sense: for any $u_2(\cdot), \tilde{u}_2(\cdot) \in \mathcal{U}_2[t, T]$, and any $\tau \in (t, T]$,

$$u_2(s) = \tilde{u}_2(s), \qquad \text{a.e. } s \in [t, \tau],$$
$$\Rightarrow \quad \alpha_1[u_2(\cdot)](s) = \alpha_1[\tilde{u}_2(\cdot)](s), \qquad \text{a.e. } s \in [t, \tau].$$

We denote $\mathcal{A}_1[t,T]$ to be the set of all E–K strategies of Player 1 on $[t,T]$. Similarly, we may define E–K strategies of Player 2 on $[t,T]$, and denote $\mathcal{A}_2[t,T]$ the set of all E–K strategies of Player 2 on $[t,T]$.

Here are some simple examples for E–K strategies of Player 1:

$$\alpha_1[u_2(\cdot)](s) = \theta(s, u_2(s)), \qquad \forall s \in [t,T], \quad u_2(\cdot) \in \mathcal{U}_2[t,T],$$

$$\alpha_1[u_2(\cdot)](s) = \int_t^s \theta(r, u_2(r))dr, \qquad \forall s \in [t,T], \quad u_2(\cdot) \in \mathcal{U}_2[t,T].$$

It is easy to cook up more examples.

Sometimes, we need to consider the differential game on $[0,\infty)$, for this, we may similarly define the admissible control sets $\mathcal{U}_i[t,\infty)$ and the E–K strategy sets $\mathcal{A}_i[t,\infty)$.

3.2 Open-Loop Pareto Optima and Nash Equilibria

For any given $\lambda = (\lambda_1, \lambda_2) \in (0,\infty)^2$, let us introduce the following functional:

$$J^\lambda(t, x; u_1(\cdot), u_2(\cdot))$$
$$= \lambda_1 J_1(t, x; u_1(\cdot), u_2(\cdot)) + \lambda_2 J_2(t, x; u_1(\cdot), u_2(\cdot))$$
$$= \int_t^T g^\lambda(s, X(s), u_1(s), u_2(s))ds + h^\lambda(X(T)),$$

where

$$\begin{cases} g^\lambda(t, x, u_1, u_2) = \lambda_1 g_1(t, x, u_1, u_2) + \lambda_2 g_2(t, x, u_1, u_2), \\ h^\lambda(x) = \lambda_1 h_1(x) + \lambda_2 h_2(x). \end{cases}$$

We pose the following optimal control problem.

Problem (DG)$^\lambda$. For any given $(t, x) \in [0, T) \times \mathbb{R}^n$, find a pair $(\bar{u}_1^\lambda(\cdot), \bar{u}_2^\lambda(\cdot)) \in \mathcal{U}_1[\tau, T] \times \mathcal{U}_2[t, T]$ such that

$$J^\lambda(t, x; \bar{u}_1^\lambda(\cdot), \bar{u}_2^\lambda(\cdot)) = \inf_{(u_1(\cdot), u_2(\cdot)) \in \mathcal{U}_1[t,T] \times \mathcal{U}_2[t,T]} J^\lambda(t, x; u_1(\cdot), u_2(\cdot))$$
$$\equiv V^\lambda(t, x).$$

Similar to Proposition 1.2.3, we have the following result.

Proposition 3.2.1. *For any* $\lambda = (\lambda_1, \lambda_2) \in (0,\infty)^2$ *and* $(t, x) \in [0, T) \times \mathbb{R}^n$, *suppose* $(\bar{u}_1^\lambda(\cdot), \bar{u}_2^\lambda(\cdot)) \in \mathcal{U}_1[t, T] \times \mathcal{U}_2[t, T]$ *is an optimal control of Problem* (DG)$^\lambda$ *for* (t, x). *Then it is an open-loop Pareto optimum of Problem* (DG) *for* (t, x).

Since Problem $(DG)^\lambda$ can be regarded as a standard optimal control problem, all the theory presented in Chapter 2 concerning optimal control problem can be applied here to study those Pareto optima that can be obtained as optimal controls of Problem $(DG)^\lambda$ for some $\lambda \in (0, \infty)^2$. For examples, we have existence of a Pareto optimum $(\bar{u}_1^\lambda(\cdot), \bar{u}_2^\lambda(\cdot))$, maximum principle for Pareto optima, and viscosity solution of Hamilton-Jacobi-Bellman equation characterization for the value function $V^\lambda(\cdot, \cdot)$. We leave the details to the readers.

For open-loop Nash equilibria of Problem (DG), we present a Pontryagin type maximum principle. To simplify the presentation, we omit the precise conditions on the involved functions.

Theorem 3.2.2. *Let* $(\bar{u}_1(\cdot), \bar{u}_2(\cdot)) \in \mathcal{U}_1[t, T] \times \mathcal{U}_2[t, T]$ *be an open-loop Nash equilibrium of Problem (DG). Then*

$$H_1(s, \bar{X}(s), \bar{u}_1(s), \bar{u}_2(s), \psi_1(s)) = \max_{u_1 \in U_1} H_1(s, \bar{X}(s), u_1, \bar{u}_2(s), \psi_1(s)),$$

$$H_2(s, \bar{X}(s), \bar{u}_1(s), \bar{u}_2(s), \psi_2(s)) = \max_{u_2 \in U_2} H_2(s, \bar{X}(s), \bar{u}_1(s), u_2, \psi_2(s)),$$

$$\text{a.e. } s \in [t, T],$$

where for $i = 1, 2$,

$$H_i(t, x, u_1, u_2, \psi) = \langle \psi, f(t, x, u_1, u_2) \rangle - g_i(t, x, u_1, u_2).$$

$$\begin{cases} \dot{\psi}_i(s) = -f_x(s, \bar{X}(s), \bar{u}_1(s), \bar{u}_2(s))^T \psi_i(s) \\ \qquad\qquad + (g_i)_x(s, \bar{X}(s), \bar{u}_1(s), \bar{u}_2(s))^T, \quad s \in [t, T], \\ \psi_i(T) = -(h_i)_x(\bar{X}(T)). \end{cases}$$

The proof is straightforward.

3.3   Two-Person Zero-Sum Differential Games

Now, we consider state equation (3.1) and cost functionals (3.4) with the property

$$\begin{cases} g_1(t, x, u_1, u_2) + g_2(t, x, u_1, u_2) = 0, \\ h_1(x) + h_2(x) = 0. \end{cases}$$

Then

$$J_1(t, x; u_1(\cdot), u_2(\cdot)) + J_2(t, x; u_1(\cdot), u_2(\cdot)) = 0.$$

Recall from Section 1.3.2, in this case, Problem (DG) is a *two-person zero-sum differential game*. For convenience, we call it Problem (Z). We denote

$$\begin{cases} g(t,x,u_1,u_2) = g_1(t,x,u_1,u_2) = -g_2(t,x,u_1,u_2), \\ h(x) = h_1(x) = -h_2(x), \end{cases}$$

and

$$J(t,x;u_1(\cdot),u_2(\cdot)) = J_1(t,x;u_1(\cdot),u_2(\cdot)) = -J_2(t,x;u_1(\cdot),u_2(\cdot)).$$

Then Player 1 wants to minimize $J(t,x;u_1(\cdot),u_2(\cdot))$ and Player 2 wants to maximize $J(t,x;u_1(\cdot),u_2(\cdot))$.

Now, for any initial pair $(t,x) \in [0,T] \times \mathbb{R}^n$, we define

$$\begin{cases} \bar{V}^+(t,x) = \displaystyle\inf_{u_1(\cdot)\in\mathcal{U}_1[t,T]} \sup_{u_2\in\mathcal{U}_2[t,T]} J(t,x;u_1(\cdot),u_2(\cdot)), \\ \bar{V}^-(t,x) = \displaystyle\sup_{u_2(\cdot)\in\mathcal{U}_2[t,T]} \inf_{u_1\in\mathcal{U}_1[t,T]} J(t,x;u_1(\cdot),u_2(\cdot)), \end{cases}$$

which are called the *open-loop upper* and *lower value functions* of Problem (Z), respectively. From the above definition, we have

$$\bar{V}^-(t,x) \leqslant \bar{V}^+(t,x), \qquad \forall (t,x) \in [0,T] \times \mathbb{R}^n.$$

In the case that

$$\bar{V}^+(t,x) = \bar{V}^-(t,x) \equiv \bar{V}(t,x), \qquad (t,x) \in [0,T] \times \mathbb{R}^n,$$

we say Problem (Z) admits an *open-loop value function* $\bar{V}(\cdot,\cdot)$. Further, if there is a pair $(\bar{u}_1(\cdot),\bar{u}_2(\cdot)) \in \mathcal{U}_1[t,T] \times \mathcal{U}_2[t,T]$ such that

$$J(t,x;\bar{u}_1(\cdot),u_2(\cdot)) \leqslant J(t,x;\bar{u}_1(\cdot),\bar{u}_2(\cdot)) \leqslant J(t,x;u_1(\cdot),\bar{u}_2(\cdot)),$$
$$\forall (u_1(\cdot),u_2(\cdot)) \in \mathcal{U}_1[t,T] \times \mathcal{U}_2[t,T],$$

then we call the pair $(\bar{u}_1(\cdot),\bar{u}_2(\cdot))$ an *open-loop saddle point* of Problem (Z) for the initial pair (t,x). The following, called a *minimax principle*, is a simple consequence of Theorem 3.2.2.

Corollary 3.3.1. *Let* $(\bar{u}_1(\cdot),\bar{u}_2(\cdot))$ *be an open-loop saddle point of Problem* (Z). *Let* $\psi(\cdot)$ *be the solution to the following adjoint equation:*

$$\begin{cases} \dot{\psi}(s) = -f_x(s,\bar{X}(s),\bar{u}_1(s),\bar{u}_2(s))^T \psi(s) + g_x(s,\bar{X}(s),\bar{u}_1(s),\bar{u}_2(s))^T, \\ \psi(T) = -h_x(\bar{X}(T)). \end{cases}$$

Then

$$\max_{u_1\in U_1} H(s,\bar{X}(s),u_1,\bar{u}_2(s),\psi(s)) = H(s,\bar{X}(s),\bar{u}_1(s),\bar{u}_2(s),\psi(s))$$
$$= \min_{u_2\in U_2} H(s,\bar{X}(s),\bar{u}_1(s),u_2,\psi(s)),$$
$$a.e. \ s \in [t,T],$$

with

$$H(t, x, u_1, u_2, \psi) = \langle \psi, f(t, x, u_1, u_2) \rangle - g(t, x, u_1, u_2).$$

Or, equivalently,

$$H(x, \bar{X}(s), u_1, \bar{u}_2(s), \psi(s)) \leqslant H(s, \bar{X}(s), \bar{u}_1(s), \bar{u}_2(s), \psi(s))$$
$$\leqslant H(s, \bar{X}(s), \bar{u}_1(s), u_2, \psi(s)),$$
$$\forall (u_1, u_2) \in U_1 \times U_2, \text{ a.e. } s \in [t, T],$$

which implies that

$$H(s, \bar{X}(s), \bar{u}_1(s), \bar{u}_2(s), \psi(s)) = \max_{u_2 \in U_2} \min_{u_1 \in U_1} H(s, \bar{X}(s), u_1, u_2, \psi(s))$$
$$= \min_{u_1 \in U_1} \max_{u_2 \in U_2} H(s, \bar{X}(s), u_1, u_2, \psi(s)).$$

We now look at some examples.

Example 3.3.2. For any $(t, x) \in [0, T) \times \mathbb{R}$, consider system

$$\begin{cases} \dot{X}(s) = [u_1(s) - u_2(s)]^2, & s \in [t, T], \\ X(t) = x, \end{cases}$$

with $U_1 = U_2 = [-1, 1]$, and with cost functional

$$J(t, x; u_1(\cdot), u_2(\cdot)) = X(T).$$

Clearly,

$$J(t, x; u_1(\cdot), u_2(\cdot)) = x + \int_t^T [u_1(s) - u_2(s)]^2 ds.$$

Therefore,

$$\bar{V}^-(t, x) = \sup_{u_2(\cdot) \in \mathcal{U}_2[t,T]} \inf_{u_1(\cdot) \in \mathcal{U}_1[t,T]} J(t, x; u_1(\cdot), u_2(\cdot))$$

$$= x + \sup_{u_2(\cdot) \in \mathcal{U}_2[t,T]} \inf_{u_1(\cdot) \in \mathcal{U}_1[t,T]} \int_t^T [u_1(s) - u_2(s)]^2 ds = x,$$

and

$$\bar{V}^+(t, x) = \inf_{u_1(\cdot) \in \mathcal{U}_1[t,T]} \sup_{u_2(\cdot) \in \mathcal{U}_2[t,T]} J(t, x; u_1(\cdot), u_2(\cdot))$$

$$= x + \inf_{u_1(\cdot) \in \mathcal{U}_1[t,T]} \sup_{u_2(\cdot) \in \mathcal{U}_2[t,T]} \int_t^T [u_1(s) - u_2(s)]^2 ds$$

$$= x + \inf_{u_1(\cdot) \in \mathcal{U}_1[t,T]} \int_t^T [1 + |u_1(s)|]^2 ds = x + T - t.$$

This implies

$$\bar{V}^-(t,x) < \bar{V}^+(t,x).$$

Hence, Problem (Z) does not admit an open-loop value function.

Example 3.3.3. For any $(t,x) \in [0,T) \times \mathbb{R}$, consider the following system:

$$\begin{cases} \dot{X}(s) = u_1(s) + u_2(s), & s \in [t,T], \\ X(t) = x, \end{cases}$$

with $U_1 = U_2 = [-1,1]$ and with

$$J(t,x;u_1(\cdot),u_2(\cdot)) = X(T).$$

Then

$$\bar{V}^-(t,x) = \sup_{u_2(\cdot)\in\mathcal{U}_2[t,T]} \inf_{u_1(\cdot)\in\mathcal{U}_1[t,T]} J(t,x;u_1(\cdot),u_2(\cdot))$$

$$= x + \inf_{u_1(\cdot)\in\mathcal{U}_1[t,T]} \int_t^T u_1(s)ds + \sup_{u_2(\cdot)\in\mathcal{U}_2[t,T]} \int_t^T u_2(s)ds$$

$$= x - (T-t) + (T-t) = x$$

$$= \inf_{u_1(\cdot)\in\mathcal{U}_1[t,T]} \sup_{u_2(\cdot)\in\mathcal{U}_2[t,T]} J(t,x;u_1(\cdot),u_2(\cdot)) = \bar{V}^+(t,x).$$

Thus, the current differential game admits an open-loop value function. Moreover, if we let

$$\bar{u}_1(s) = -1, \qquad \bar{u}_2(s) = 1, \qquad s \in [t,T],$$

then it is a saddle point of Problem (Z) for the initial pair (t,x).

Next, for any $(t,x) \in [0,T] \times \mathbb{R}^n$, we define

$$\begin{cases} V^+(t,x) = \sup_{\alpha_2\in\mathcal{A}_2[t,T]} \inf_{u_1(\cdot)\in\mathcal{U}_1[t,T]} J(t,x;u_1(\cdot),\alpha_2[u_1(\cdot)]), \\ V^-(t,x) = \inf_{\alpha_1\in\mathcal{A}_1[t,T]} \sup_{u_2(\cdot)\in\mathcal{U}_2[t,T]} J(t,x;\alpha_1[u_2(\cdot)],u_2(\cdot)), \end{cases}$$

which are called *Elliott–Kalton upper* and *lower value functions* of Problem (Z), respectively. We will show that under mild conditions,

$$V^-(t,x) \leqslant V^+(t,x), \qquad \forall (t,x) \in [0,T] \times \mathbb{R}^n.$$

If the following holds:

$$V^+(t,x) = V^-(t,x) \equiv V(t,x), \qquad \forall (t,x) \in [0,T] \times \mathbb{R}^n,$$

we say that Problem (Z) admits an *Elliott–Kalton value function* $V(\cdot,\cdot)$.

Similarly to Problem (OC), we have the following result concerning some basic properties of the upper and lower value functions.

Theorem 3.3.4. *Let* (DG1)–(DG2) *hold. Then the upper and lower value functions* $V^{\pm}(\cdot,\cdot)$ *are continuous on* $[0,T] \times \mathbb{R}^n$.

Proof. First, the same as the proof of Proposition 2.4.1, we have

$$|J(t,x_1;u_1(\cdot),u_2(\cdot)) - J(t,x_2;u_1(\cdot),u_2(\cdot))| \leqslant \bar{\omega}(|x_1| \vee |x_2|, |x_1 - x_2|),$$
$$\forall (t,u_1(\cdot),u_2(\cdot)) \in [0,T] \times \mathcal{U}_1[t,T] \times \mathcal{U}_2[t,T], x_1, x_2 \in \mathbb{R}^n,$$

for some local modulus of continuity $\bar{\omega}(\cdot,\cdot)$. Hence, one has

$$|V^{\pm}(t,x_1) - V^{\pm}(t,x_2)| \leqslant \bar{\omega}(|x_1| \vee |x_1|, |x_1 - x_2|),$$
$$t \in [0,T], x_1, x_2 \in \mathbb{R}^n.$$

We now prove the continuity of $t \mapsto V^+(t,x)$. The proof for $t \mapsto V^-(t,x)$ is similar.

Let $0 \leqslant t < \tau \leqslant T$. Then for any $(u_1(\cdot),u_2(\cdot)) \in \mathcal{U}_1[t,T] \times \mathcal{U}_2[t,T]$, by Proposition 3.1.1,

$$|X(s;t,x,u_1(\cdot),u_2(\cdot)) - X(s;\tau,x,u_1(\cdot),u_2(\cdot))|$$
$$\leqslant e^{L(s-\tau)}|X(\tau;t,x,u_1(\cdot),u_2(\cdot)) - x|$$
$$\leqslant e^{L(s-\tau)}[e^{L(\tau-t)} - 1](1+|x|).$$

Now, for any $\alpha_2 \in \mathcal{A}_2[\tau,T]$, fix a $u_2^* \in U_2$ and define $u_2^* \oplus \alpha_2 \in \mathcal{A}_2[t,T]$ as follows: For any $u_1(\cdot) \in \mathcal{U}_1[t,T]$,

$$(u_2^* \oplus \alpha_2)[u_1(\cdot)](s) = \begin{cases} u_2^*, & s \in [t,\tau), \\ \alpha_2[u_1(\cdot)|_{[\tau,T]}](s), & s \in [\tau,T], \end{cases}$$

where $u_1(\cdot)|_{[\tau,T]}$ is the restriction of $u_1(\cdot) \in \mathcal{U}_1[t,T]$ on $[\tau,T]$. Then

$$J(t,x;u_1(\cdot),(u_2^* \oplus \alpha_2)[u_1(\cdot)])$$
$$= \int_t^{\tau} g(s,X(s),u_1(s),u_2^*)ds + J(\tau,X(\tau);u_1(\cdot)|_{[\tau,T]},\alpha_2[u_1(\cdot)|_{[\tau,T]}])$$
$$\geqslant -\int_t^{\tau} [g_0 + \omega(|X(s)|,|X(s)|)]ds - \bar{\omega}(|x| \vee |X(\tau)|; |X(\tau) - x|)$$
$$\quad + J(\tau,x;u_1(\cdot)|_{[\tau,T]},\alpha_2[u_1(\cdot)|_{[\tau,T]}])$$
$$\geqslant -[g_0 + \omega(e^{LT}(1+|x|),e^{LT}(1+|x|))](\tau - t)$$
$$\quad -\bar{\omega}(e^{LT}(1+|x|); (e^{L(\tau-t)} - 1)(1+|x|))$$
$$\quad + J(\tau,x;u_1(\cdot)|_{[\tau,T]},\alpha_2[u_1(\cdot)|_{[\tau,T]}])$$
$$\equiv -\tilde{\omega}(|x|,\tau - t) + J(\tau,x;u_1(\cdot)|_{[\tau,T]},\alpha_2[u_1(\cdot)|_{[\tau,T]}]),$$

with $\widetilde{\omega}(\cdot,\cdot)$ being some local modulus of continuity. Now, we first take infimum in $u_1(\cdot) \in \mathcal{U}_1[t,T]$, then taking supremum in $\alpha_2 \in \mathcal{A}_2[\tau,T]$ we have

$$
\begin{aligned}
V^+(t,x) &= \sup_{\alpha_2 \in \mathcal{A}_2[t,T]} \inf_{u_1(\cdot) \in \mathcal{U}_1[t,T]} J\big(t,x;u_1(\cdot),\alpha_2[u_1(\cdot)]\big) \\
&\geqslant \sup_{\alpha_2 \in \mathcal{A}_2[\tau,T]} \inf_{u_1(\cdot) \in \mathcal{U}_1[t,T]} J\big(t,x;u_1(\cdot),(u_2^* \oplus \alpha_2)[u_1(\cdot)]\big) \\
&\geqslant V^+(\tau,x) - \widetilde{\omega}(|x|,\tau - t).
\end{aligned}
$$

Here, we should note that

$$
\Big\{ u_1(\cdot)\big|_{[\tau,T]} \ \big| \ u_1(\cdot) \in \mathcal{U}_1[t,T] \Big\} = \mathcal{U}_1[\tau,T],
$$

and

$$
\Big\{ u_2^* \oplus \alpha_2 \ \big| \ \alpha_2 \in \mathcal{A}_2[\tau,T] \Big\} \subseteq \mathcal{A}_2[t,T],
$$

in which the equality does not hold, in general.

On the other hand, for any $u_1(\cdot) \in \mathcal{U}_1[\tau,T]$, fix a $u_1^* \in U_1$ and we let

$$
u_1^* \oplus u_1(s) = \begin{cases} u_1^*, & s \in [t,\tau), \\ u_1(s), & s \in [\tau,T]. \end{cases}
$$

Also, for any $\alpha_2 \in \mathcal{A}_2[t,T]$, we define a restriction $\alpha_2\big|_{[\tau,T]} \in \mathcal{A}_2[\tau,T]$ (depending on u_1^*) as follows:

$$
\alpha_2\big|_{[\tau,T]}[u_1(\cdot)] = \alpha_2[u_1^* \oplus u_1(\cdot)], \qquad \forall u_1(\cdot) \in \mathcal{U}_1[\tau,T].
$$

Then for any $u_1(\cdot) \in \mathcal{U}_1[\tau,T]$,

$$
\begin{aligned}
&J\big(t,x;u_1^* \oplus u_1(\cdot),\alpha_2[u_1^* \oplus u_1(\cdot)]\big) \\
&= \int_t^\tau g\big(s,X(s),u_1^*,\alpha_2[u_1^* \oplus u_1(\cdot)]\big)ds + J\big(\tau,X(\tau);u_1(\cdot),\alpha_2\big|_{[\tau,T]}[u_1(\cdot)]\big) \\
&\leqslant \big[g_0 + \omega\big(e^{LT}(1+|x|);e^{LT}(1+|x|)\big)(\tau - t) \\
&\quad + \bar{\omega}\big(e^{LT}(1+|x|);(e^{L(\tau-t)}-1)(1+|x|)\big) + J\big(\tau,x;u_1(\cdot),\alpha_2\big|_{[\tau,T]}[u_1(\cdot)]\big) \\
&\equiv \widetilde{\omega}(|x|;\tau - t) + J\big(\tau,x;u_1(\cdot),\alpha_2\big|_{[\tau,T]}[u_1(\cdot)]\big).
\end{aligned}
$$

Hence,

$$
\begin{aligned}
&\inf_{u_1(\cdot) \in \mathcal{U}_1[t,T]} J\big(t,x;u_1(\cdot),\alpha_2[u_1(\cdot)]\big) \\
&\leqslant \inf_{u_1(\cdot) \in \mathcal{U}_1[\tau,T]} J\big(t,x;u_1^* \oplus u_1(\cdot),\alpha_2[u_1^* \oplus u_1(\cdot)]\big) \\
&\leqslant \widetilde{\omega}(|x|;\tau - t) + \inf_{u_1(\cdot) \in \mathcal{U}_1[\tau,T]} J\big(\tau,x;u_1(\cdot),\alpha_2\big|_{[\tau,T]}[u_1(\cdot)]\big) \\
&\leqslant \widetilde{\omega}(|x|;\tau - t) + V^+(\tau,x).
\end{aligned}
$$

Taking supremum in $\alpha_2 \in \mathcal{A}_2[t,T]$, we finally obtain

$$V^+(t,x) \leqslant \widetilde{\omega}(|x|; \tau - t) + V^+(\tau, x).$$

This completes the proof. $\square$

The following is the optimality principle for the upper and lower value functions.

Theorem 3.3.5. *Let* (DG1)–(DG2) *hold. Then for any* $(t,x) \in [0,T) \times \mathbb{R}^n$ *and* $\tau \in [t,T]$,

$$V^+(t,x) = \sup_{\alpha_2 \in \mathcal{A}_2[t,T]} \inf_{u_1(\cdot) \in \mathcal{U}_1[t,T]} \left\{ \int_t^\tau g(s, X(s), u_1(s), \alpha_2[u_1(\cdot)](s)) ds \right. \tag{3.5}$$
$$\left. + V^+(\tau, X(\tau)) \right\},$$

and

$$V^-(t,x) = \inf_{u_1(\cdot) \in \mathcal{U}_1[t,T]} \sup_{\alpha_2 \in \mathcal{A}_2[t,T]} \left\{ \int_t^\tau g(s, X(s), u_1(s), \alpha_2[u_1(\cdot)](s)) ds \right. \tag{3.6}$$
$$\left. + V^-(\tau, X(\tau)) \right\}.$$

Proof. First, the same as the proof of Proposition 2.4.1, we have

$$|V^\pm(t,x_1) - V^\pm(t,x_2)| \leqslant \bar{\omega}(|x_1| \vee |x_1|, |x_1 - x_2|).$$

We now prove the continuity of $t \mapsto V^\pm(t,x)$.

We prove (3.5) only (the proof of (3.6) is similar). Denote the right-hand side of (3.5) by $\widehat{V}(t,x)$. For any $\varepsilon > 0$, there exists an $\alpha_2^\varepsilon \in \mathcal{A}_2[t,T]$ such that

$$\widehat{V}(t,x) - \varepsilon \leqslant \inf_{u_1(\cdot) \in \mathcal{U}_1[t,T]} \left\{ \int_t^\tau g(s, X^\varepsilon(s), u_1(s), \alpha_2^\varepsilon[u_1(\cdot)](s)) ds \right.$$
$$\left. + V^+(\tau, X^\varepsilon(\tau)) \right\},$$

where $X^\varepsilon(\cdot) \equiv X(\cdot; t, x, u_1(\cdot), \alpha_2^\varepsilon[u_1(\cdot)])$. By the definition of $V^+(\tau, X^\varepsilon(\tau))$, there exists an $\bar{\alpha}_2^\varepsilon \in \mathcal{A}_2[\tau, T]$, depending on $(\tau, X^\varepsilon(\tau))$ (thus also depends on $\{u_1(s), \alpha_2^\varepsilon[u_1(\cdot)](s), s \in [t,\tau]\}$) such that

$$V^+(\tau, X^\varepsilon(\tau)) - \varepsilon \leqslant \inf_{u_1(\cdot) \in \mathcal{U}_1[\tau,T]} J(\tau, X^\varepsilon(\tau); u_1(\cdot), \bar{\alpha}_2^\varepsilon[u_1(\cdot)]).$$

Now, we define $\widetilde{\alpha}_2^\varepsilon \in \mathcal{A}_2[t,T]$ as follows: For any $u_1(\cdot) \in \mathcal{U}_1[t,T]$,

$$\widetilde{\alpha}_2^\varepsilon[u_1(\cdot)](s) = \begin{cases} \alpha_2^\varepsilon[u_1(\cdot)](s), & s \in [t,\tau), \\ \bar{\alpha}_2^\varepsilon[u(\cdot)|_{\tau,T}](s), & s \in [\tau, T]. \end{cases}$$

Then,

$$V^+(t,x) \geqslant \inf_{u_1(\cdot)\in\mathcal{U}_1[t,T]} J(t,x;u_1(\cdot),\tilde{\alpha}_2^\varepsilon[u_1(\cdot)])$$

$$= \inf_{u_1(\cdot)\in\mathcal{U}_1[t,T]} \left\{ \int_t^\tau g(s,X^\varepsilon(s),u_1(s),\alpha_2^\varepsilon[u_1(\cdot)](s))ds \right.$$

$$\left. +J\big(\tau,X^\varepsilon(\tau);u_1(\cdot)\big|_{[\tau,T]},\tilde{\alpha}_2^\varepsilon[u_1(\cdot)]\big|_{[\tau,T]}\big) \right\}$$

$$\geqslant \inf_{u_1(\cdot)\in\mathcal{U}_1[t,T]} \left\{ \int_t^\tau g(s,X^\varepsilon(s),u_1(s),\alpha_2^\varepsilon[u_1(\cdot)](s))ds \right.$$

$$\left. +V^+(\tau,X^\varepsilon(\tau)) \right\} - \varepsilon \geqslant \widehat{V}(t,x) - 2\varepsilon.$$

Since $\varepsilon > 0$ is arbitrary, we obtain

$$V^+(t,x) \geqslant \widehat{V}(t,x).$$

On the other hand, for any $\varepsilon > 0$, there exists an $\alpha_2^\varepsilon \in \mathcal{A}_2[t,T]$, such that

$$V^+(t,x) - \varepsilon \leqslant \inf_{u_1(\cdot)\in\mathcal{U}_1[t,T]} J\big(t,x;u_1(\cdot),\alpha_2^\varepsilon[u_1(\cdot)]\big).$$

By the definition of $\widehat{V}(t,x)$, one has

$$\widehat{V}(t,x) \geqslant \inf_{u_1(\cdot)\in\mathcal{U}_1[t,T]} \left\{ \int_t^\tau g\big(s,X^\varepsilon(s),u_1(s),\alpha_2^\varepsilon[u_1(\cdot)](s)\big)ds + V^+(\tau,X^\varepsilon(\tau)) \right\},$$

where $X^\varepsilon(\cdot) \equiv X(\cdot;t,x,u_1(\cdot),\alpha_2^\varepsilon[u_1(\cdot)])$. Then for $\varepsilon > 0$, there exists a $u_1^\varepsilon(\cdot) \in \mathcal{U}_1[t,T]$ such that

$$\widehat{V}(t,x) + \varepsilon \geqslant \int_t^\tau g\big(s,X^\varepsilon(s),u_1^\varepsilon(s),\alpha_2^\varepsilon[u_1^\varepsilon(\cdot)](s)\big)ds + V^+(\tau,X^\varepsilon(\tau)).$$

Here, $X^\varepsilon(\cdot) = X(\cdot;t,x,u_1^\varepsilon(\cdot),\alpha_2^\varepsilon[u_1^\varepsilon(\cdot)])$. Now, for any $u_1(\cdot) \in \mathcal{U}_1[\tau,T]$, let

$$[u_1^\varepsilon(\cdot)\oplus u_1(\cdot)](s) = \begin{cases} u_1^\varepsilon(s), & s\in[t,\tau), \\ u_1(s), & s\in[\tau,T]. \end{cases}$$

We define $\bar{\alpha}_2^\varepsilon \in \mathcal{A}_2[\tau,T]$ as follows (note $\alpha_2^\varepsilon \in \mathcal{A}_2[t,T]$)

$$\bar{\alpha}_2^\varepsilon[u_1(\cdot)] = \alpha_2^\varepsilon[u_1^\varepsilon(\cdot)\oplus u_1(\cdot)], \qquad \forall u_1(\cdot)\in\mathcal{U}_1[\tau,T].$$

Thus, $u_1^\varepsilon(\cdot)\oplus u_1(\cdot)$ is an extension of $u_1(\cdot)$ to $[t,T]$, and $\bar{\alpha}_2^\varepsilon$ is a restriction of α_2^ε on $[\tau,T]$. Then

$$V^+(\tau,X^\varepsilon(\tau)) \geqslant \inf_{u_1(\cdot)\in\mathcal{U}_1[\tau,T]} J\big(\tau,X^\varepsilon(\tau);u_1(\cdot),\bar{\alpha}_2^\varepsilon[u_1(\cdot)]\big).$$

Hence, there exists a $\bar{u}_1^\varepsilon(\cdot) \in \mathcal{U}_1[\tau,T]$ such that

$$V^+(\tau,X^\varepsilon(\tau)) + \varepsilon \geqslant J\big(\tau,X^\varepsilon(\tau);\bar{u}_1^\varepsilon(\cdot),\alpha_2^\varepsilon[u_1^\varepsilon(\cdot)]\big).$$

Note that $u_1^\varepsilon(\cdot) \oplus \bar{u}_1^\varepsilon(\cdot) \in \mathcal{U}_1[t, T]$. Combining the above, we have

$$\widehat{V}(t, x) + \varepsilon \geqslant \int_t^\tau g\big(s, X^\varepsilon(s), u_1^\varepsilon(s), \alpha_2^\varepsilon[u_1^\varepsilon(\cdot)](s)\big) ds + V^+(\tau, X^\varepsilon(\tau))$$

$$\geqslant \int_t^\tau g\big(s, X^\varepsilon(s), u_1^\varepsilon(s), \alpha_2^\varepsilon[u_1^\varepsilon(\cdot)](s)\big) ds + J\big(\tau, X^\varepsilon(\tau); \bar{u}_1^\varepsilon(\cdot), \alpha_2^\varepsilon[u_1^\varepsilon(\cdot)]\big) - \varepsilon$$

$$= J\big(t, x; u_1^\varepsilon(\cdot) \oplus \bar{u}_1^\varepsilon(\cdot), \alpha_2^\varepsilon[u_1^\varepsilon(\cdot) \oplus \bar{u}_1^\varepsilon(\cdot)]\big) - \varepsilon$$

$$\geqslant \inf_{u_1(\cdot)\in\mathcal{U}_1[t,T]} J\big(t, x; u_1(\cdot), \alpha_2^\varepsilon[u_1(\cdot)]\big) - \varepsilon \geqslant V^+(t, x) - 2\varepsilon.$$

Since $\varepsilon > 0$ is arbitrary, we have

$$\widehat{V}(t, x) \geqslant V^+(t, x).$$

This completes the proof. $\qquad\qquad\qquad\qquad\qquad\qquad\qquad\qquad\square$

Having the above, we now state the following result.

Theorem 3.3.6. *Let* (DG1)–(DG2) *hold. Then* $V^\pm(\cdot, \cdot)$ *are respectively the unique viscosity solutions to the following:*

$$\begin{cases} V_t^\pm(t, x) + H^\pm(t, x, V_x^\pm(t, x)) = 0, & (t, x) \in [0, T] \times \mathbb{R}^n, \\ V^\pm(T, x) = h(x), \end{cases} \quad (3.7)$$

where

$$\begin{cases} H^+(t, x, p) = \displaystyle\inf_{u_1\in U_1} \sup_{u_2\in U_2} \Big[\langle p, f(t, x, u_1, u_2) \rangle + g(t, x, u_1, u_2) \Big], \\ H^-(t, x, p) = \displaystyle\sup_{u_2\in U_2} \inf_{u_1\in U_1} \Big[\langle p, f(t, x, u_1, u_2) \rangle + g(t, x, u_1, u_2) \Big], \end{cases}$$

and

$$V^-(t, x) \leqslant V^+(t, x), \qquad (t, x) \in [0, T] \times \mathbb{R}^n.$$

In the case that the following Isaacs condition holds:

$$H^-(t, x, p) = H^+(t, x, p), \qquad \forall(t, x, p) \in [0, T] \times \mathbb{R}^n \times \mathbb{R}^n,$$

then

$$V^-(t, x) = V^+(t, x), \qquad \forall(t, x) \in [0, T] \times \mathbb{R}^n.$$

Namely, Problem (Z) admits a value function.

In the above, $H^+(\cdot, \cdot, \cdot)$ and $H^-(\cdot, \cdot, \cdot)$ are called *upper* and *lower Hamiltonians*, respectively. Equations (3.7) are called *upper* and *lower Hamilton-Jacobi-Isaacs* (HJI, for short) *equations*, respectively.

Proof. First of all, under (DG1)–(DG2), we can check that

$$
\begin{cases}
|H^\pm(t,x,p) - H^\pm(t,x,q)| \leqslant L(1+|x|)|p-q|, \\
\qquad \forall t \in [0,T], \ x,p,q \in \mathbb{R}^n, \\
|H^\pm(t,x,p) - H^\pm(t,y,p)| \leqslant L|p|\,|x-y| + \omega(|x| \vee |y|, |x-y|), \\
\qquad \forall t \in [0,T], \ x,y,p \in \mathbb{R}^n.
\end{cases}
$$

This implies that (2.20) holds for $H^\pm(\cdot,\cdot,\cdot)$. Hence, by Theorem 2.5.3, viscosity solutions of (3.7) are unique. Consequently, it suffices to show that $V^\pm(\cdot,\cdot)$ are viscosity solutions to (3.7).

We now prove that $V^+(\cdot,\cdot)$ is a viscosity solution to the H.JI equation (3.7). The proof for $V^-(\cdot,\cdot)$ is similar. To prove that $V^+(\cdot,\cdot)$ is a viscosity sub-solution to (3.7), let $\varphi(\cdot,\cdot) \in C^1([0,T] \times \mathbb{R}^n)$ such that $V^+(\cdot,\cdot) - \varphi(\cdot,\cdot)$ attains a local maximum at $(t,x) \in [0,T) \times \mathbb{R}^n$. Then as long as $0 < \tau - t$ is small enough, for any $(u_1(\cdot), \alpha_2[\,\cdot\,]) \in \mathcal{U}_1[t,T] \times \mathcal{A}_2[t,T]$, one has

$$
V^+(t,x) - \varphi(t,x) \geqslant V^+(\tau, X(\tau)) - \varphi(\tau, X(\tau)),
$$

where $X(\cdot) \equiv X(\cdot\,; t, x, u_1(\cdot), \alpha_2[u_1(\cdot)])$. Then by (3.5), one has

$$
\begin{aligned}
0 = \ & \sup_{\alpha_2 \in \mathcal{A}_2[t,T]} \inf_{u_1(\cdot) \in \mathcal{U}_1[t,T]} \left\{ \int_t^\tau g\big(s, X(s), u_1(s), \alpha_2[u_1(\cdot)](s)\big)\, ds \right. \\
& \left. \qquad\qquad + V^+(\tau, X(\tau)) - V^+(t,x) \right\} \\
\leqslant \ & \sup_{\alpha_2 \in \mathcal{A}_2[t,T]} \inf_{u_1(\cdot) \in \mathcal{U}_1[t,T]} \left\{ \int_t^\tau g\big(s, X(s), u_1(s), \alpha_2[u_1(\cdot)](s)\big)\, ds \right. \\
& \left. \qquad\qquad + \varphi(\tau, X(\tau)) - \varphi(t,x) \right\} \\
= \ & \sup_{\alpha_2 \in \mathcal{A}_2[t,T]} \inf_{u_1(\cdot) \in \mathcal{U}_1[t,T]} \left\{ \int_t^\tau \Big[g\big(s, X(s), u_1(s), \alpha_2[u_1(\cdot)](s)\big) + \varphi_s(s, X(s)) \right. \\
& \left. \qquad\qquad + \langle \varphi_x(s, X(s)), f(s, X(s), u_1(s), \alpha_2[u_1(\cdot)](s)) \rangle \Big] ds \right\} \\
= \ & \sup_{\alpha_2 \in \mathcal{A}_2[t,T]} \inf_{u_1(\cdot) \in \mathcal{U}_1[t,T]} \left\{ \int_t^\tau \Big[g\big(t, x, u_1(s), \alpha_2[u_1(\cdot)](s)\big) + \varphi_t(t,x) \right. \\
& \left. \qquad\qquad + \langle \varphi_x(t,x), f(t, x, u_1(s), \alpha_2[u_1(\cdot)](s)) \rangle \Big] ds \right\} + o(\tau - t).
\end{aligned}
$$

Thus, for any $\varepsilon > 0$ and given $\tau \in (t,T]$ with $\tau - t > 0$ small, there exists an $\alpha_2^{\varepsilon,\tau} \in \mathcal{A}_2[t,T]$ such that

$$-\varepsilon(\tau - t) \leqslant \varphi_t(t,x)(\tau - t) + \inf_{u_1(\cdot) \in \mathcal{U}_1[t,T]} \left\{ \int_t^\tau \left[g(t,x,u_1(s),\alpha_2^{\varepsilon,\tau}[u_1(\cdot)](s)) \right. \right.$$
$$\left. + \langle \varphi_x(t,x), f(t,x,u_1(s),\alpha_2^{\varepsilon,\tau}[u_1(\cdot)](s)) \rangle \right] ds + o(\tau - t) \Big\}$$
$$\leqslant \varphi_t(t,x)(\tau - t) + \inf_{u_1(\cdot) \in \mathcal{U}_1[t,T]} \left\{ \int_t^\tau \sup_{u_2 \in U_2} \left[g(t,x,u_1(s),u_2) \right. \right.$$
$$\left. + \langle \varphi_x(t,x), f(t,x,u_1(s),u_2) \rangle \right] ds + o(\tau - t) \Big\}$$
$$\leqslant (\tau - t) \Big\{ \varphi_t(t,x) + \sup_{u_2 \in U_2} \left[g(t,x,u_1,u_2) + \langle \varphi_x(t,x), f(t,x,u_1,u_2) \rangle + o(1) \right] \Big\},$$

where $u_1 \in U_1$ is arbitrary. Dividing by $\tau - t$ and sending $\tau - t \to 0$, then sending $\varepsilon \to 0$, we obtain

$$0 \leqslant \varphi_t(t,x) + \sup_{u_2 \in U_2} \left[\langle \varphi_x(t,x), f(t,x,u_1,u_2) \rangle + g(t,x,u_1,u_2) \right].$$

Finally, taking infimum over $u_1 \in U_1$, we arrive at

$$\varphi_t(t,x) + H^+(t,x,\varphi_x(t,x)) \geqslant 0.$$

Thus, $V^+(\cdot,\cdot)$ is a viscosity sub-solution of upper HJI equation (3.7).

Next, we prove that $V^+(\cdot,\cdot)$ is a viscosity super-solution to upper HJI equation (3.7). Let $\varphi(\cdot,\cdot) \in C^1([0,T] \times \mathbb{R}^n)$ such that $V(\cdot,\cdot) - \varphi(\cdot,\cdot)$ attains a local minimum at (t,x). Then provided $\tau \in (t,T)$ with $\tau - \tau > 0$ small enough, for any $(u_1(\cdot), \alpha_2[\cdot]) \in \mathcal{U}_1[t,T] \times \mathcal{A}_2[t,T]$,

$$V^+(t,x) - \varphi(t,x) \leqslant V^+(\tau,X(\tau)) - \varphi(\tau,X(\tau)),$$

where $X(\cdot) \equiv X(\cdot; t, x, u_1(\cdot), \alpha_2[u_1(\cdot)])$. Then by (3.5) again

$$0 = \sup_{\alpha_2 \in \mathcal{A}_2[t,T]} \inf_{u_1(\cdot) \in \mathcal{U}_1[t,T]} \left\{ \int_t^\tau g(s,X(s),u_1(s),\alpha_2[u_1(\cdot)](s)) ds \right.$$
$$\left. + V^+(\tau,X(\tau)) - V^+(t,x) \right\}$$
$$\geqslant \sup_{\alpha_2 \in \mathcal{A}_2[t,T]} \inf_{u_1(\cdot) \in \mathcal{U}_1[t,T]} \left\{ \int_t^\tau g(s,X(s),u_1(s),\alpha_2[u_1(\cdot)](s)) ds \right.$$
$$\left. + \varphi(\tau,X(\tau)) - \varphi(t,x) \right\}$$
$$= \sup_{\alpha_2 \in \mathcal{A}_2[t,T]} \inf_{u_1(\cdot) \in \mathcal{U}_1[t,T]} \left\{ \int_t^\tau \left[g(s,X(s),u_1(s),\alpha_2[u_1(\cdot)](s)) + \varphi_s(s,X(s)) \right. \right.$$
$$\left. + \langle \varphi_x(s,X(s)), f(s,X(s),u_1(s),\alpha_2[u_1(\cdot)](s)) \rangle \right] ds \Big\}$$
$$= \sup_{\alpha_2 \in \mathcal{A}_2[t,T]} \inf_{u_1(\cdot) \in \mathcal{U}_1[t,T]} \left\{ \int_t^\tau \left[g(t,x,u_1(s),\alpha_2[u_1(\cdot)](s)) + \varphi_t(t,x) \right. \right.$$
$$\left. + \langle \varphi_x(t,x), f(t,x,u_1(s),\alpha_2[u_1(\cdot)](s)) \rangle \right] ds \Big\} + o(\tau - t).$$

We claim that for any $\varepsilon > 0$, there exists an $\alpha_2^\varepsilon \in \mathcal{A}_2[t,T]$ such that for any $u_1(\cdot) \in \mathcal{U}_1[t,T]$ and $s \in [t,T]$,

$$g\big(t,x,u_1(s),\alpha_2^\varepsilon[u_1(\cdot)](s)\big) + \big\langle \varphi_x(t,x), f\big(t,x,u_1(s),\alpha_2^\varepsilon[u_1(\cdot)](s)\big) \big\rangle$$
$$\geqslant \sup_{u_2 \in U_2} \Big(g\big(t,x,u_1(s),u_2\big) + \big\langle \varphi_x(t,x), f\big(t,x,u_1(s),u_2\big) \big\rangle \Big) - \varepsilon. \tag{3.8}$$

To show this, let us introduce the following:

$$g^\varepsilon(s,u_2) = \Big[\sup_{\bar{u}_2 \in U_2} \Big(g\big(t,x,u_1(s),\bar{u}_2\big) + \big\langle \varphi_x(t,x), f\big(t,x,u_1(s),\bar{u}_2\big) \big\rangle \Big)$$
$$- \Big(g\big(t,x,u_1(s),u_2\big) + \big\langle \varphi_x(t,x), f\big(,x,u_1(s),u_2\big) \big\rangle \Big) - \varepsilon \Big]^+.$$

Clearly, $s \mapsto g^\varepsilon(s,u_2)$ is measurable and $u_2 \mapsto g^\varepsilon(s,u_2)$ is continuous. Further, by the definition of supremum, we have

$$0 \in g^\varepsilon(s,U_2).$$

Thus, by Lemma 1.4.4 (Filippov's lemma), there exists a $u_2(\cdot) \in \mathcal{U}_2[t,T]$ such that

$$g^\varepsilon(s,u_2(s)) = 0, \qquad s \in [t,T].$$

From this, we can define $\alpha_2^\varepsilon[\,\cdot\,]$ and our claim (3.8) follows. Then

$$0 \geqslant \sup_{\alpha_2 \in \mathcal{A}_2[t,T]} \inf_{u_1(\cdot) \in \mathcal{U}_1[t,T]} \Big\{ \int_t^\tau \Big[g\big(t,x,u_1(s),\alpha_2[u_1(\cdot)](s)\big) + \varphi_t(t,x)$$
$$+ \big\langle \varphi_x(t,x), f\big(t,x,u_1(s),\alpha_2[u_1(\cdot)](s)\big) \big\rangle \Big] ds \Big\} + o(\tau - t)$$

$$\geqslant \varphi_t(t,x)(\tau - t) + \inf_{u_1(\cdot) \in \mathcal{U}_1[t,T]} \Big\{ \int_t^\tau \sup_{u_2 \in U_2} \Big[g\big(t,x,u_1(s),u_2\big)$$
$$+ \big\langle \varphi_x(t,x), f\big(t,x,u_1(s),u_2\big) \big\rangle \Big] ds \Big\} + o(\tau - t)$$

$$\geqslant \varphi_t(t,x)(\tau - t) + \int_t^\tau H^+(t,x,\varphi_x(t,x)) ds + o(\tau - t).$$

Consequently, dividing $\tau - t$ and then sending $\tau \to t$, we obtain that

$$\varphi_t(t,x) + H^+(t,x,\varphi_x(t,x)) \leqslant 0.$$

This means that $V^+(\cdot,\cdot)$ is a viscosity super-solution of the upper HJI equation (3.7), $\qquad \qquad \qquad \qquad \qquad \qquad \qquad \qquad \qquad \qquad \square$

Now, let us take a further look at Examples 3.3.2–3.3.3. For Example 3.3.2, we have

$$H^+(t,x,p) = \inf_{u_1 \in U_1} \sup_{u_2 \in U_2} p(u_1 - u_2)^2 = p^+,$$

and

$$H^-(t,x,p) = \sup_{u_2 \in U_2} \inf_{u_1 \in U_1} p(u_1 - u_2)^2 = -p^-.$$

Hence, the Isaacs condition fails. In the current case, the upper and lower HJI equations take the following forms:

$$\begin{cases} V_t^\pm(t,x) \pm \left(V_x^\pm(t,x)\right)^\pm = 0, & (t,x) \in [0,T] \times \mathbb{R}^n, \\ V^\pm(T,x) = x. \end{cases}$$

Clearly,

$$V^-(t,x) = x, \qquad V^+(t,x) = x + T - t$$

are unique classical solutions to the corresponding upper and lower HJI equations, and they are different. In this case, open-loop upper and lower value functions coincide with E-K upper and lower value functions:

$$\bar{V}^\pm(t,x) = V^\pm(t,x), \qquad \forall(t,x) \in [0,T] \times \mathbb{R}.$$

For Examples 3.3.3, we have

$$H^\pm(t,x,p) = \inf_{u_1 \in U_1} pu_1 + \sup_{u_2 \in U_2} pu_2 = -|p| + |p| = 0,$$

and Isaacs condition holds. Hence, the HJI equation is

$$\begin{cases} V_t(t,x) = 0, & (t,x) \in [0,T] \times \mathbb{R}, \\ V(T,x) = x, & x \in \mathbb{R}, \end{cases}$$

and the unique viscosity solution is given by

$$V(t,x) = x, \qquad \forall(t,x) \in [0,T] \times \mathbb{R},$$

which coincides with $\bar{V}(\cdot\,,\cdot) \equiv \bar{V}^\pm(\cdot\,,\cdot)$.

3.4   Brief Historic Remarks

In 1951, Isaacs ([57]) initiated the study of pursuit and evasion differential games and two-person zero sum differential games (see also [58] and [59]). In the mid 1950s, Berkovitz and Fleming got involved in the study ([44], [14], [45], see also [46], [47], [8]), "tried to make Isaacs' ideas more rigorous". See Breitner [20] for many interesting remarks on the history of differential games.

In later 1960s and early 1970s, Berkovitz ([8], [9], see also [11], [12], [13]), Krasovskii–Subbotin ([64], see also [65]), Friedman ([50], [51]), Elliott–Kalton ([40]) also made contributions. It is worthy of mentioning that in [40], Elliott and Kalton introduced a concept of strategy, named Elliott–Kalton strategy in the current book, is a very convenient notion. Such a strategy allows us to make the theory of two-person zero sum differential games rigorous later on ([41], [71], [72], [61]).

For two-person zero-sum differential games, although the works of Berkovitz, Fleming, Krasovskii–Subbotin, Friedman, Elliott–Kalton, and so on, made Isaacs' ideas mathematically rigorous, the results were not very satisfactory, since the proof of the existence of the value function was very complicated and the characterization of the value function was very vague. The reasons mainly are the following: On one hand, the formally derived HJI equation, which is a first order partial differential equation, might not have a classical solution. On the other hand, in general, the upper and lower value functions are not necessarily differentiable. Hence, the theory remains unsatisfactory for a couple of decays. In 1983, Crandall and Lions introduced the notion of viscosity solution for the first order partial differential equations ([30]). This leads to a very satisfactory characterization of the value function for the two-person zero-sum differential games, as the unique viscosity solution to the corresponding HJI equation ([71], [72]).

Chapter 4

Differential Games with Unbounded Controls

4.1 Unbounded Controls

To motivate the study in this chapter, let us look at the following one-dimensional linear state equation

$$\begin{cases} \dot{X}(s) = X(s) + u_1(s) + u_2(s), & s \in [t, T], \\ X(t) = x, \end{cases}$$

with performance functional

$$J(t, x; u_1(\cdot), u_2(\cdot)) = \int_t^T \Big[|X(s)|^\mu + u_1(s)^2 - u_2(s)^2 \Big] ds + |X(T)|^\mu,$$

for some $\mu > 0$, and with $U_1 = U_2 = \mathbb{R}$. Note that with

$$f(t, x, u_1, u_2) = x + u_1 + u_2,$$

the inequality

$$|f(s, 0, u_1, u_2)| = |u_1 + u_2| \leqslant L, \qquad \forall u_1, u_2 \in \mathbb{R},$$

does not hold, due to the control variables u_1 and u_2 taking values in unbounded sets. Thus, the theory presented in the previous chapter does not apply to the problem with the above state equation and performance functional. In this chapter, we are going to develop a theory for two-person zero-sum differential games with unbounded controls, which can cover the above situation.

We consider the following control system:

$$\begin{cases} \dot{X}(s) = f(s, X(s), u_1(s), u_2(s)), & s \in [t, T], \\ X(t) = x. \end{cases} \tag{4.1}$$

where, for $i = 1, 2$, $U_i \subseteq \mathbb{R}^{m_i}$ are nonempty closed sets which could be unbounded (and could even be $U_i = \mathbb{R}^{m_i}$), $f : [0,T] \times \mathbb{R}^n \times U_1 \times U_2 \to \mathbb{R}^n$ is a given map. In the current case, the control pair $(u_1(\cdot), u_2(\cdot))$ is taken from the set $\mathcal{U}_1[t,T] \times \mathcal{U}_2[t,T]$ of *admissible controls*, defined by the following:

$$\mathcal{U}_i[t,T] = \left\{ u_i : [t,T] \to U_i \;\middle|\; \|u_i(\cdot)\|_{L^2(t,T)} \equiv \left[\int_t^T |u_i(s)|^2 ds \right]^{\frac{1}{2}} < \infty \right\},$$

$$i = 1, 2.$$

The *performance functional* associated with (4.1) is the following:

$$J(t,x;u_1(\cdot), u_2(\cdot)) = \int_t^T g(s, X(s), u_1(s), u_2(s)) ds + h(X(T)), \qquad (4.2)$$

with $g : [0,T] \times \mathbb{R}^n \times U_1 \times U_2 \to \mathbb{R}$ and $h : \mathbb{R}^n \to \mathbb{R}$ being some given maps.

Let us formally look at the above problem. First of all, under some mild conditions, for any *initial pair* $(t,x) \in [0,T] \times \mathbb{R}^n$ and any control pair $(u_1(\cdot), u_2(\cdot)) \in \mathcal{U}_1[t,T] \times \mathcal{U}_2[t,T]$, the state equation (4.1) admits a unique solution $X(\cdot) \equiv X(\cdot\,;t,x,u_1(\cdot), u_2(\cdot))$, and the performance functional $J(t,x;u_1(\cdot), u_2(\cdot))$ is well-defined. By adopting the notion of Elliott–Kalton strategies, we can define the *Elliott–Kalton upper* and *lower value functions* $V^\pm : [0,T] \times \mathbb{R}^n \to \mathbb{R}$. Further, $V^\pm(\cdot, \cdot)$ should be the unique viscosity solutions to the following upper and lower HJI equations, respectively:

$$\begin{cases} V_t^\pm(t,x) + H^\pm(t,x,V_x^\pm(t,x)) = 0, & (t,x) \in [0,T] \times \mathbb{R}^n, \\ V^\pm(T,x) = h(x), & x \in \mathbb{R}^n, \end{cases}$$

where $H^\pm(t,x,p)$ are the *upper* and *lower Hamiltonians* defined by the following, respectively:

$$\begin{cases} H^+(t,x,p) = \inf_{u_1 \in U_1} \sup_{u_2 \in U_2} \Big[\langle p, f(t,x,u_1,u_2) \rangle + g(t,x,u_1,u_2) \Big], \\ H^-(t,x,p) = \sup_{u_2 \in U_2} \inf_{u_1 \in U_1} \Big[\langle p, f(t,x,u_1,u_2) \rangle + g(t,x,u_1,u_2) \Big], \\ \qquad\qquad\qquad\qquad (t,x,p) \in [0,T] \times \mathbb{R}^n \times \mathbb{R}^n. \end{cases}$$

Consequently, in the case that the following *Isaacs condition*:

$$H^+(t,x,p) = H^-(t,x,p), \qquad \forall (t,x,p) \in [0,T] \times \mathbb{R}^n \times \mathbb{R}^n,$$

holds, the upper and lower value functions should coincide and the corresponding two-person zero-sum differential game admits the value function

$$V(t,x) = V^+(t,x) = V^-(t,x), \qquad (t,x) \in [0,T] \times \mathbb{R}^n.$$

We have seen from previous chapter that in the case that both U_1 and U_2 are bounded, the above procedure goes through. Now, when either U_1 or U_2 is unbounded, we first need some coercivity conditions to guarantee the upper and lower Hamiltonians $H^{\pm}(t, x, p)$ to be defined. Second, when either U_1 or U_2 is unbounded, one does not have condition (2.20) which plays a crucial role in proving the uniqueness of viscosity solution to the HJ equation. Thus, we need to re-develop the whole theory.

4.2 Upper and Lower Hamiltonians

Let us introduce the following standing assumptions.

(H0) For $i = 1, 2$, the set $U_i \subseteq \mathbb{R}^{m_i}$ is closed and

$$0 \in U_i, \qquad i = 1, 2. \tag{4.3}$$

The time horizon $T > 0$ is fixed.

Note that both U_1 and U_2 could be unbounded and may even be equal to $\mathbb{R}^{m_1}$ and $\mathbb{R}^{m_2}$, respectively. Condition (4.3) is for convenience. We may make a translation of the control domains and make corresponding changes in the control systems and performance functional to achieve this.

Let us now introduce the following assumptions for the involved functions f and g in the state equation (4.1) and the performance functional (4.2). We denote $\langle x \rangle = \sqrt{1 + |x|^2}$.

(H1) Map $f : [0, T] \times \mathbb{R}^n \times U_1 \times U_2 \to \mathbb{R}^n$ is continuous and

$$|f(t, x, u_1, u_2)| \leqslant L(\langle x \rangle + |u_1| + |u_2|),$$
$$\forall (t, x, u_1, u_2) \in [0, T] \times \mathbb{R}^n \times U_1 \times U_2.$$

(H2) Map $g : [0, T] \times \mathbb{R}^n \times U_1 \times U_2 \to \mathbb{R}$ is continuous and there exist constants $L, c > 0$ and $\mu \in [0, 2)$ such that

$$c|u_1|^2 - L(\langle x \rangle^{\mu} + |u_2|^2) \leqslant g(t, x, u_1, u_2) \leqslant L(\langle x \rangle^{\mu} + |u_1|^2) - c|u_2|^2,$$
$$\forall (t, x, u_1, u_2) \in [0, T] \times \mathbb{R}^n \times U_1 \times U_2.$$

Note that the case $\mu = 2$ is excluded. Such a case is critical, and a special case of that will be studied separately in a later chapter.

Now, we let

$$\mathbb{H}(t, x, p, u_1, u_2) = \langle p, f(t, x, u_1, u_2) \rangle + g(t, x, u_1, u_2),$$
$$(t, x, u_1, u_2) \in [0, T] \times \mathbb{R}^n \times U_1 \times U_2.$$

The *upper* and *lower Hamiltonians* are defined as follows:

$$\begin{cases} H^+(t,x,p) = \inf_{u_1 \in U_1} \sup_{u_2 \in U_2} \mathbb{H}(t,x,p,u_1,u_2), \\ H^-(t,x,p) = \sup_{u_2 \in U_2} \inf_{u_1 \in U_1} \mathbb{H}(t,x,p,u_1,u_2), \\ (t,x,p) \in [0,T] \times \mathbb{R}^n \times \mathbb{R}^n, \end{cases} \tag{4.4}$$

provided the involved infimum and supremum are finite. Note that the upper and lower Hamiltonians have nothing to do with the function $h(\cdot)$ (appears as the terminal cost/payoff in (4.2)). The main result of this section is the following.

Theorem 4.2.1. *Under* (H0)–(H2), *the upper and lower Hamitonians* $H^\pm(\cdot,\cdot,\cdot)$ *are well-defined and continuous. Moreover, there is a constant* $K > 0$ *such that*

$$|H^\pm(t,x,p)| \leqslant L\big(\langle x \rangle^\mu + \langle x \rangle \,|p|\big) + \frac{L^2}{4c}|p|^2, \tag{4.5}$$

$$\forall (t,x,p) \in [0,T] \times \mathbb{R}^n \times \mathbb{R}^n,$$

and

$$|H^\pm(t,x,p) - H^\pm(t,x,q)| \leqslant K\big(\langle x \rangle + |p| \vee |q|\big)|p-q|, \tag{4.6}$$

$$\forall (t,x) \in [0,T] \times \mathbb{R}^n, \; p,q \in \mathbb{R}^n.$$

Proof. Let us look at $H^+(t,x,p)$ carefully ($H^-(t,x,p)$ can be treated similarly). First, by our assumption, we have

$$\begin{aligned} \mathbb{H}(t,x,p,u_1,u_2) &\leqslant |p|\,|f(t,x,u_1,u_2)| + g(t,x,u_1,u_2) \\ &\leqslant L\big(\langle x \rangle + |u_1| + |u_2|\big)|p| + L\big(\langle x \rangle^\mu + |u_1|^2\big) - c|u_2|^2 \\ &= L\big(\langle x \rangle^\mu + \langle x \rangle \,|p| + |p|\,|u_1| + |u_1|^2\big) + L|p|\,|u_2| - c|u_2|^2 \\ &= L\big(\langle x \rangle^\mu + \langle x \rangle \,|p| + |p|\,|u_1| + |u_1|^2\big) + \frac{L^2}{4c}|p|^2 - c\Big(|u_2| - \frac{L|p|}{2c}\Big)^2 \\ &\leqslant L\big(\langle x \rangle^\mu + \langle x \rangle \,|p| + |p|\,|u_1| + |u_1|^2\big) + \frac{L^2}{4c}|p|^2, \end{aligned} \tag{4.7}$$

and

$$\begin{aligned} \mathbb{H}(t,x,p,u_1,u_2) &\geqslant -|p|\,|f(t,x,u_1,u_2)| + g(t,x,u_1,u_2) \\ &\geqslant -L\big(\langle x \rangle + |u_1| + |u_2|\big)|p| - L\big(\langle x \rangle^\mu + |u_2|^2\big) + c|u_1|^2 \\ &= -L\big(\langle x \rangle^\mu + \langle x \rangle \,|p| + |p|\,|u_2| + |u_2|^2\big) - \frac{L^2}{4c}|p|^2 + c\Big(|u_1| - \frac{L|p|}{2c}\Big)^2 \\ &\geqslant -L\big(\langle x \rangle^\mu + \langle x \rangle \,|p| + |p|\,|u_2| + |u_2|^2\big) - \frac{L^2}{4c}|p|^2. \end{aligned} \tag{4.8}$$

From (4.7), we see that $u_2 \mapsto \mathbb{H}(t, x, p, u_1, u_2)$ is coercive from above. Thus, by the closeness of U_2, for any $(t, x, p, u_1) \in [0, T] \times \mathbb{R}^n \times \mathbb{R}^n \times U_1$, there exists a $\bar{u}_2 \equiv \bar{u}_2(t, x, p, u_1) \in U_2$ such that

$$
\begin{aligned}
\mathcal{H}^+(t, x, p, u_1) &\equiv \sup_{u_2 \in U_2} \mathbb{H}(t, x, p, u_1, u_2) \\
&= \sup_{u_2 \in U_2, |u_2| \leqslant |\bar{u}_2|} \mathbb{H}(t, x, p, u_1, u_2) = \mathbb{H}(t, x, p, u_1, \bar{u}_2) \\
&\leqslant L\big(\langle x \rangle^\mu + \langle x \rangle \, |p| + |p| \, |u_1| + |u_1|^2\big) + L|p| \, |\bar{u}_2| - c|\bar{u}_2|^2 \\
&\leqslant L\big(\langle x \rangle^\mu + \langle x \rangle \, |p| + |p| \, |u_1| + |u_1|^2\big) + \frac{L^2}{4c}|p|^2.
\end{aligned}
\tag{4.9}
$$

On the other hand, from (4.8), for any $(t, x, p, u_1) \in [0, T] \times \mathbb{R}^n \times \mathbb{R}^n \times U_1$, noting $0 \in U_2$, we have

$$
\begin{aligned}
\mathcal{H}^+(t, x, p, u_1) &= \sup_{u_2 \in U_2} \mathbb{H}(t, x, p, u_1, u_2) \geqslant \mathbb{H}(t, x, p, u_1, 0) \\
&\geqslant -L\big(\langle x \rangle^\mu + \langle x \rangle \, |p|\big) - L|p| \, |u_1| + c|u_1|^2 \\
&\geqslant -L\big(\langle x \rangle^\mu + \langle x \rangle \, |p|\big) - \frac{L^2}{4c}|p|^2.
\end{aligned}
\tag{4.10}
$$

Combining (4.9) and (4.10) yields

$$
\begin{aligned}
\frac{c}{2}|\bar{u}_2|^2 &\leqslant L\big(\langle x \rangle^\mu + \langle x \rangle \, |p| + |p| \, |u_1| + |u_1|^2\big) \\
&\quad - \mathcal{H}^+(t, x, p, u_1) + L|p||\bar{u}_2| - \frac{c}{2}|\bar{u}_2|^2 \\
&\leqslant L\big(2\langle x \rangle^\mu + 2\langle x \rangle \, |p| + |p| \, |u_1| + |u_1|^2\big) + \frac{L^2}{4c}|p|^2 \\
&\quad - \frac{c}{2}\Big(|\bar{u}_2| - \frac{L|p|}{c}\Big)^2 + \frac{L^2}{2c}|p|^2 \\
&\leqslant L\big(2\langle x \rangle^\mu + 2\langle x \rangle \, |p| + |p| \, |u_1| + |u_1|^2\big) + \frac{3L^2}{4c}|p|^2.
\end{aligned}
\tag{4.11}
$$

The above implies that for any compact set $G \subseteq [0, T] \times \mathbb{R}^n \times \mathbb{R}^n \times U_1$, there exists a compact set $\widehat{U}_2(G) \subseteq U_2$ (depending on G) such that

$$
\mathcal{H}^+(t, x, p, u_1) = \sup_{u_2 \in \widehat{U}_2(G)} \mathbb{H}(t, x, p, u_1, u_2), \qquad \forall (t, x, p, u_1) \in G.
$$

Hence, $\mathcal{H}^+(\cdot, \cdot, \cdot, \cdot)$ is continuous on $[0, T] \times \mathbb{R}^n \times \mathbb{R}^n \times U_1$.

Noting $0 \in U_1$, one has (see (4.9))

$$
\begin{aligned}
H^+(t, x, p) &= \inf_{u_1 \in U_1} \mathcal{H}^+(t, x, p, u_1) \leqslant \mathcal{H}^+(t, x, p, 0) \\
&\leqslant L\big(\langle x \rangle^\mu + \langle x \rangle \, |p|\big) + \frac{L^2}{4c}|p|^2,
\end{aligned}
\tag{4.12}
$$

and noting $0 \in U_2$, we obtain (see (4.8))

$$H^+(t,x,p) = \inf_{u_1 \in U_1} \sup_{u_2 \in U_2} \mathbb{H}(t,x,p,u_1,u_2) \geqslant \inf_{u_1 \in U_1} \mathbb{H}(t,x,p,u_1,0)$$

$$\geqslant -L\big(\langle x \rangle^\mu + \langle x \rangle \, |p|\big) - \frac{L^2}{4c}|p|^2. \tag{4.13}$$

Thus, $H^+(\cdot,\cdot,\cdot)$ is well-defined and (4.5) holds.

Next, we want to show that $H^+(\cdot,\cdot,\cdot)$ is continuous and $p \mapsto H^+(t,x,p)$ is locally Lipschitz. To this end, we introduce

$$U_1(|x|,|p|) = \Big\{ u_1 \in U_1 \mid \frac{c}{2}|u_1|^2 \leqslant 2L\big(\langle x \rangle^\mu + \langle x \rangle \, |p|\big) + \frac{3L^2}{4c}|p|^2 + 1 \Big\},$$

$$\forall x, p \in \mathbb{R}^n,$$

which, for any given $x, p \in \mathbb{R}^n$, is a compact set. Clearly, for any $u_1 \in U_1 \setminus U_1(|x|,|p|)$, one has

$$c|u_1|^2 - L|p|\,|u_1| = \frac{c}{2}|u_1|^2 + \frac{c}{2}\Big(|u_1| - \frac{L}{c}|p|\Big)^2 - \frac{L^2}{2c}|p|^2$$

$$\geqslant \frac{c}{2}|u_1|^2 - \frac{L^2}{2c}|p|^2 > 2L\big(\langle x \rangle^\mu + \langle x \rangle \, |p|\big) + \frac{L^2}{4c}|p|^2 + 1.$$

Thus, for such a u_1, by (4.10) and (4.12),

$$\mathcal{H}^+(t,x,p,u_1) \geqslant -L\big(\langle x \rangle^\mu + \langle x \rangle \, |p|\big) - L|p|\,|u_1| + c|u_1|^2$$

$$> L\big(\langle x \rangle^\mu + \langle x \rangle \, |p|\big) + \tfrac{L^2}{4c}|p|^2 + 1 \tag{4.14}$$

$$\geqslant H^+(t,x,p) + 1 = \inf_{u_1 \in U_1} \mathcal{H}^+(t,x,p,u_1) + 1.$$

Hence,

$$\inf_{u_1 \in U_1} \mathcal{H}^+(t,x,p,u_1) = \inf_{u_1 \in U_1(|x|,|p|)} \mathcal{H}^+(t,x,p,u_1). \tag{4.15}$$

Now, by the definition of $U_1(|x|,|p|)$, we have

$$|u_1| \leqslant K_1\big(\langle x \rangle + |p|\big), \qquad \forall u_1 \in U_1(|x|,|p|),$$

for some absolute constant $K_1 > 0$ only depending on L and c. Hence, it follows from (4.11) that

$$|\bar{u}_2| \leqslant K_2\big(\langle x \rangle + |p|\big), \tag{4.16}$$

for some absolute constant K_2 only depending on L and c. Hence, if we let

$$U_2(|x|,|p|) = \Big\{ u_2 \in U_2 \mid |u_2| \leqslant K_2\big(\langle x \rangle + |p|\big) \Big\},$$

which is a compact set (for any given $x, p \in \mathbb{R}^n$), then for any $(t, x, p) \in [0, T] \times \mathbb{R}^n \times \mathbb{R}^n$,

$$H^+(t, x, p) = \inf_{u_1 \in U_1(|x|, |p|)} \sup_{u_2 \in U_2(|x|, |p|)} \mathbb{H}(t, x, p, u_1, u_2).$$

This implies that $H^+(\cdot, \cdot, \cdot)$ is continuous. Next, for any $(t, x) \in [0, T] \times \mathbb{R}^n$, $p, q \in \mathbb{R}^n$ and $u_i \in U_i(|x|, |p| \vee |q|)$ $(i = 1, 2)$, we have (without loss of generality, let $|q| \leqslant |p|$)

$$\begin{aligned}
&|\mathbb{H}(t, x, p, u_1, u_2) - \mathbb{H}(t, x, q, u_1, u_2)| \\
&\leqslant |p - q| \, |f(t, x, u_1, u_2)| \leqslant L(\langle x \rangle + |u_1| + |u_2|) |p - q| \\
&\leqslant K(\langle x \rangle + |p| \vee |q|) |p - q|.
\end{aligned}$$

This proves (4.6). $\qquad\qquad\qquad\qquad\qquad\qquad\qquad\qquad\qquad\qquad\square$

4.3 Uniqueness of Viscosity Solution

Having well-defined upper and lower Hamiltonians, we can write down the corresponding upper and lower HJI equations. Since (2.20) might not hold for $H^\pm(\cdot, \cdot, \cdot)$, the uniqueness of viscosity solution to such kind of HJ equation has to be re-established. To this end, we consider the following *HJ inequalities*:

$$\begin{cases} V_t(t, x) + H(t, x, V_x(t, x)) \geqslant 0, & (t, x) \in [0, T] \times \mathbb{R}^n, \\ V(T, x) \leqslant h(x), & x \in \mathbb{R}^n, \end{cases} \qquad (4.17)$$

and

$$\begin{cases} V_t(t, x) + H(t, x, V_x(t, x)) \leqslant 0, & (t, x) \in [0, T] \times \mathbb{R}^n, \\ V(T, x) \geqslant h(x), & x \in \mathbb{R}^n, \end{cases} \qquad (4.18)$$

as well as the following HJ equation:

$$\begin{cases} V_t(t, x) + H(t, x, V_x(t, x)) = 0, & (t, x) \in [0, T] \times \mathbb{R}^n, \\ V(T, x) = h(x), & x \in \mathbb{R}^n. \end{cases} \qquad (4.19)$$

We introduce the following definition (compare with Definition 2.5.1).

Definition 4.3.1. (i) A continuous function $V(\cdot, \cdot)$ is called a *viscosity sub-solution* of (4.17) if

$$V(T, x) \leqslant h(x), \qquad \forall x \in \mathbb{R}^n,$$

and for any continuous differentiable function $\varphi(\cdot,\cdot)$, if $(t_0, x_0) \in (0, T) \times \mathbb{R}^n$ is a local maximum of $(t, x) \mapsto V(t, x) - \varphi(t, x)$, then

$$\varphi_t(t_0, x_0) + H(t_0, x_0, \varphi_x(t_0, x_0)) \geqslant 0.$$

In this case, we also say that $V(\cdot,\cdot)$ satisfies (4.17) in the viscosity sense.

(ii) A continuous function $V(\cdot,\cdot)$ is called a *viscosity super-solution* of (4.18) if

$$V(T, x) \geqslant h(x), \qquad \forall x \in \mathbb{R}^n,$$

and for any continuous differentiable function $\varphi(\cdot,\cdot)$, if $(t_0, x_0) \in [0, T) \times \mathbb{R}^n$ is a local minimum of $(t, x) \mapsto V(t, x) - \varphi(t, x)$, then

$$\varphi_t(t_0, x_0) + H(t_0, x_0, \varphi_x(t_0, x_0)) \leqslant 0.$$

In this case, we also say that $V(\cdot,\cdot)$ satisfies (4.18) in the viscosity sense.

(iii) A continuous function $V(\cdot,\cdot)$ is called a *viscosity solution* of (4.19) if it is a viscosity sub-solution of (4.17) and a viscosity super-solution of (4.18).

We first present the following lemma.

Proposition 4.3.2. (i) *In the above definition, the local maximum and local minimum can be replaced by strict local maximum and strict local minimum, respectively.*

(ii) *In the above definition, the local maximum, and local minimum* (t_0, x_0) *are allowed to have* $t_0 = 0$.

Proof. (i) If, say, $V(\cdot,\cdot) - \varphi(\cdot,\cdot)$ attains a local maximum at (t_0, x_0). Then replacing $\varphi(\cdot,\cdot)$ by $\widehat{\varphi}(\cdot,\cdot)$ with

$$\widehat{\varphi}(t, x) = \varphi(t, x) + (t - t_0)^2 + |x - x_0|^2,$$

we see that

$$\begin{aligned}
V(t, x) - \widehat{\varphi}(t, x) &= V(t, x) - \varphi(t, x) - |t - t_0|^2 - |x - x_0|^2 \\
&\leqslant V(t_0, x_0) - \varphi(t_0, x_0) - |t - t_0|^2 - |x - x_0|^2 < V(t_0, x_0) - \widehat{\varphi}(t_0, x_0), \\
&\qquad\qquad \forall (t, x) \neq (t_0, x_0),
\end{aligned}$$

and

$$\varphi_t(t_0, x_0) = \widehat{\varphi}_t(t_0, x_0), \qquad \varphi_x(t_0, x_0) = \widehat{\varphi}_x(t_0, x_0).$$

Thus, in the definition, $\varphi(\cdot,\cdot)$ can be replaced by $\widehat{\varphi}(\cdot,\cdot)$, for which $V(\cdot,\cdot) - \widehat{\varphi}(\cdot,\cdot)$ attains a strict local maximum at (t_0, x_0).

(ii) Let $V(\cdot,\cdot) - \varphi(\cdot,\cdot)$ attain a strict local maximum at $(0, x_0)$, i.e.,

$$V(t,x) - \varphi(t,x) < V(0,x_0) - \varphi(0,x_0), \qquad \text{for } (t,x) \text{ near } (0,x_0).$$

then there exist $r > 0$ and $\mu = \mu(r) > 0$, with $\mu(r) \to 0$ as $r \to 0$, such that

$$V(t,x) - \varphi(t,x) \leqslant V(0,x_0) - \varphi(0,x_0) - \mu,$$
$$\forall(t,x) \in \partial\Gamma_r \setminus [\{0\} \times \overline{B_r(x_0)}],$$

where

$$\Gamma_r = (0,r) \times B_r(x_0), \quad B_r(x_0) = \Big\{x \in \mathbb{R}^n \mid |x - x_0| < r\Big\}.$$

By the continuity of $V(\cdot,\cdot)$ and $\varphi(\cdot,\cdot)$ on $\overline{\Gamma}_r$, we have some $t_0 \in (0,r)$ such that

$$|V(t_0,x_0) - V(0,x_0)| + |\varphi(t_0,x_0) - \varphi(0,x_0)| \leqslant \frac{\mu}{2}.$$

Now, for any $\varepsilon \in (0, \frac{\mu t_0}{2})$, let

$$\psi^\varepsilon(t,x) = \varphi(t,x) + \frac{\varepsilon}{t}, \qquad (t,x) \in \Gamma_r.$$

Then

$$\lim_{t\downarrow 0}[V(t,x)) - \psi^\varepsilon(t,x)] = -\infty, \qquad x \in \bar{B}_r(x_0),$$

and for $(t,x) \in \partial\Gamma_r \setminus [\{0\} \times \overline{B_r(x_0)}]$, we have

$$V(t,x) - \psi^\varepsilon(t,x) = V(t,x) - \varphi(t,x) - \frac{\varepsilon}{t}$$
$$\leqslant V(0,x_0) - \varphi(0,x_0) - \mu - \frac{\varepsilon}{t}$$
$$\leqslant V(t_0,x_0) - \varphi(t_0,x_0) - \frac{\varepsilon}{t_0} + \frac{\mu}{2} - \mu + \frac{\varepsilon}{t_0} - \frac{\varepsilon}{t}$$
$$\leqslant V(t_0,x_0) - \psi^\varepsilon(t_0,x_0) - \frac{\mu}{2} + \frac{\varepsilon}{t_0} \leqslant V(t_0,x_0) - \psi^\varepsilon(t_0,x_0).$$

Hence, $V(\cdot,\cdot) - \psi^\varepsilon(\cdot,\cdot)$ attains a local maximum at some $(t_\varepsilon, x_\varepsilon) \in \Gamma_r$. Then

$$0 \leqslant \psi_t^\varepsilon(t_\varepsilon, x_\varepsilon) + H(t_\varepsilon, x_\varepsilon, \psi_x^\varepsilon(t_\varepsilon, x_\varepsilon))$$
$$= \varphi_t(t_\varepsilon, x_\varepsilon) - \frac{\varepsilon}{t_\varepsilon^2} + H(t_\varepsilon, x_\varepsilon, \varphi_x(t_\varepsilon, x_\varepsilon))$$
$$\leqslant \varphi_t(t_\varepsilon, x_\varepsilon) + H(t_\varepsilon, x_\varepsilon, \varphi_x(t_\varepsilon, x_\varepsilon)).$$

Sending $\varepsilon \to 0$, we may assume that $(t_\varepsilon, x_\varepsilon) \to (t_r, x_r) \in \overline{\Gamma}_r$, and

$$0 \leqslant \varphi_t(t_r, x_r) + H(t_r, x_r, \varphi_x(t_r, x_r)).$$

Finally, sending $r \to 0$, we get

$$0 \leqslant \varphi_t(0, x_0) + H(0, x_0, \varphi_x(0, x_0)).$$

This proves (ii) for sub-solution case. The super-solution case can be proved similarly. □

Lemma 4.3.3. *Suppose* $H : [0, T] \times \mathbb{R}^n \times \mathbb{R} \times \mathbb{R}^n \to \mathbb{R}$ *is continuous and*

$$\begin{cases} V_t(t, x) + H(t, x, V_x(t, x)) \geqslant 0, \\ \widehat{V}_t(t, x) + H(t, x, \widehat{V}_x(t, x)) \leqslant 0, \end{cases}$$

in the viscosity sense. Then

$$W(t, x, y) = V(t, x) - \widehat{V}(t, y)$$

satisfies the following in the viscosity sense:

$$W_t(t, x, y) + H(t, x, W_x(t, x, y)) - H(t, y, -W_x(t, x, y)) \geqslant 0.$$

Proof. Let $\varphi : (0, T) \times \mathbb{R}^n \times \mathbb{R}^n \to \mathbb{R}$ be C^1 such that $W(t, x, y) - \varphi(t, x, y)$ attains a strict local maximum at $(t_0, x_0, y_0) \in (0, T) \times \mathbb{R}^n \times \mathbb{R}^n$. Let

$$d(t, x, s, y) = \max \left\{ |t - t_0|, |s - t_0|, |x - x_0|, |y - y_0| \right\},$$

and let

$$\Gamma_r \equiv \left\{ (t, x, s, y) \in (0, T) \times \mathbb{R}^n \times (0, T) \times \mathbb{R}^n, \ d(t, x, s, y) < r \right\}.$$

Then there exists a small enough $r_0 > 0$ and a continuous function $\rho : [0, r_0] \to [0, \infty)$ with

$$\Gamma_{r_0} \subseteq (0, T) \times \mathbb{R}^n \times (0, T) \times \mathbb{R}^n,$$
$$\rho(0) = 0, \quad \rho(r) > 0, \quad r \in (0, r_0],$$

such that

$$V(t, x) - \widehat{V}(t, y) - \varphi(t, x, y)$$
$$\leqslant V(t_0, x_0) - \widehat{V}(t_0, y_0) - \varphi(t_0, x_0, y_0) - \rho\big(d(t, x, t, y)\big),$$
$$(t, x, t, y) \in \Gamma_{r_0}.$$

We fix an $r \in (0, r_0)$. Then from $\rho(r) > 0$, by the continuity of $V(\cdot, \cdot)$, $\widehat{V}(\cdot, \cdot)$, and $\varphi(\cdot, \cdot, \cdot)$, we have a $\delta > 0$ such that

$$|V(t, x) - V(s, x)| + |\widehat{V}(t, y) - \widehat{V}(s, y)| + |\varphi(t, x, y) - \varphi(s, x, y)| < \frac{\rho(r)}{2},$$
$$(t, x, s, y) \in \overline{\Gamma}_{r_0}, \quad |t - s| \leqslant \delta,$$

and by choosing

$$0 < \varepsilon < \frac{\delta^2}{M},$$

with

$$M = \sup_{(t,x,s,y)\in\overline{\Gamma}_{r_0}} \Big\{ |V(t,x) - V(s,y)| + |\widehat{V}(t,x) - \widehat{V}(s,y)| $$
$$+ |\varphi(t,s,x) - \varphi(s,x,y)| \Big\},$$

we have

$$|V(t,x) - V(s,x)| + |\widehat{V}(t,y) - \widehat{V}(s,y)| + |\varphi(t,x,y) - \varphi(s,x,y)|$$
$$\leqslant M < \frac{(t-s)^2}{\varepsilon}, \qquad (t,x,s,y) \in \overline{\Gamma}_{r_0}, \quad |t-s| \geqslant \delta.$$

Hence,

$$|V(t,x) - V(s,x)| + |\widehat{V}(t,y) - \widehat{V}(s,y)| + |\varphi(t,x,y) - \varphi(s,x,y)|$$
$$\leqslant M < \frac{(t-s)^2}{\varepsilon} + \frac{\rho(r)}{2}, \qquad (t,x,s,y) \in \overline{\Gamma}_{r_0}.$$

Now, we let

$$\Psi(t,x,s,y) = V(t,x) - \widehat{V}(s,y) - \varphi(t,x,y) - \frac{(t-s)^2}{\varepsilon},$$
$$(t,x,s,y) \in \overline{\Gamma}_r.$$

If $(t,x,s,y) \in \partial\Gamma_r$, then

$$d(t,x,s,y) \equiv \max\{|t-t_0|, |s-t_0|, |x-x_0|, |y-y_0|\} = r. \qquad (4.20)$$

In this case, we claim that

$$d(t,x,t,y) = r, \quad \text{or} \quad d(s,x,s,y) = r. \qquad (4.21)$$

In fact, when (4.20) holds, if $|t-t_0| = r$, then

$$|s-t_0|, |x-x_0|, |y-y_0| \leqslant r,$$

and thus

$$d(t,x,t,y) = \max\{|t-t_0|, |t-t_0|, |x-x_0|, |y-y_0|\} = |t-t_0| = r.$$

Similarly, in the case $|s-t_0| = r$, we have

$$d(s,x,s,y) = \max\{|s-t_0|, |s-t_0|, |x-x_0|, |y-y_0|\} = |s-t_0| = r.$$

Finally, in the case, $|t-t_0|, |s-t_0| < r$, $\max\{|x-x_0|, |y-y_0|\} = r$. Thus,

$$d(t,x,t,y) = \max\{|x-x_0|, |y-y_0|\} = r = d(s,x,s,y).$$

This means that (4.21) holds. Consequently, when (4.21) holds, we have

$$\Psi(t,x,s,y) = V(t,x) - \widehat{V}(t,x) - \varphi(t,x,y) - \left(\widehat{V}(s,y) - \widehat{V}(t,x) + \frac{(t-s)^2}{\varepsilon}\right)$$

$$\leqslant V(t_0,x_0) - \widehat{V}(t_0,x_0) - \varphi(t_0,x_0,y_0) - \rho\big(d(t,x,t,y)\big)$$

$$+ |\widehat{V}(s,y) - \widehat{V}(t,y)| - \frac{(t-s)^2}{\varepsilon}$$

$$\leqslant \Psi(t_0,x_0,t_0,y_0) - \rho(r) + \frac{(t-s)^2}{\varepsilon} + \frac{\rho(r)}{2} - \frac{(t-s)^2}{\varepsilon}$$

$$= \Psi(t_0,x_0,t_0,y_0) - \frac{\rho(r)}{2}.$$

Therefore, $\Psi(\cdot,\cdot,\cdot,\cdot)$ admits a local maximum $(t_r,x_r,s_r,y_r) \in \Gamma_r$. This implies

$$(t,x) \mapsto V(t,x) - \left[\widehat{V}(s_r,y_r) + \varphi(t,x,y_r) + \frac{(t-s_r)^2}{\varepsilon}\right]$$

attains a local maximum at (t_r,x_r), we have

$$\varphi_t(t_r,x_r,y_r) + \frac{2(t_r-s_r)}{\varepsilon} + H(t_r,x_r,\varphi_x(t_r,x_r,y_r)) \geqslant 0. \qquad (4.22)$$

Also,

$$(t,y) \mapsto \widehat{V}(s,y) - \left[V(t_r,x_r) - \varphi(t_r,x_r,y) - \frac{(s-t_r)^2}{\varepsilon}\right]$$

attains a local minimum at (s_r,y_r). Then

$$-\frac{2(s_r-t_r)}{\varepsilon} + H(s_r,y_r,-\varphi_y(t_r,x_r,y_r)) \leqslant 0. \qquad (4.23)$$

Combining (4.22) and (4.23), we have

$$\varphi_t(t_r,x_r,y_r) + H(t_r,x_r,\varphi_x(t_r,x_r,y_r)) - H(s_r,y_r,-\varphi_y(t_r,x_r,y_r)) \geqslant 0.$$

Sending $r \to 0$, we obtain

$$\varphi_t(t_0,x_0,y_0) + H(t_0,x_0,\varphi_x(t_0,x_0,y_0)) - H(t_0,y_0,-\varphi_y(t_0,x_0,y_0)) \geqslant 0.$$

This proves what we have claimed. $\qquad\qquad\qquad\qquad\qquad\qquad\qquad\square$

Now we assume the following (compare with (2.20)).

(HJ) The maps $H : [0,T] \times \mathbb{R}^n \times \mathbb{R}^n \to \mathbb{R}$ and $h : \mathbb{R}^n \to \mathbb{R}$ are continuous and there is a constant $K_0 > 0$ and a continuous function $\omega : [0,\infty)^3 \to [0,\infty)$ with property $\omega(r,s,0) = 0$, such that

$$\begin{cases} |H(t,x,p) - H(t,y,p)| \leqslant \omega\big(|x| + |y|, |p|, |x-y|\big), \\ |H(t,x,p) - H(t,x,q)| \leqslant K_0(\langle x \rangle + |p| \vee |q|)|p-q|, \\ \qquad\qquad \forall t \in [0,T], \ x,y,p,q \in \mathbb{R}^n. \end{cases}$$

and

$$|h(x) - h(y)| \leqslant K_0 (\langle x \rangle + \langle y \rangle) |x - y|, \qquad \forall x, y \in \mathbb{R}^n.$$

Our main result of this section is the following comparison theorem.

Theorem 4.3.4. *Let* (HJ) *hold. Suppose* $V(\cdot, \cdot)$ *and* $\widehat{V}(\cdot, \cdot)$ *are the viscosity sub- and super-solution of* (4.17) *and* (4.18), *respectively. Moreover, let*

$$\begin{aligned} |V(t, x) - V(t, y)|, |\widehat{V}(t, x) - \widehat{V}(t, y)| \\ \leqslant K (\langle x \rangle + \langle y \rangle) |x - y|, \quad \forall t \in [0, T], \; x, y \in \mathbb{R}^n, \end{aligned}$$

for some $K > 0$. *Then*

$$V(t, x) \leqslant \widehat{V}(t, x), \qquad \forall (t, x) \in [0, T] \times \mathbb{R}^n.$$

Proof. Suppose there exists a $(\bar{t}, \bar{x}) \in [0, T) \times \mathbb{R}^n$ such that

$$V(\bar{t}, \bar{x}) - \widehat{V}(\bar{t}, \bar{x}) > 0.$$

Let $\alpha, \beta > 0$ be undetermined. Define

$$Q \equiv Q(\alpha, \beta) = \Big\{ (t, x) \in [0, T] \times \mathbb{R}^n \mid \langle x \rangle \leqslant \langle \bar{x} \rangle \, e^{\alpha + \beta (t - \bar{t})} \Big\},$$

and

$$G \equiv G(\alpha, \beta) = \Big\{ (t, x, y) \in [0, T] \times \mathbb{R}^n \times \mathbb{R}^n \mid (t, x), (t, y) \in Q \Big\}.$$

Now, for $\delta > 0$ small, define

$$\psi(t, x) \equiv \psi^{\beta, \delta}(t, x) = \Big[\frac{\langle x \rangle}{\langle \bar{x} \rangle} \, e^{\beta (\bar{t} - t)} \Big]^{\frac{1}{\delta}} \equiv e^{\frac{1}{\delta} \left[\log \frac{\langle x \rangle}{\langle \bar{x} \rangle} + \beta (\bar{t} - t) \right]}.$$

Then

$$\psi(\bar{t}, \bar{x}) = 1, \qquad \psi(T, x) = \Big[\frac{\langle x \rangle}{\langle \bar{x} \rangle} \, e^{\beta (\bar{t} - T)} \Big]^{\frac{1}{\delta}},$$

and

$$\psi_t(t, x) = -\frac{\beta \psi(t, x)}{\delta}, \qquad \psi_x(t, x) = \frac{x \psi(t, x)}{\delta \langle x \rangle^2}.$$

For any $(t, x) \in \bar{Q}$, we have

$$\langle x \rangle \leqslant \langle \bar{x} \rangle \, e^{\alpha + \beta (t - \bar{t})} \leqslant \langle \bar{x} \rangle \, e^{\alpha + \beta (T - \bar{t})}.$$

Thus, Q is bounded and $\bar{G}$ is compact. We introduce

$$\Psi(t, x, y) = V(t, x) - \widehat{V}(t, y) - \frac{|x - y|^2}{\varepsilon} - \sigma \psi(t, x) - \frac{\sigma(T - t)}{T - \bar{t}},$$
$$(t, x, y) \in \bar{G},$$

where $\varepsilon > 0$ small and

$$0 < \sigma \leqslant \frac{V(\bar{t}, \bar{x}) - \widehat{V}(\bar{t}, \bar{x})}{3}.$$

Clearly,

$$\Psi(\bar{t}, \bar{x}, \bar{x}) = V(\bar{t}, \bar{x}) - \widehat{V}(\bar{t}, \bar{x}) - \sigma - \sigma \geqslant 3\sigma - 2\sigma = \sigma > 0.$$

Since $\Psi(\cdot, \cdot, \cdot)$ is continuous on the compact set $\bar{G}$, we may let $(t_0, x_0, y_0) \in \bar{G}$ be a maximum of $\Psi(\cdot, \cdot, \cdot)$ over $\bar{G}$. By the optimality of (t_0, x_0, y_0), we have

$$V(t_0, x_0) - \widehat{V}(t_0, x_0) - \sigma\psi(t_0, x_0) - \frac{\sigma(T - t_0)}{T - \bar{t}}$$

$$= \Psi(t_0, x_0, x_0) \leqslant \Psi(t_0, x_0, y_0)$$

$$= V(t_0, x_0) - \widehat{V}(t_0, y_0) - \frac{|x_0 - y_0|^2}{\varepsilon} - \sigma\psi(t_0, x_0) - \frac{\sigma(T - t_0)}{T - \bar{t}},$$

which implies

$$\frac{|x_0 - y_0|^2}{\varepsilon} \leqslant \widehat{V}(t_0, x_0) - \widehat{V}(t_0, y_0) \leqslant K(\langle x_0 \rangle + \langle y_0 \rangle)|x_0 - y_0|.$$

Thus,

$$|x_0 - y_0| \leqslant K(\langle x_0 \rangle + \langle y_0 \rangle)\varepsilon.$$

Now, if $t_0 = T$, then

$$\langle x_0 \rangle, \ \langle y_0 \rangle \leqslant \langle \bar{x} \rangle e^{\alpha + \beta(T - \bar{t})}.$$

Hence,

$$\Psi(T, x_0, y_0) = h(x_0) - h(y_0) - \frac{|x_0 - y_0|^2}{\varepsilon} - \sigma\psi(T, x_0)$$

$$\leqslant K_0(\langle x_0 \rangle + \langle y_0 \rangle)|x_0 - y_0| \leqslant K_0 K(\langle x_0 \rangle + \langle y_0 \rangle)^2 \varepsilon$$

$$\leqslant K_0 K \big[2\langle \bar{x} \rangle e^{[\alpha + \beta(T - \bar{t})]}\big]^2 \varepsilon.$$

Thus, for $\varepsilon > 0$ small enough, the following holds:

$$\Psi(T, x_0, y_0) < \sigma \leqslant \Psi(\bar{t}, \bar{x}, \bar{x}) \leqslant \Psi(t_0, x_0, y_0),$$

which means that $t_0 \in [0, T)$. Next, we note that for $(t, x) \in (\partial Q) \cap [(0, T) \times \mathbb{R}^n]$, one has

$$\log \frac{\langle x \rangle}{\langle \bar{x} \rangle} + \beta(\bar{t} - t) = \alpha, \qquad \text{and} \qquad 0 < t < T,$$

which implies

$$\psi(t, x) = e^{\frac{\alpha}{\delta}} \to \infty, \qquad \delta \to 0,$$

uniformly in $(t, x) \in (\partial Q) \cap [(0, T) \times \mathbb{R}^n]$. This implies that for $\delta > 0$ small (only depending on α),

$$(t_0, x_0, y_0) \in G \cup \Big[\{0\} \times \mathbb{R}^n \times \mathbb{R}^n \Big].$$

By Lemma 4.3.3, we have

$$0 \leqslant \sigma \psi_t(t_0, x_0) - \frac{\sigma}{T - \bar{t}} + H\Big(t_0, x_0, \frac{2(x_0 - y_0)}{\varepsilon} + \sigma \psi_x(t_0, x_0) \Big)$$
$$- H\Big(t_0, y_0, -\frac{2(y_0 - x_0)}{\varepsilon} \Big)$$

$$= \sigma \psi_t(t_0, x_0) - \frac{\sigma}{T - \bar{t}} + H\Big(t_0, x_0, \frac{2(x_0 - y_0)}{\varepsilon} + \sigma \psi_x(t_0, x_0) \Big)$$
$$- H\Big(t_0, x_0, \frac{2(x_0 - y_0)}{\varepsilon} \Big) + H\Big(t_0, x_0, \frac{2(x_0 - y_0)}{\varepsilon} \Big)$$
$$- H\Big(t_0, y_0, \frac{2(x_0 - y_0)}{\varepsilon} \Big)$$

$$\leqslant \sigma \psi_t(t_0, x_0) - \frac{\sigma}{T - \bar{t}}$$
$$+ K_0 \Big[\langle x_0 \rangle + \frac{2|x_0 - y_0|}{\varepsilon} + \sigma |\psi_x(t_0, x_0)| \Big] \sigma |\psi_x(t_0, x_0)|$$
$$+ \omega \Big(|x_0| + |y_0|, \frac{2|x_0 - y_0|}{\varepsilon}, |x_0 - y_0| \Big)$$

$$\leqslant -\sigma \frac{\alpha}{\delta} \psi(t_0, x_0) - \frac{\sigma}{T - \bar{t}}$$
$$+ \sigma K_0 \Big[\langle x_0 \rangle + 2K(\langle x_0 \rangle + \langle y_0 \rangle) + \frac{\sigma \psi(t_0, x_0)}{\delta \langle x_0 \rangle} \Big] \frac{\psi(t_0, x_0)}{\delta \langle x_0 \rangle}$$
$$+ \omega \Big(|x_0| + |y_0|, \frac{2|x_0 - y_0|}{\varepsilon}, |x_0 - y_0| \Big).$$

Note that $(t_0, x_0, y_0) \equiv (t_{0,\varepsilon}, x_{0,\varepsilon}, y_{0,\varepsilon}) \in \bar{G}(\alpha, \beta)$ (a fixed compact set). Let $\varepsilon \to 0$ along a suitable sequence, we have $|x_{0,\varepsilon} - y_{0,\varepsilon}| \to 0$, and $\frac{|x_{0,\varepsilon} - y_{0,\varepsilon}|}{\varepsilon}$ stays bounded. For notational simplicity, we denote $(t_{0,\varepsilon}, x_{0,\varepsilon}, y_{0,\varepsilon}) \to (t_0, x_0, x_0)$. Thus, after sending $\varepsilon \to 0$, one has

$$0 \leqslant -\sigma \frac{\alpha}{\delta} \psi(t_0, x_0) - \frac{\sigma}{T - \bar{t}}$$
$$+ \sigma K_0 \Big[(4K + 1) \langle x_0 \rangle + \frac{\sigma \psi(t_0, x_0)}{\delta \langle x_0 \rangle} \Big] \frac{\psi(t_0, x_0)}{\delta \langle x_0 \rangle}.$$

Then cancel σ and send $\sigma \to 0$, one obtains

$$\frac{1}{T - \bar{t}} \leqslant -\frac{\alpha \psi(t_0, x_0)}{\delta} + K_0(4K + 1) \langle x_0 \rangle \frac{\psi(t_0, x_0)}{\delta \langle x_0 \rangle}$$
$$= -\big(\alpha - K_0(4K + 1)\big) \frac{\psi(t_0, x_0)}{\delta}.$$

Thus, by taking $\alpha > K_0(4K + 1)$, we obtain a contradiction, proving our conclusion. $\square$

The following corollary is clear.

Corollary 4.3.5. *Let* (H1)–(H2) *hold, and let* $h : \mathbb{R}^n \to \mathbb{R}$ *be continuous. Let* $H^{\pm}(\cdot,\cdot,\cdot)$ *be upper and lower Hamiltonians defined by* (4.4). *Then each of the following upper and lower HJI equations*

$$\begin{cases} V_t^{\pm}(t,x) + H^{\pm}(t,x,V^{\pm}(t,x)) = 0, & (t,x) \in [0,T] \times \mathbb{R}^n, \\ V^{\pm}(T,x) = h(x), & x \in \mathbb{R}^n \end{cases} \qquad (4.24)$$

has at most one viscosity solution.

4.4 Upper and Lower Value Functions

In this section, we are going to look at the upper and lower value functions defined via the Elliottt–Kalton strategies. Some basic properties of upper and lower value functions will be established carefully.

4.4.1 *State trajectories and Elliott–Kalton strategies*

Let us introduce the following hypotheses, in addition to (H1)–(H2), some local Lipschitz continuity is added for the involved maps.

(H1)$'$ Map $f : [0,T] \times \mathbb{R}^n \times U_1 \times U_2 \to \mathbb{R}^n$ satisfies (H1). Moreover,

$$|f(t,x,u_1,u_2) - f(t,y,u_1,u_2)| \leqslant L\big[\langle x \rangle \vee \langle y \rangle + |u_1| + |u_2|\big]|x-y|,$$
$$\forall (t,u_1,u_2) \in [0,T] \times U_1 \times U_2, \ x,y \in \mathbb{R}^n,$$

and

$$\langle f(t,x,u_1,u_2) - f(t,y,u_1,u_2), x - y \rangle \leqslant L|x-y|^2,$$
$$\forall (t,u_1,u_2) \in [0,T] \times U_1 \times U_2, \ x,y \in \mathbb{R}^n. \qquad (4.25)$$

(H2)$'$ Map $g : [0,T] \times \mathbb{R}^n \times U_1 \times U_2 \to \mathbb{R}$ satisfies (H2). Moreover,

$$|g(t,x,u_1,u_2) - g(t,y,u_1,u_2)|$$
$$\leqslant L\big[(\langle x \rangle \vee \langle y \rangle)^{\mu-1} + |u_1| + |u_2|\big]|x-y|,$$
$$\forall (t,u_1,u_2) \in [0,T] \times U_1 \times U_2, \ x,y \in \mathbb{R}^n.$$

Also, map $h : \mathbb{R}^n \to \mathbb{R}$ is continuous and

$$\begin{cases} |h(x) - h(y)| \leqslant L(\langle x \rangle \vee \langle y \rangle)^{\mu-1}|x-y|, & \forall x,y \in \mathbb{R}^n, \\ |h(0)| \leqslant L. \end{cases}$$

Let us present the following Gronwall type inequality.

Lemma 4.4.1. *Let* $\theta, \alpha, \beta : [t, T] \to \mathbb{R}_+$ *and* $\theta_0 \geqslant 0$ *satisfy*

$$\theta(s)^2 \leqslant \theta_0^2 + \int_t^s \left[\alpha(r)\theta(r)^2 + \beta(r)\theta(r)\right] dr, \qquad s \in [t, T].$$

Then

$$\theta(s) \leqslant e^{\frac{1}{2}\int_t^s \alpha(\tau)d\tau}\theta_0 + \frac{1}{2}\int_t^s e^{\frac{1}{2}\int_r^s \alpha(\tau)d\tau}\beta(r)dr, \qquad s \in [t, T].$$

Proof. First, by the usual Gronwall's inequality, we have

$$\theta(s)^2 \leqslant e^{\int_t^s \alpha(\tau)d\tau}\theta_0^2 + \int_t^s e^{\int_r^s \alpha(\tau)d\tau}\beta(r)\theta(r)dr.$$

This implies

$$e^{-\int_t^s \alpha(\tau)d\tau}\theta(s)^2 \leqslant \theta_0^2 + \int_t^s e^{-\int_t^r \alpha(\tau)d\tau}\beta(r)\theta(r)dr \equiv \Theta(s).$$

Then

$$\frac{d}{ds}\sqrt{\Theta(s)} = \frac{1}{2}\Theta(s)^{-\frac{1}{2}}\dot{\Theta}(s) = \frac{1}{2}\Theta(s)^{-\frac{1}{2}}e^{-\int_t^s \alpha(\tau)d\tau}\beta(s)\theta(s)$$
$$\leqslant \frac{1}{2}e^{-\frac{1}{2}\int_t^s \alpha(\tau)d\tau}\beta(s).$$

Consequently,

$$\theta(s) \leqslant e^{\frac{1}{2}\int_t^s \alpha(\tau)d\tau}\sqrt{\Theta(s)}$$
$$\leqslant e^{\frac{1}{2}\int_t^s \alpha(\tau)d\tau}\left[\sqrt{\Theta(t)} + \int_t^s \frac{1}{2}e^{-\frac{1}{2}\int_t^r \alpha(\tau)d\tau}\beta(r)dr\right]$$
$$= e^{\frac{1}{2}\int_t^s \alpha(\tau)d\tau}\theta_0 + \frac{1}{2}\int_t^s e^{\frac{1}{2}\int_r^s \alpha(\tau)d\tau}\beta(r)dr,$$

proving our conclusion. □

We now prove the following result concerning the state trajectories.

Proposition 4.4.2. *Let* (H1)′ *hold. Then, for any* $(t, x) \in [0, T) \times \mathbb{R}^n$, $(u_1(\cdot), u_2(\cdot)) \in \mathcal{U}_1[t, T] \times \mathcal{U}_2[t, T]$, *state equation* (4.1) *admits a unique solution* $X(\cdot) \equiv X(\cdot; t, x, u_1(\cdot), u_2(\cdot)) \equiv X_{t,x}(\cdot)$. *Moreover, there exists a constant* $K_0 > 0$ *only depends on* L, T, t *such that*

$$\langle X_{t,x}(s) \rangle \leqslant K_0\left\{\langle x \rangle + \int_t^s \left(|u_1(r)| + |u_2(r)|\right)dr\right\}, \qquad s \in [t, T], \quad (4.26)$$

$$|X_{t,x}(s) - x| \leqslant K_0\left\{\langle x \rangle(s - t) + \int_t^s \left(|u_1(r)| + |u_2(r)|\right)dr\right\}, \quad (4.27)$$
$$s \in [t, T],$$

and for $(\bar{t}, \bar{x}) \in [0, T] \times \mathbb{R}^n$ with $\bar{t} \in [t, T]$,

$$|X_{t,x}(s) - X_{\bar{t},\bar{x}}(s)| \leqslant K_0 \Big\{ |x - \bar{x}| + \langle x \rangle \vee \langle \bar{x} \rangle (\bar{t} - t) \tag{4.28}$$
$$+ \int_t^{\bar{t}} \big(|u_1(r)| + |u_2(r)| \big) dr \Big\}, \quad s \in [t, T].$$

Proof. First, under (H1)′, for any $(t, x) \in [0, T) \times \mathbb{R}^n$, and any $(u_1(\cdot), u_2(\cdot)) \in \mathcal{U}_1[t, T] \times \mathcal{U}_2[t, T]$, the map $y \mapsto f(s, y, u_1(s), u_2(s))$ is locally Lipschitz continuous. Thus, state equation (4.1) admits a unique local solution $X(\cdot) = X(\cdot \,; t, x, u_1(\cdot), u_2(\cdot))$. Next, by (4.25) and (H1), we have

$$\langle x, f(t, x, u_1, u_2) \rangle$$
$$= \langle x, f(t, x, u_1, u_2) - f(t, 0, u_1, u_2) \rangle + \langle x, f(t, 0, u_1, u_2) \rangle$$
$$\leqslant L \Big[|x|^2 + |x| \big(1 + |u_1| + |u_2| \big) \Big], \quad \forall (t, x, u_1, u_2) \in [0, T] \times \mathbb{R}^n \times U_1 \times U_2.$$

Thus,

$$\langle X(s) \rangle^2 = \langle x \rangle^2 + 2 \int_t^s \langle X(r), f(r, X(r), u_1(r), u_2(r)) \rangle \, dr$$
$$\leqslant \langle x \rangle^2 + 2L \int_t^s \Big[\langle X(r) \rangle^2 + \langle X(r) \rangle \big(1 + |u_1(r)| + |u_2(r)| \big) \Big] dr.$$

Then, it follows from Lemma 4.4.1 that

$$\langle X(s) \rangle \leqslant e^{L(T-t)} \langle x \rangle + L \int_t^s e^{L(s-r)} \big(1 + |u_1(r)| + |u_2(r)| \big) dr$$
$$= e^{L(T-t)} \langle x \rangle + e^{L(s-t)} - 1 + L \int_t^s e^{L(s-r)} \big(|u_1(r)| + |u_2(r)| \big) dr.$$

This implies that the solution $X(\cdot)$ of the state equation (4.1) globally exists on $[t, T]$ and (4.26) holds with any

$$K_0 \geqslant (2 \vee L) e^{L(T-t)}.$$

Also, we have

$$|X(s) - x|^2 = 2 \int_t^s \langle X(r) - x, f(r, X(r), u_1(r), u_2(r)) \rangle \, dr$$
$$\leqslant 2 \int_t^s \Big(L|X(r) - x|^2 + \langle X(r) - x, f(r, x, u_1(r), u_2(r)) \rangle \Big) dr$$
$$\leqslant 2L \int_t^s \Big(|X(r) - x|^2 + |X(r) - x| \big(\langle x \rangle + |u_1(r)| + |u_2(r)| \big) \Big) dr.$$

Thus, by Lemma 4.4.1 again, we obtain

$$|X(s) - x| \leqslant L \int_t^s e^{L(s-r)} \Big(\langle x \rangle + |u_1(r)| + |u_2(r)| \Big) dr$$

$$\leqslant Le^{L(s-t)} \Big[\langle x \rangle (s-t) + \int_t^s \big(|u_1(r)| + |u_2(r)| \big) dr \Big].$$

Therefore, (4.27) holds with any

$$K_0 \geqslant Le^{L(T-t)}.$$

Now, for any $(t,x), (\bar{t}, \bar{x}) \in [0,T] \times \mathbb{R}^n$, with $0 \leqslant t \leqslant \bar{t} < T$, we have

$$|X_{t,x}(s) - X_{\bar{t},\bar{x}}(s)|^2$$
$$= |X_{t,x}(\bar{t}) - \bar{x}|^2 + 2 \int_{\bar{t}}^s \langle X_{t,x}(r) - X_{\bar{t},\bar{x}}(r), f(r, X_{t,x}(r), u_1(r), u_2(r))$$
$$- f(r, X_{\bar{t},\bar{x}}(r), u_1(r), u_2(r)) \rangle \, dr$$
$$\leqslant |X_{t,x}(\bar{t}) - x|^2 + 2L \int_{\bar{t}}^s |X_{t,x}(r) - X_{\bar{t},\bar{x}}(r)|^2 dr.$$

Thus, it follows from the Gronwall's inequality that

$$|X_{t,x}(s) - X_{\bar{t},\bar{x}}(s)|$$
$$\leqslant e^{L(s-\bar{t})} |X_{t,x}(\bar{t}) - \bar{x}| \leqslant e^{L(s-\bar{t})} \big(|x - \bar{x}| + |X_{t,x}(\bar{t}) - x| \big)$$
$$\leqslant e^{L(s-\bar{t})} \Big[|x - \bar{x}| + Le^{L(\bar{t}-t)} \Big(\langle x \rangle (\bar{t}-t) + \int_t^{\bar{t}} \big(|u_1(r)| + |u_2(r)| \big) dr \Big) \Big]$$
$$\leqslant e^{L(s-\bar{t})} |x - \bar{x}| + Le^{L(s-t)} \Big[\langle x \rangle (\bar{t}-t) + \int_t^{\bar{t}} \big(|u_1(r)| + |u_2(r)| \big) dr \Big].$$

Hence, (4.28) holds with any

$$K_0 \geqslant (1 \vee L) e^{L(T-t)}.$$

This completes the proof. $\qquad\blacksquare$

From the above proposition, together with (H2)$'$, we see that for any $u_i(\cdot) \in \mathcal{U}_i[t,T]$, $i = 1, 2$, the performance functional $J(t, x; u_1(\cdot), u_2(\cdot))$ is well-defined. Similar to those in the previous chapter, we may define Elliott–Kalton (E-K, for short) strategies for Players 1 and 2, respectively. The sets of all E-K strategies for Player i on $[t, T]$ is denoted by $\mathcal{A}_i[t,T]$.

Recall that $0 \in U_i$ $(i = 1, 2)$. For later convenience, we hereafter let $u_1^0(\cdot) \in \mathcal{U}_1[t,T]$ and $u_2^0(\cdot) \in \mathcal{U}_2[t,T]$ be defined by

$$u_1^0(s) = 0, \quad u_2^0(s) = 0, \qquad \forall s \in [t, T],$$

and let $\alpha_1^0 \in \mathcal{A}_1[t, T]$ be the E-K strategy that

$$\alpha_1^0[u_2(\cdot)](s) = 0, \qquad \forall s \in [t, T], \quad u_2(\cdot) \in \mathcal{U}_2[t, T].$$

We call such an α_1^0 the *zero E-K strategy* for Player 1. Similarly, we define zero E-K strategy $\alpha_2^0 \in \mathcal{A}_2[t, T]$ for Player 2.

Now, we define

$$\begin{cases} V^+(t, x) = \displaystyle\sup_{\alpha_2 \in \mathcal{A}_2[t,T]} \inf_{u_1(\cdot) \in \mathcal{U}_1[t,T]} J(t, x; u_1(\cdot), \alpha_2[u_1(\cdot)]), \\ V^-(t, x) = \displaystyle\inf_{\alpha_1 \in \mathcal{A}_1[t,T]} \sup_{u_2(\cdot) \in \mathcal{U}_2[t,T]} J(t, x; \alpha_1[u_2(\cdot)], u_2(\cdot)), \qquad (4.29) \\ \qquad\qquad (t, x) \in [0, T] \times \mathbb{R}^n, \end{cases}$$

which are called *Elliott–Kalton upper* and *lower value functions* of our two-person zero-sum differential game.

4.4.2   Upper and lower value functions, and optimality principle

Although the upper and lower value functions are formally defined in (4.29), there seems to be no guarantee that they are well-defined. The following result states that under suitable conditions, $V^\pm(\cdot, \cdot)$ are indeed well-defined.

Theorem 4.4.3. *Let* (H1)′–(H2)′ *hold. Then the upper and lower value functions* $V^\pm(\cdot, \cdot)$ *are well-defined and there exists a constant* $K > 0$ *such that*

$$|V^\pm(t, x)| \leqslant K \langle x \rangle^\mu, \qquad (t, x) \in [0, T] \times \mathbb{R}^n. \qquad (4.30)$$

Moreover,

$$\begin{cases} V^+(t, x) = \displaystyle\sup_{\alpha_2 \in \mathcal{A}_2[t,T;N(|x|)]} \inf_{u_1(\cdot) \in \mathcal{U}_1[t,T;N(|x|)]} J(t, x; u_1(\cdot), \alpha_2[u_1(\cdot)]), \\ V^-(t, x) = \displaystyle\inf_{\alpha_1 \in \mathcal{A}_1[t,T;N(|x|)]} \sup_{u_2(\cdot) \in \mathcal{U}_2[t,T;N(|x|)]} J(t, x; \alpha_1[u_2(\cdot)], u_2(\cdot)), \end{cases} \qquad (4.31)$$

where $N : [0, \infty) \to [0, \infty)$ *is some nondecreasing continuous function,*

$$\mathcal{U}_i[t, T; r] = \Big\{ u_i \in \mathcal{U}_i[t, T] \ \Big| \ \|u_i(\cdot)\|^2_{L^2(t,T)} \leqslant r \Big\}, \qquad i = 1, 2, \qquad (4.32)$$

and

$$\begin{cases} \mathcal{A}_1[t, T; r] = \Big\{ \alpha_1 : \mathcal{U}_2[t, T] \to \mathcal{U}_1[t, T; r] \ \big| \ \alpha_1 \in \mathcal{A}_1[t, T] \Big\}, \\ \mathcal{A}_2[t, T; r] = \Big\{ \alpha_2 : \mathcal{U}_1[t, T] \to \mathcal{U}_2[t, T; r] \ \big| \ \alpha_2 \in \mathcal{A}_2[t, T] \Big\}. \end{cases} \qquad (4.33)$$

Proof. First of all, for any $(t, x) \in [0, T] \times \mathbb{R}^n$ and $(u_1(\cdot), u_2(\cdot)) \in \mathcal{U}_1[t, T] \times \mathcal{U}_2[t, T]$, by Proposition 4.5.2, we have

$$\langle X(s) \rangle \leqslant C_0 \Big[\langle x \rangle + \int_t^s \big(|u_1(r)| + |u_2(r)| \big) dr \Big].$$

Thus,

$$J(t, x; u_1(\cdot), u_2(\cdot)) = \int_t^T g(s, X(s), u_1(s), u_2(s)) ds + h(X(T))$$

$$\geqslant \int_t^T \Big[c|u_1(s)|^2 - L\big(\langle X(s) \rangle^\mu + |u_2(s)|^2 \big) \Big] ds - L \langle X(T) \rangle^\mu$$

$$\geqslant c\|u_1(\cdot)\|_{L^2(t,T)}^2 - L\|u_2(\cdot)\|_{L^2(t,T)}^2$$
$$- L(T+1) \Big[K_0 \big(\langle x \rangle + \|u_1(\cdot)\|_{L^1(t,T)} + \|u_2(\cdot)\|_{L^1(t,T)} \big) \Big]^\mu.$$

Hence, in the case $1 < \mu < 2$, we have (note $3^{\mu-1} \leqslant 3$)

$$J(t, x; u_1(\cdot), u_2(\cdot)) \geqslant c\|u_1(\cdot)\|_{L^2(t,T)}^2 - L\|u_2(\cdot)\|_{L^2(t,T)}^2$$
$$- 3^{\mu-1} L(T+1) K_0^\mu \Big[\langle x \rangle^\mu + \|u_1(\cdot)\|_{L^1(t,T)}^\mu + \|u_2(\cdot)\|_{L^1(t,T)}^\mu \Big]$$

$$\geqslant -3L(T+1) K_0^\mu \langle x \rangle^\mu$$
$$- L\|u_2(\cdot)\|_{L^2(t,T)}^2 - 3L(T+1) K_0^\mu \|u_2(\cdot)\|_{L^1(t,T)}^\mu \qquad (4.34)$$
$$+ c\|u_1(\cdot)\|_{L^2(t,T)}^2 - 3L(T+1) K_0^\mu (T-t)^{\frac{\mu}{2}} \|u_1(\cdot)\|_{L^2(t,T)}^\mu$$

$$\geqslant -K \langle x \rangle^\mu - K\|u_2(\cdot)\|_{L^2(t,T)} + \frac{c}{2}\|u_1(\cdot)\|_{L^2(t,T)}^2,$$

and in the case $\mu \in [0, 1]$,

$$J(t, x; u_1(\cdot), u_2(\cdot)) \geqslant c\|u_1(\cdot)\|_{L^2(t,T)}^2 - L\|u_2(\cdot)\|_{L^2(t,T)}^2$$
$$- L(T+1) K_0^\mu \Big[\langle x \rangle^\mu + \|u_1(\cdot)\|_{L^1(t,T)}^\mu + \|u_2(\cdot)\|_{L^1(t,T)}^\mu \Big]$$

$$\geqslant -L(T+1) K_0^\mu \langle x \rangle^\mu$$
$$- L\|u_2(\cdot)\|_{L^2(t,T)}^2 - L(T+1) K_0^\mu \|u_2(\cdot)\|_{L^1(t,T)}^\mu \qquad (4.35)$$
$$+ c\|u_1(\cdot)\|_{L^2(t,T)}^2 - L(T+1) K_0^\mu (T-t)^{\frac{\mu}{2}} \|u_1(\cdot)\|_{L^2(t,T)}^\mu$$

$$\geqslant -K \langle x \rangle^\mu - K\|u_2(\cdot)\|_{L^2(t,T)} + \frac{c}{2}\|u_1(\cdot)\|_{L^2(t,T)}^2.$$

From the above, we see that

$$V^+(t, x) = \sup_{\alpha_2 \in \mathcal{A}_2[t,T]} \inf_{u_1(\cdot) \in \mathcal{U}_1[t,T]} J(t, x; u_1(\cdot), \alpha_2[u_1(\cdot)])$$

$$\geqslant \inf_{u_1(\cdot) \in \mathcal{U}_1[t,T]} J(t, x; u_1(\cdot), \alpha_2^0[u_1(\cdot)]) \geqslant -K \langle x \rangle^\mu.$$

Likewise, for any $(u_1(\cdot), u_2(\cdot)) \in \mathcal{U}_1[t,T] \times \mathcal{U}_2[t,T]$, we have

$$J(t,x;u_1(\cdot),u_2(\cdot)) = \int_t^T g(s,X(s),u_1(s),u_2(s))ds + h(X(T))$$

$$\leqslant K \langle x \rangle^\mu + K\|u_1(\cdot)\|^2_{L^2(t,T)} - \frac{c}{2}\|u_2(\cdot)\|^2_{L^2(t,T)}$$

$$\leqslant K \langle x \rangle^\mu + K\|u_1(\cdot)\|^2_{L^2(t,T)}.$$

Thus,

$$V^+(t,x) = \sup_{\alpha_2 \in \mathcal{A}_2[0,T]} \inf_{u_1(\cdot) \in \mathcal{U}_1[t,T]} J(t,x;u_1(\cdot),\alpha_2[u_1(\cdot)])$$

$$\leqslant \sup_{\alpha_2 \in \mathcal{A}_2[t,T]} J(t,x;u_1^0(\cdot),\alpha_2[u_1^0(\cdot)]) \leqslant K \langle x \rangle^\mu.$$

Similar results also hold for the lower value function $V^-(\cdot,\cdot)$. Therefore, we obtain that $V^\pm(t,x)$ are well-defined for all $(t,x) \in [0,T] \times \mathbb{R}^n$ and (4.30) holds.

Next, for any $u_1(\cdot) \in \mathcal{U}_1[t,T] \setminus \mathcal{U}_1[t,T; \frac{4K}{c} \langle x \rangle^\mu]$, where $K > 0$ is the constant appears in (4.30) and $\mathcal{U}_1[t,T;r]$ is defined by (4.32). From (4.34)–(4.35), we see that

$$J(t,x;u_1(\cdot),\alpha_2^0[u_1(\cdot)]) \geqslant -K \langle x \rangle^\mu + \frac{c}{2}\|u_1(\cdot)\|^2_{L^2(t,T)} > K \langle x \rangle^\mu$$

$$\geqslant V^+(t,x) = \sup_{\alpha_2 \in \mathcal{A}_2[t,T]} \inf_{u_1(\cdot) \in \mathcal{U}_1[t,T]} J(t,x;u_1(\cdot),\alpha_2[u_1(\cdot)]).$$

Thus,

$$V^+(t,x) = \sup_{\alpha_2 \in \mathcal{A}_2[t,T]} \inf_{u_1(\cdot) \in \mathcal{U}_1[t,T; \frac{4K}{c} \langle x \rangle^\mu]} J(t,x;u_1(\cdot),\alpha_2[u_1(\cdot)]).$$

Consequently, from (4.30), for any $u_1(\cdot) \in \mathcal{U}_1[t,T; \frac{4K}{c} \langle x \rangle^\mu]$, we have

$$-K \langle x \rangle^\mu \leqslant V^+(t,x) \leqslant \sup_{\alpha_2 \in \mathcal{A}_2[t,T]} J(t,x;u_1(\cdot),\alpha_2[u_1(\cdot)])$$

$$\leqslant K \langle x \rangle^\mu + K\|u_1(s)\|^2_{L^2(t,T)} - \frac{c}{2}\|\alpha_2[u_1(\cdot)]\|^2_{L^2(t,T)}$$

$$\leqslant K \langle x \rangle^\mu + \frac{4K^2}{c} \langle x \rangle^\mu - \frac{c}{2}\|\alpha_2[u_1(\cdot)]\|^2_{L^2(t,T)}.$$

This implies that

$$\frac{c}{2}\|\alpha_2[u_1(\cdot)]\|^2_{L^2(t,T)} \leqslant \widetilde{K} \langle x \rangle^\mu, \quad \forall u_1(\cdot) \in \mathcal{U}_1[t,T; \frac{4K}{c} \langle x \rangle^\mu],$$

with $\widetilde{K} > 0$ being another absolute constant. Hence, if we replace the original $N(r)$ by the following:

$$N(r) = \frac{2\widetilde{K}}{c} r^\mu,$$

and let $\mathcal{A}_2[t, T; r]$ be defined by (4.33), then the first relation in (4.31) holds.

The second relation in (4.31) can be proved similarly. ☐

Next, we want to establish a modified Bellman's principle of optimality. To this end, we introduce some sets. For any $(t, x) \in [0, T) \times \mathbb{R}^n$ and $\bar{t} \in (t, T]$, let

$$\mathcal{U}_i[t, \bar{t}; r] = \left\{ u_i(\cdot) \in \mathcal{U}_i[t, T] \mid \int_t^{\bar{t}} |u_i(s)|^2 ds \leqslant r \right\}, \qquad i = 1, 2,$$

and

$$\begin{cases} \mathcal{A}_1[t, \bar{t}; r] = \left\{ \alpha_1 : \mathcal{U}_2[t, T] \to \mathcal{U}_1[t, \bar{t}; r] \mid \alpha_1 \in \mathcal{A}_1[t, T] \right\}, \\ \mathcal{A}_2[t, \bar{t}; r] = \left\{ \alpha_2 : \mathcal{U}_1[t, T] \to \mathcal{U}_2[t, \bar{t}; r] \mid \alpha_2 \in \mathcal{A}_2[t, T] \right\}. \end{cases}$$

It is clear that

$$\begin{cases} \mathcal{U}_i[t, T; r] \subseteq \mathcal{U}_i[t, \bar{t}; r] \subseteq \mathcal{U}_i[t, T], \\ \mathcal{A}_i[t, T; r] \subseteq \mathcal{A}_i[t, \bar{t}; r] \subseteq \mathcal{A}_i[t, T], \end{cases} \qquad i = 1, 2.$$

Thus, from the proof of Theorem 4.4.3, we see that for a suitable choice of $N(\cdot)$, say, $N(r) = Kr^\mu$ for some large $K > 0$, the following holds:

$$\begin{cases} V^+(t, x) = \sup_{\alpha_2 \in \mathcal{A}_2[t, \bar{t}; N(|x|)]} \inf_{u_1(\cdot) \in \mathcal{U}_1[t, T; N(|x|)]} J(t, x; u_1(\cdot), \alpha_2[u_1(\cdot)]), \\ V^-(t, x) = \inf_{\alpha_1 \in \mathcal{A}_1[t, \bar{t}; N(|x|)]} \sup_{u_2(\cdot) \in \mathcal{U}_2[t, \bar{t}; N(|x|)]} J(t, x; \alpha_1[u_2(\cdot)], u_2(\cdot)). \end{cases} \tag{4.36}$$

The following is a modified Bellman's principle of optimality.

Theorem 4.4.4. *Let* (H1)′–(H2)′ *hold. Let* $(t, x) \in [0, T) \times \mathbb{R}^n$ *and* $\bar{t} \in (t, T]$. *Let* $N : [0, \infty) \to [0, \infty)$ *be a nondecreasing continuous function such that* (4.36) *holds. Then*

$$V^+(t, x) = \sup_{\alpha_2 \in \mathcal{A}_2[t, \bar{t}; N(|x|)]} \inf_{u_1(\cdot) \in \mathcal{U}_1[t, \bar{t}; N(|x|)]} \left\{ V^+(\bar{t}, X(\bar{t})) \right. \tag{4.37}$$
$$\left. + \int_t^{\bar{t}} g(s, X(s), u_1(s), \alpha_2[u_1(\cdot)](s)) ds \right\},$$

and

$$V^-(t, x) = \inf_{\alpha_1 \in \mathcal{A}_1[t, \bar{t}; N(|x|)]} \sup_{u_2(\cdot) \in \mathcal{U}_2[t, \bar{t}; N(|x|)]} \left\{ V^-(\bar{t}, X(\bar{t})) \right. \tag{4.38}$$
$$\left. + \int_t^{\bar{t}} g(s, X(s), \alpha_1[u_2(\cdot)](s), u_2(s)) ds \right\}.$$

We note that if in (4.37) and (4.38), $\mathcal{A}_i[t, \bar{t}; N(|x|)]$ and $\mathcal{U}_i[t, \bar{t}; N(|x|)]$ are replaced by $\mathcal{A}_i[t, T]$ and $\mathcal{U}_i[t, T]$, respectively, the result is standard and the proof is routine. However, for the above case, some careful modification is necessary.

Proof. We only prove (4.37). The other can be proved similarly. Since $N(|x|)$ and $\bar{t}$ are fixed, for notational simplicity, we denote below that

$$\widetilde{\mathcal{U}}_1 = \mathcal{U}_1[t, \bar{t}; N(|x|)], \qquad \widetilde{\mathcal{A}}_2 = \mathcal{A}_2[t, \bar{t}; N(|x|)].$$

Denote the right-hand side of (4.37) by $\widehat{V}^+(t, x)$. For any $\varepsilon > 0$, there exists an $\alpha_2^\varepsilon \in \widetilde{\mathcal{A}}_2$ such that

$$\widehat{V}^+(t, x) - \varepsilon < \inf_{u_1(\cdot) \in \widetilde{\mathcal{U}}_1} \left\{ \int_t^{\bar{t}} g(s, X(s), u_1(s), \alpha_2^\varepsilon[u_1(\cdot)](s)) ds + V^+(\bar{t}, X(\bar{t})) \right\}.$$

By the definition of $V^+(\bar{t}, X(\bar{t}))$, there exists an $\bar{\alpha}_2^\varepsilon \in \mathcal{A}_2[\bar{t}, T]$ such that

$$V^+(\bar{t}, X(\bar{t})) - \varepsilon < \inf_{\bar{u}_1(\cdot) \in \mathcal{U}_1[\bar{t}, T]} J(\bar{t}, X(\bar{t}); \bar{u}_1(\cdot), \bar{\alpha}_2^\varepsilon[\bar{u}_1(\cdot)]).$$

Now, we define an extension $\widehat{\alpha}_2^\varepsilon \in \mathcal{A}_2[t, T]$ of $\alpha_2^\varepsilon \in \mathcal{A}_2[\bar{t}, T]$ as follows: For any $u_1(\cdot) \in \mathcal{U}_1[t, T]$,

$$\widehat{\alpha}_2^\varepsilon[u_1(\cdot)](s) = \begin{cases} \alpha_2^\varepsilon[u_1(\cdot)](s), & s \in [t, \bar{t}), \\ \bar{\alpha}_2^\varepsilon[u_1(\cdot)|_{[\bar{t}, T]}](s), & s \in [\bar{t}, T]. \end{cases}$$

Since $\alpha_2^\varepsilon \in \widetilde{\mathcal{A}}_2$, we have

$$\int_t^{\bar{t}} |\widehat{\alpha}^\varepsilon[u_1(\cdot)](s)|^2 ds = \int_t^{\bar{t}} |\alpha_2^\varepsilon[u_1(\cdot)](s)|^2 ds \leqslant N(|x|).$$

This means that $\widehat{\alpha}_2^\varepsilon \in \widetilde{\mathcal{A}}_2$. Consequently,

$$V^+(t, x) \geqslant \inf_{u_1(\cdot) \in \widetilde{\mathcal{U}}_1} J(t, x; u_1(\cdot), \widehat{\alpha}_2^\varepsilon[u_1(\cdot)])$$

$$= \inf_{u_1(\cdot) \in \widetilde{\mathcal{U}}_1} \left\{ \int_t^{\bar{t}} g(s, X(s), u_1(s), \alpha_2^\varepsilon[u_1(\cdot)](s)) ds \right.$$

$$\left. + J(\bar{t}, X(\bar{t}); u_1(\cdot)|_{[\bar{t}, T]}, \bar{\alpha}_2^\varepsilon[u_1(\cdot)|_{[\bar{t}, T]}]) \right\}$$

$$\geqslant \inf_{u_1(\cdot) \in \widetilde{\mathcal{U}}_1} \left\{ \int_t^{\bar{t}} g(s, X(s), u_1(s), \alpha_2^\varepsilon[u_1(\cdot)](s)) ds \right.$$

$$\left. + \inf_{\bar{u}_1(\cdot) \in \mathcal{U}_1[\bar{t}, T]} J(\bar{t}, X(\bar{t}); \bar{u}_1(\cdot), \bar{\alpha}_2^\varepsilon[\bar{u}_1(\cdot)]) \right\}$$

$$\geqslant \inf_{u_1(\cdot) \in \widetilde{\mathcal{U}}_1} \left\{ \int_t^{\bar{t}} g(s, X(s), u_1(s), \alpha_2^\varepsilon[u_1(\cdot)](s)) ds + V^+(\bar{t}, X(\bar{t})) \right\} - \varepsilon$$

$$\geqslant \widehat{V}^+(t, x) - 2\varepsilon.$$

Since $\varepsilon > 0$ is arbitrary, we obtain

$$\widehat{V}^+(t,x) \leqslant V^+(t,x).$$

On the other hand, for any $\varepsilon > 0$, there exists an $\alpha_2^\varepsilon \in \widetilde{\mathcal{A}}_2$ such that

$$V^+(t,x) - \varepsilon < \inf_{u_1(\cdot) \in \widetilde{\mathcal{U}}_1} J(t,x;u_1(\cdot),\alpha_2^\varepsilon[u_1(\cdot)]).$$

Also, by definition of $\widehat{V}^+(t,x)$,

$$\widehat{V}^+(t,x) \geqslant \inf_{u_1(\cdot) \in \widetilde{\mathcal{U}}_1} \left\{ \int_t^{\bar{t}} g(s,X(s),u_1(s),\alpha_2^\varepsilon[u_1(\cdot)](s))ds + V^+(\bar{t},X(\bar{t})) \right\}.$$

Thus, there exists a $u_1^\varepsilon(\cdot) \in \widetilde{\mathcal{U}}_1$ such that

$$\widehat{V}^+(t,x) + \varepsilon \geqslant \int_t^{\bar{t}} g(s,X(s),u_1^\varepsilon(s),\alpha_2^\varepsilon[u_1^\varepsilon(\cdot)](s))ds + V^+(\bar{t},X(\bar{t})).$$

Now, for any $\bar{u}_1(\cdot) \in \mathcal{U}_1[\bar{t},T]$, define a particular extension $\widetilde{u}_1(\cdot) \in \mathcal{U}_1[t,T]$ by the following:

$$\widetilde{u}_1^\varepsilon(s) = \begin{cases} u_1^\varepsilon(s), & s \in [t,\bar{t}), \\ \bar{u}_1(s), & s \in [\bar{t},T]. \end{cases}$$

Namely, we patch $u_1^\varepsilon(\cdot)$ to $\bar{u}_1(\cdot)$ on $[t,\bar{t})$. Since

$$\int_t^{\bar{t}} |\widetilde{u}_1^\varepsilon(s)|^2 ds = \int_t^{\bar{t}} |u_1^\varepsilon(s)|^2 ds \leqslant N(|x|),$$

we see that $\widetilde{u}_1^\varepsilon(\cdot) \in \widetilde{\mathcal{U}}_1$. Next, we define a restriction $\bar{\alpha}_2^\varepsilon \in \mathcal{A}[\bar{t},T]$ of $\alpha_2^\varepsilon \in \widetilde{\mathcal{A}}_2$, as follows:

$$\bar{\alpha}_2^\varepsilon[\bar{u}_1(\cdot)] = \alpha_2^\varepsilon[\widetilde{u}_1^\varepsilon(\cdot)].$$

For such an $\bar{\alpha}_2^\varepsilon$, we have

$$V^+(\bar{t},X(\bar{t})) \geqslant \inf_{\bar{u}_1(\cdot) \in \mathcal{U}_1[\bar{t},T]} J(\bar{t},X(\bar{t}),\bar{u}_1(\cdot),\bar{\alpha}_2^\varepsilon[\bar{u}_1(\cdot)]).$$

Hence, there exists a $\bar{u}_1^\varepsilon(\cdot) \in \mathcal{U}_1[\bar{t},T]$ such that

$$V^+(\bar{t},X(\bar{t})) + \varepsilon > J(\bar{t},X(\bar{t}),\bar{u}_1^\varepsilon(\cdot),\bar{\alpha}_2^\varepsilon[\bar{u}_1^\varepsilon(\cdot)]).$$

Then we further let

$$\widehat{u}_1^\varepsilon(s) = \begin{cases} u_1^\varepsilon(s), & s \in [t,\bar{t}), \\ \bar{u}_1^\varepsilon(s), & s \in [\bar{t},T]. \end{cases}$$

Again, $\widehat{u}_1^\varepsilon(\cdot) \in \widetilde{\mathcal{U}}_1$, and therefore,

$$\widehat{V}^+(t,x) + \varepsilon \geqslant \int_t^{\bar{t}} g(s, X(s), u_1^\varepsilon(s), \alpha_2^\varepsilon[u_1^\varepsilon(\cdot)](s))ds + V^+(\bar{t}, X(\bar{t}))$$

$$\geqslant \int_t^{\bar{t}} g(s, X(s), u_1^\varepsilon(s), \alpha_2^\varepsilon[u_1^\varepsilon(\cdot)](s))ds + J(\bar{t}, X(\bar{t}), \bar{u}_1^\varepsilon(\cdot), \bar{\alpha}_2^\varepsilon[\bar{u}_1^\varepsilon(\cdot)]) - \varepsilon$$

$$= J(t, x; \widetilde{u}_1^\varepsilon(\cdot), \alpha_2^\varepsilon[\widetilde{u}_1^\varepsilon(\cdot)]) - \varepsilon$$

$$\geqslant \inf_{u_1(\cdot) \in \widetilde{\mathcal{U}}_1} J(t, x; u_1(\cdot), \alpha_2^\varepsilon[u_1(\cdot)]) - \varepsilon > V^+(t,x) - 2\varepsilon.$$

Since $\varepsilon > 0$ is arbitrary, we obtain

$$\widehat{V}^+(t,x) \geqslant V^+(t,x).$$

This completes the proof. □

4.4.3   *Continuity of upper and lower value functions*

In this subsection, we are going to establish the continuity of the upper and lower value functions. Let us state the main results now.

Theorem 4.4.5. *Let* (H1)$'$–(H2)$'$ *hold. Then* $V^\pm(\cdot,\cdot)$ *are continuous. Moreover, there exists a nondecreasing continuous function* $N : [0,\infty) \to [0,\infty)$ *such that the following estimates hold:*

$$|V^\pm(t,x) - V^\pm(t,\bar{x})| \leqslant N(|x| \vee |\bar{x}|)|x - \bar{x}|, \quad t \in [0,T], \ x, \bar{x} \in \mathbb{R}^n, \quad (4.39)$$

and

$$|V^\pm(t,x) - V^\pm(\bar{t},x)| \leqslant N(|x|)|t - \bar{t}|^{\frac{1}{2}}, \quad \forall t, \bar{t} \in [0,T], \ x \in \mathbb{R}^n. \quad (4.40)$$

Proof. We will only prove the conclusions for $V^+(\cdot,\cdot)$. The conclusions for $V^-(\cdot,\cdot)$ can be proved similarly.

First, let $0 \leqslant t \leqslant T$, $x, \bar{x} \in \mathbb{R}^n$, and let $N : [0,\infty) \to [0,\infty)$ be nondecreasing and continuous such that (4.31) holds. Take

$$u_1(\cdot) \in \mathcal{U}_1[t, T; N(|x| \vee |\bar{x}|)], \quad \alpha_2 \in \mathcal{A}_2[t, T; N(|x| \vee |\bar{x}|)].$$

Denote $u_2(\cdot) = \alpha_2[u_1(\cdot)]$. For the simplicity of notations, in what follows, we will let $N(\cdot)$ be a generic nondecreasing function which can be different line by line. Making use of Proposition 4.4.2, we have

$$|X_{t,x}(s)|, |X_{t,\bar{x}}(s)| \leqslant K_0\Big[\langle x \rangle \vee \langle \bar{x} \rangle + \int_t^T \Big(|u_1(r)| + |u_2(r)|\Big)dr\Big]$$

$$\leqslant N(\langle x \rangle \vee \langle \bar{x} \rangle), \quad s \in [t, T],$$

and

$$|X_{t,x}(s) - X_{t,\bar{x}}(s)| \leqslant K \langle x \rangle \vee \langle \bar{x} \rangle |x - \bar{x}|$$
$$\leqslant N(|x| \vee |\bar{x}|)|x - \bar{x}|, \qquad s \in [t, T].$$

Consequently,

$$|J(t, x; u_1(\cdot), u_2(\cdot)) - J(t, \bar{x}; u_1(\cdot), u_2(\cdot))|$$

$$\leqslant \int_t^T |g(s, X_{t,x}(s), u_1(s), u_2(s)) - g(s, X_{t,\bar{x}}(s), u_1(s), u_2(s))|ds$$

$$+ |h(X_{t,x}(T)) - h(X_{t,\bar{x}}(T))|$$

$$\leqslant \int_t^T L\Big[(\langle X_{t,x}(s) \rangle \vee \langle X_{t,\bar{x}}(s) \rangle)^{\mu-1} + |u_1(s)| + |u_2(s)|\Big]$$

$$\cdot |X_{t,x}(s) - X_{t,\bar{x}}(s)|ds$$

$$+ L\Big(\langle X_{t,x}(T) \rangle \vee \langle X_{t,\bar{x}}(T) \rangle\Big)^{\mu-1} |X_{t,x}(T) - X_{t,\bar{x}}(T)|$$

$$\leqslant N(|x| \vee |\bar{x}|)|x - \bar{x}|.$$

Since the above estimate is uniform in $(u_1(\cdot), \alpha_2[\cdot])$, we obtain (4.39) for $V^+(\cdot, \cdot)$.

We now prove the continuity in t. From the modified principle of optimality, we see that for any $\varepsilon > 0$, there exists an $\alpha_2^\varepsilon \in \mathcal{A}_2[t, \bar{t}; N(|x|)]$ such that

$$V^+(t, x) - \varepsilon$$

$$\leqslant \inf_{u_1(\cdot) \in \mathcal{U}_1[t, \bar{t}; N(|x|)]} \Big\{ \int_t^{\bar{t}} g(s, X(s), u_1(\cdot), \alpha_2^\varepsilon[u_1(\cdot)](s))ds + V^+(\bar{t}, X(\bar{t})) \Big\}$$

$$\leqslant \int_t^{\bar{t}} g(s, X(s), 0, \alpha_2^\varepsilon[u_1^0(\cdot)](s))ds + V^+(\bar{t}, X(\bar{t}))$$

$$\leqslant \int_t^{\bar{t}} L\Big(\langle X(s) \rangle^\mu - c|\alpha_2[u_1^0(\cdot)](s))|^2\Big)ds + V^+(\bar{t}, x)$$

$$+ |V^+(\bar{t}, X(\bar{t})) - V^+(\bar{t}, x)|$$

$$\leqslant \int_t^{\bar{t}} L\langle X(s) \rangle^\mu ds + V^+(\bar{t}, x) + |V^+(\bar{t}, X(\bar{t})) - V^+(\bar{t}, x)|.$$

By Proposition 4.4.2, we have (denote $u_2^\varepsilon(\cdot) = \alpha_2^\varepsilon[u_1^0(\cdot)]$)

$$|X(\bar{t}) - x| \leqslant K\Big[\langle x \rangle(\bar{t} - t) + \int_t^{\bar{t}} |u_2^\varepsilon(s)|ds\Big]$$

$$\leqslant K\Big[\langle x \rangle(\bar{t} - t) + \Big(\int_t^{\bar{t}} |u_2^\varepsilon(s)|^2 ds\Big)^{\frac{1}{2}}(\bar{t} - t)^{\frac{1}{2}}\Big]$$

$$\leqslant K\Big[\langle x \rangle(\bar{t} - t) + N(|x|)(\bar{t} - t)^{\frac{1}{2}}\Big].$$

Hereafter, $N(\cdot)$ is understood as a generic nondecreasing continuous function which could be different from line to line. Also,

$$|X(s)| \leqslant K\Big[\langle x \rangle + \int_t^{\bar{t}} |u_2^\varepsilon(s)|ds\Big] \leqslant N(|x|), \qquad s \in [t,\bar{t}].$$

Hence, by the proved (4.39), we obtain

$$|V^+(\bar{t}, X(\bar{t})) - V^+(\bar{t}, x)| \leqslant N(|x| \vee |X(\bar{t})|)|X(\bar{t}) - x|$$
$$\leqslant N(|x|)(\bar{t} - t)^{\frac{1}{2}}.$$

Consequently,

$$V^+(t, x) - V^+(\bar{t}, x) \leqslant N(|x|)(\bar{t} - t)^{\frac{1}{2}} + \varepsilon,$$

which yields

$$V^+(t, x) - V^+(\bar{t}, x) \leqslant N(|x|)(\bar{t} - t)^{\frac{1}{2}}.$$

On the other hand,

$$V^+(t, x) \geqslant \inf_{u_1(\cdot) \in \mathcal{U}_1[t,T;N(|x|)]} \Big\{ \int_t^{\bar{t}} g(s, X(s), u_1(s), 0)ds + V^+(\bar{t}, X(\bar{t})) \Big\}.$$

Hence, for any $\varepsilon > 0$, there exists a $u_1^\varepsilon(\cdot) \in \mathcal{U}_1[t,T;N(|x|)]$ such that

$$V^+(t, x) + \varepsilon \geqslant \int_t^{\bar{t}} g(s, X(s), u_1^\varepsilon(s), 0)ds + V^+(\bar{t}, X(\bar{t}))$$
$$\geqslant -\int_t^{\bar{t}} L\langle X(s)\rangle^\mu ds + c\int_t^{\bar{t}} |u_1^\varepsilon(s)|^2 ds + V^+(\bar{t}, x)$$
$$-|V^+(\bar{t}, X(\bar{t})) - V^+(\bar{t}, x)|$$
$$\geqslant -\int_t^{\bar{t}} L\langle X(s)\rangle^\mu ds + V^+(\bar{t}, x) - |V^+(\bar{t}, X(\bar{t})) - V^+(\bar{t}, x)|.$$

Now, in the current case, we have

$$|X(\bar{t}) - x| \leqslant K\Big[\langle x \rangle(\bar{t} - t) + \int_t^{\bar{t}} |u_1^\varepsilon(s)|ds\Big]$$
$$\leqslant K\Big[\langle x \rangle(\bar{t} - t) + \Big(\int_t^{\bar{t}} |u_1^\varepsilon(s)|^2 ds\Big)^{\frac{1}{2}}(\bar{t} - t)^{\frac{1}{2}}\Big]$$
$$\leqslant K\Big[\langle x \rangle(\bar{t} - t) + N(|x|)(\bar{t} - t)^{\frac{1}{2}}\Big].$$

Also,

$$|X(s)| \leqslant K\Big[\langle x \rangle + \int_t^{\bar{t}} |u_1^\varepsilon(s)|ds\Big] \leqslant N(|x|), \qquad s \in [t,\bar{t}].$$

Hence, by the proved (4.39), we obtain

$$|V^+(\bar{t}, X(\bar{t})) - V^+(\bar{t}, x)| \leqslant N(|x| \vee |X(\bar{t})|)|X(\bar{t}) - x| \leqslant N(|x|)(\bar{t} - t)^{\frac{1}{2}}.$$

Consequently,

$$V^+(t, x) - V^+(\bar{t}, x) \geqslant -N(|x|)(\bar{t} - t)^{\frac{1}{2}} - \varepsilon,$$

which yields

$$V^+(t, x) - V^+(\bar{t}, x) \geqslant -N(|x|)(\bar{t} - t)^{\frac{1}{2}}.$$

Hence, we obtain the estimate (4.40) for $V^+(\cdot, \cdot)$. □

Once we have the continuity, we are able to routinely prove the following result (see Corollary 4.3.5).

Theorem 4.4.6. *Let (H1)′–(H2)′ hold. Then $V^\pm(\cdot, \cdot)$ are the unique viscosity solution to the upper and lower HJI equations (4.24), respectively. Further, if the Isaacs' condition holds:*

$$H^+(t, x, p) = H^-(t, x, p), \qquad \forall (t, x, p) \in [0, T] \times \mathbb{R}^n \times \mathbb{R}^n,$$

then

$$V^+(t, x) = V^-(t, x), \qquad \forall (t, x) \in [0, T] \times \mathbb{R}^n.$$

We have seen that (H1)–(H2) enable us to defined the upper and lower Hamiltonians so that the upper and lower HJI equations can be well-formulated. Moreover, under some even weaker conditions, we can proved the uniqueness of the viscosity solutions to the upper and lower HJI equations. On the other hand, we have assumed much stronger hypotheses (H1)′–(H2)′ to obtain the upper and lower value functions $V^\pm(\cdot, \cdot)$ being well-defined so that the corresponding upper and lower HJI equations have viscosity solutions. In other words, weaker conditions ensure the uniqueness of viscosity solutions to the upper and lower HJI equations, and stronger conditions seem to be needed for the existence.

4.5 Brief Historic Remarks

In 1997, Bardi–Da Lio ([2]) studied nonlinear optimal control problem for which the control domains are allowed to be unbounded and the functions involved in the state equation and the cost functional have certain growth by means of viscosity solutions (see also [33], [53], [34]). Two-person zero-sum differential games with only one player is allowed to have unbounded control domain were studied by Rampazzo ([103]), McEneaney ([75]), and Soravia ([109]). The case that both players are allowed to take unbounded controls was studied by Qiu–Yong ([102]). This chapter is essentially based on [102].

Chapter 5

Differential Games of Pursuit and Evasion

In this chapter, we consider the following autonomous state equation:

$$\begin{cases} \dot{X}(s) = f(X(s), u_1(s), u_2(s)), & s \in [0, \infty), \\ X(0) = x, \end{cases} \tag{5.1}$$

for some $f : \mathbb{R}^n \times U_1 \times U_2 \to \mathbb{R}^n$. We modify (DG1) correspondingly.

(PE1) The map $f : \mathbb{R}^n \times U_1 \times U_2 \to \mathbb{R}^n$ is measurable and there exists a constant $L > 0$ such that

$$\begin{cases} |f(x_1, u_1, u_2) - f(x_2, u_1, u_2)| \leqslant L|x_1 - x_2|, \\ \qquad\qquad (u_1, u_2) \in U_1 \times U_2, \ x_1, x_2 \in \mathbb{R}^n, \\ |f(0, u_1, u_2)| \leqslant L, \qquad (u_1, u_2) \in U_1 \times U_2. \end{cases}$$

Under (PE1), for any $x \in \mathbb{R}^n$ and $(u_1(\cdot), u_2(\cdot)) \in \mathcal{U}_1[0, \infty) \times \mathcal{U}_2[0, \infty)$, there exists a unique solution $X(\cdot) \equiv X(\cdot\,; x, u_1(\cdot), u_2(\cdot))$ to (5.1). Moreover, the conclusion of Proposition 2.1.1 holds. Throughout this chapter, we let $M \subseteq \mathbb{R}^n$ be a non-empty closed target set and $M \neq \mathbb{R}^n$. We consider the following game situation. Player 1 wants to choose a control $u_1(\cdot) \in \mathcal{U}_1[0, \infty)$ such that

$$X(\tau; x, u_1(\cdot), u_2(\cdot)) \in M,$$

for some $\tau \geqslant 0$, and Player 2 wants to choose a control $u_2(\cdot) \in \mathcal{U}_2[0, \infty)$ such that

$$X(s; x, u_1(\cdot), u_2(\cdot)) \notin M, \qquad \forall s \in [0, \infty),$$

or even better, the following holds:

$$d(X(s; x, u_1(\cdot), u_2(\cdot)), M) \equiv \inf_{z \in M} |X(s) - z| \geqslant \delta, \quad \forall s \in [0, \infty),$$

for some $\delta > 0$. The above described is called a *differential game* of *pursuit* and *evasion*. In terms of control problem terminology, Player 1 faces a controllability problem, whereas, Player 2 faces a viability problem. Because of the above game situation, we call Player 1 the *pursuer* and Player 2 the *evader*.

5.1   Differential Pursuit Games

Under condition (PE1), for any $x \in \mathbb{R}^n$, any $(\alpha_1[\,\cdot\,], u_2(\cdot)) \in \mathcal{A}_1[0, \infty) \times \mathcal{U}_2[0, \infty)$, the following problem

$$\begin{cases} \dot{X}(s) = f(X(s), \alpha_1[u_2(\cdot)](s), u_2(s)), & s \in [0, \infty), \\ X(0) = x, \end{cases} \tag{5.2}$$

admits a unique solution $X(\cdot) \equiv X(\cdot\,; x, \alpha_1[u_2(\cdot)], u_2(\cdot))$. We now state the following problem.

Problem (P). For any $x \in \mathbb{R}^n \setminus M$, Find an Elliott–Kalton strategy $\alpha_1[\,\cdot\,] \in \mathcal{A}_1[0, \infty)$ and a $\mathcal{T}(x, \alpha_1) \in [0, \infty)$ such that for any $u_2(\cdot) \in \mathcal{U}_2[0, \infty)$,

$$X(\tau; x, \alpha_1[u_2(\cdot)], u_2(\cdot)) \in M, \tag{5.3}$$

for some $\tau \in [0, \mathcal{T}(x, \alpha_1)]$.

Any $\tau > 0$ and $\alpha_1[\,\cdot\,] \in \mathcal{A}_1[0, \infty)$ satisfying (5.3) are called a *capturing time* and a *capturing strategy*, respectively. We point out that in the above, $\mathcal{T}(x, \alpha_1)$ is independent of $u_2(\cdot) \in \mathcal{U}[0, \infty)$.

5.1.1   *Capturability*

We now introduce the following definition.

Definition 5.1.1. (i) System (5.2) is said to be *capturable* from $x \in \mathbb{R}^n$ to M if there exists an $\alpha_1[\,\cdot\,] \in \mathcal{A}_1[0, \infty)$ and a $\mathcal{T}(x, \alpha_1) \in (0, \infty)$ such that for any $u_2(\cdot) \in \mathcal{U}_2[0, \infty)$, (5.3) holds for some $\tau \in [0, \mathcal{T}(x, \alpha_1)]$, possibly depending on $(x, u_2(\cdot))$.

(ii) System (5.2) is said to be *locally capturable* to M if there exists an open neighborhood $\mathcal{O}(M)$ of M such that for any $x \in \mathcal{O}(M)$, System (5.2) is capturable from x to M.

(iii) System (5.2) is said to be *globally capturable* to M if for any $x \in \mathbb{R}^n$, system (5.2) is capturable from x to M.

(iv) System (5.2) is said to be *small time locally capturable* (STLC, for short) to M if for any $\varepsilon > 0$, there exists an (open) neighborhood $\mathcal{O}(M)$ of M such that for any $x \in \mathcal{O}(M)$, System (5.2) is capturable from x to M with $\mathcal{T}(x, \alpha_1) \in [0, \varepsilon]$.

(v) System (5.2) is said to be *small time globally capturable* (STGC, for short) to M if for any $\varepsilon > 0$ and any $x \in \mathbb{R}^n$, system (5.2) is capturable from x to M with $\mathcal{T}(x, \alpha_1) \in [0, \varepsilon]$.

Note that the capturability can be regarded as the controllability of systems with disturbance $u_2(\cdot)$. Therefore, it is acceptable that the abbreviation STLC (originally for small time local controllability) is also used for small time local capturability hereafter.

We define

$$
\begin{cases}
\mathcal{P}(s; M) = \Big\{ x \in \mathbb{R}^n \;\Big|\; \exists \alpha_1 \in \mathcal{A}_1[0, \infty),\; \mathcal{T}(x, \alpha_1) \leqslant s \Big\}, \\[2mm]
\mathcal{P}(M) \equiv \mathcal{P}(\infty; M) = \bigcup_{s \geqslant 0} \mathcal{P}(s; M) \\[2mm]
\qquad = \Big\{ x \in \mathbb{R}^n \;\Big|\; \exists \alpha_1 \in \mathcal{A}_1[0, \infty),\; \mathcal{T}(x, \alpha_1) < \infty \Big\}.
\end{cases}
$$

We call $\mathcal{P}(M)$ the *capturable set* of Problem (P). Note that by definition, one has

$$
M \subseteq \mathcal{P}(M).
$$

Also, for Problem (P), we have

$$
\begin{cases}
\text{locally capturable} & \Longleftrightarrow & \exists \mathcal{O}(M) \subseteq \mathcal{P}(\varepsilon; M), \\
\text{globally capturable} & \Longleftrightarrow & \mathbb{R}^n \subseteq \mathcal{P}(M), \\
\text{STLC} & \Longleftrightarrow & \forall \varepsilon > 0,\; \exists \mathcal{O}(M) \subseteq \mathcal{P}(\varepsilon; M), \\
\text{STGC} & \Longleftrightarrow & \forall \varepsilon > 0,\; \mathbb{R}^n \subseteq \mathcal{P}(\varepsilon; M).
\end{cases}
$$

The following result is comparable with Theorem 2.6.6.

Theorem 5.1.2. *Let* (DG1)$'$ *hold. Let* $M \subseteq \mathbb{R}^n$ *be the closure of a domain with a* C^1 *boundary* ∂M. *Suppose the following holds:*

$$
\sup_{u_2 \in U_2} \inf_{u_1 \in U_1} \langle \nu(x), f(x, u_1, u_2) \rangle \leqslant -\mu, \qquad \forall x \in \partial M,
$$

for some $\mu > 0$, *where* $\nu : \partial M \to \partial B_1(0)$ *is the outward normal map of* M. *Then system* (5.2) *is STLC to* M. *Further, if in addition,* ∂M *is compact and there exists a differentiable function* $\psi : M^c \to (0, \infty)$ *such that*

$$
\begin{cases}
\psi(x) \geqslant d(x, M) \equiv \displaystyle\inf_{z \in M} |x - z|, & \forall x \in M^c, \\[2mm]
\displaystyle\sup_{u_2 \in U_2} \inf_{u_1 \in U_1} \langle \psi_x(x), f(x, u_1, u_2) \rangle \leqslant -\beta d(x, M), & x \in M^c,
\end{cases}
\tag{5.4}
$$

for some $\beta > 0$. *Then system* (5.2) *is globally capturable to* M.

The idea of the proof is very similar to that of Theorem 2.6.6. But due to the appearance of $u_2(\cdot)$, some suitable modifications are necessary.

Proof. Since ∂M is C^1, for any $\bar{x} \in \partial M$, there exists a $\delta = \delta(\bar{x}) > 0$, a C^1 map $\varphi : B_\delta(\bar{x}) \to \mathbb{R}$ such that

$$\begin{cases} M \cap B_\delta(\bar{x}) = \{x \in B_\delta(\bar{x}) \mid \varphi(x) \leqslant 0\}, \\ \partial M \cap B_\delta(\bar{x}) = \{x \in B_\delta(\bar{x}) \mid \varphi(x) = 0\}, \end{cases}$$

and

$$0 < \sigma \leqslant |\varphi_x(x)| \leqslant \bar{\sigma}, \qquad \forall x \in B_\delta(\bar{x}),$$

for some constants $\bar{\sigma} > \sigma > 0$. Since

$$\inf_{u_1 \in U_1} \langle \varphi_x(\bar{x}), f(\bar{x}, u_1, u_2) \rangle$$
$$= |\varphi_x(\bar{x})| \inf_{u_1 \in U_1} \langle \nu(\bar{x}), f(\bar{x}, u_1, u_2) \rangle \leqslant -\mu\sigma, \qquad \forall u_2 \in U_2,$$

for any $u_2(\cdot) \in \mathcal{U}_2[0, \infty)$, by defining

$$\eta(s, u_1) = \left[\langle \varphi_x(\bar{x}), f(\bar{x}, u_1, u_2(s)) \rangle + \frac{\mu\sigma}{2} \right]^+,$$

we see that $s \mapsto \eta(s, u_1)$ is measurable, $u_1 \mapsto \eta(s, u_1)$ is continuous, and

$$0 \in \eta(s, U_1), \qquad s \in [0, T].$$

Therefore, by Filippov's Lemma, we can find a $u_1(\cdot) \in \mathcal{U}_1[0, \infty)$ such that

$$\eta(s, u_1(s)) = 0, \qquad s \in [0, \infty).$$

This defines an $\bar{\alpha}_1 \in \mathcal{A}_1[0, \infty)$ such that

$$\langle \varphi_x(\bar{x}), f(\bar{x}, \bar{\alpha}_1[u_2(\cdot)](s), u_2(s)) \rangle \leqslant -\frac{\mu\sigma}{2}, \qquad s \in [0, \infty).$$

Now, for any $x \in B_\delta(\bar{x}) \setminus M$ and $u_2(\cdot) \in \mathcal{U}_2[0, \infty)$, let the corresponding trajectory be $X(\cdot) \equiv X(\cdot\,; x, \bar{\alpha}_1[u_2(\cdot)], u_2(\cdot))$. Under (PE1), by Proposition 2.1.1, we have

$$|X(s) - \bar{x}| \leqslant |X(s) - x| + |x - \bar{x}| \leqslant (e^{Ls} - 1)(1 + |x|) + |x - \bar{x}|$$
$$\leqslant (e^{Ls} - 1)(1 + |\bar{x}|) + e^{Ls}|x - \bar{x}|, \qquad s \in [0, \bar{s}].$$

Next, under (PE1), we have

$$|\langle \varphi_x(x), f(x, u_1, u_2) \rangle - \langle \varphi_x(y), f(y, u_1, u_2) \rangle| \leqslant \omega(|x - y|),$$
$$\forall x, y \in B_\delta(\bar{x}), \ (u_1, u_2) \in U_1 \times U_2,$$

for some modulus of continuity $\omega(\cdot)$. Then (note $\varphi(\bar{x}) = 0$)

$$\varphi(X(s)) = \varphi(x) + \int_0^s \langle \varphi_x(X(r)), f(X(r), \bar{\alpha}_1[u_2(\cdot)](r), u_2(r)) \rangle \, dr$$

$$\leqslant \bar{\sigma}|x - \bar{x}| + \int_0^s \langle \varphi_x(\bar{x}), f(\bar{x}, \bar{\alpha}_1[u_2(\cdot)](r), u_2(r)) \rangle \, dr$$

$$+ \int_0^s \omega(|X(r) - \bar{x}|) \, dr$$

$$\leqslant \bar{\sigma}|x - \bar{x}| - \left[\frac{\mu\sigma}{2} - \omega\left((e^{Ls} - 1)(1 + |\bar{x}|) + e^{Ls}|x - \bar{x}|\right)\right]s, \quad s \in [0, \bar{s}].$$

Now, we take $\varepsilon > 0$ small so that

$$\varepsilon < \frac{\mu\sigma}{8\bar{\sigma}} \wedge \sqrt{\delta}, \quad \omega\left((e^{L\varepsilon} - 1)(1 + |\bar{x}|) + e^{L\varepsilon}\varepsilon^2\right) < \frac{\mu\sigma}{4}.$$

Then for any $x \in B_{\varepsilon^2}(\bar{x}) \subseteq B_\delta(\bar{x})$, one has

$$\varphi(X(\varepsilon)) \leqslant \bar{\sigma}|x - \bar{x}| - \left[\frac{\mu\sigma}{2} - \omega\left((e^{L\varepsilon} - 1)(1 + |\bar{x}|) + e^{L\varepsilon}|x - \bar{x}|\right)\right]\varepsilon$$

$$\leqslant \bar{\sigma}\varepsilon^2 - \frac{\mu\sigma\varepsilon}{4} = -\bar{\sigma}\varepsilon\left(\frac{\mu\sigma}{4\bar{\sigma}} - \varepsilon\right) < -\frac{\mu\sigma\varepsilon}{8} < 0.$$

Thus, there must be some $\tau \in [0, \varepsilon)$ such that

$$\varphi(X(\tau)) = 0,$$

which is equivalent to $X(\tau; x, \alpha_1[u_2(\cdot)], u_2(\cdot)) \in M$. This proves that the system is STLC.

Now, if (5.4) holds, then we may find an $\alpha_1 \in \mathcal{A}_1[0, \infty)$ such that for any $u_2(\cdot) \in \mathcal{U}_2[0, \infty)$,

$$d(X(s), M) \leqslant \psi(X(s))$$

$$= \psi(x) + \int_0^s \langle \psi_x(X(r)), f(X(r), \alpha_1[u_1(\cdot)](r), u_2(r)) \rangle \, dr$$

$$\leqslant \psi(x) - \int_0^s \beta d(X(r), M) \, dr.$$

Hence,

$$d(X(s), M) \leqslant \psi(x)e^{-\beta s}, \quad \forall s \geqslant 0.$$

Then by the compactness of ∂M and the proved STLC of the system to the target set M, we see that for some $\delta > 0$, $B_\delta(M) \subseteq \mathcal{P}(M)$. Therefore, system (5.1) is globally capturable to M. $\square$

We have seen that the above result gives a sufficient condition for the capturability of the pursuit game when the target set M has a non-empty interior. We now look at the following control system:

$$\dot{X}(t) = AX(t) + B_1 u_1(t) + B_2 u_2(t), \quad t \geqslant 0, \tag{5.5}$$

with the target set being a linear manifold of the following form:

$$M = M_0 + x_0, \tag{5.6}$$

for some $x_0 \in \mathbb{R}^n$ and some subspace M_0 of $\mathbb{R}^n$ with

$$\dim M_0 \leqslant n - 1. \tag{5.7}$$

Clearly, in this case, M has an empty interior. Let us denote $\Pi = \Pi_{M_0^{\perp}} : \mathbb{R}^n \to M_0^{\perp}$ to be the orthogonal projection onto $M_0^{\perp}$. We have the following result which is comparable with Theorem 2.6.10.

Theorem 5.1.3. *Let* (5.6)–(5.7) *hold for the target set* M, $U_1 = \mathbb{R}^{m_1}$ *and the following hold:*

$$\Pi\Big[\mathcal{R}\big((B_1, AB_1, \cdots, A^{n-1}B_1)\big)\Big] = M_0^{\perp}. \tag{5.8}$$

Then system (5.5) *is STGC.*

Proof. For any $(x, u_1(\cdot), u_2(\cdot)) \in \mathbb{R}^n \times \mathcal{U}_1[0, \infty) \times \mathcal{U}_2[0, \infty)$, let $X(\cdot; x, u_1(\cdot), u_2(\cdot))$ be the unique solution of (5.5). By Theorem 2.6.10, we see that for any $T > 0$, and $x \in \mathbb{R}^n \setminus M$, one can find a control $v_1(\cdot) \in \mathcal{U}_1[0, \infty)$ such that

$$d_M\big(X(T; x, v_1(\cdot), 0)\big) = \big|\Pi(X(T; x, v_1(\cdot), 0) - x_0)\big| = 0.$$

Next, we observe that

$$\mathcal{R}\big(\Pi e^{At} B_1\big) = \Pi\Big[\mathcal{R}\big((B_1, AB_1, \cdots, A^{n-1}B_1)\big)\Big]$$
$$= M_0^{\perp} \supseteq \mathcal{R}\big(\Pi e^{At} B_2\big).$$

Thus, we can find an $\alpha_1[\cdot] \in \mathcal{A}_1[0, \infty)$ such that

$$\Pi e^{A(T-s)} B_1 \alpha_1[u_2(\cdot)](s) = -\Pi e^{A(T-s)} B_2 u_2(s) + \Pi e^{A(T-s)} B_1 v_1(s),$$
$$s \in [0, T], \ \forall u_2(\cdot) \in \mathcal{U}_2[0, \infty).$$

Hence,

$$d_M\big(X(T; x, \alpha_1[u_2(\cdot)], u_2(\cdot))\big)$$
$$= \Big|\Pi\big[e^{AT} x - x_0\big] + \int_0^T \Pi e^{A(t-s)} B_1 v_1(s) ds\Big|$$
$$= \big|\Pi(X(T; x, v_1(\cdot), 0) - x_0)\big| = 0.$$

This proves our conclusion. □

We now want to relax the condition $U_1 = \mathbb{R}^{m_1}$. From Theorem 2.6.10, we know that if one defines (still denote $\Pi = \Pi_{M_0^{\perp}}$)

$$\mathbf{W}_1(t) = \int_0^t \Pi e^{A(t-s)} B_1 B_1^T e^{A^T(t-s)} \Pi ds,$$

then under (5.8), for any $t > 0$, $\mathbf{W}_1(t)$ is invertible on $M_0^\perp$. We now have the following result.

Theorem 5.1.4. *Let* (5.6)–(5.7) *hold for the target set* M, *and let* (5.8) *hold. Suppose*

$$\begin{cases} \inf_{t>0} \sup_{0 \leqslant s \leqslant t} \left| B_1^T e^{A^T(t-s)} \Pi \mathbf{W}_1(t)^\dagger \Pi e^{At} \right| = 0, \\ \inf_{t>0} \sup_{0 \leqslant s \leqslant t} \left| B_1^T e^{A^T(t-s)} \Pi \mathbf{W}_1(t)^\dagger \Pi \right| = 0, \end{cases} \tag{5.9}$$

and there exists a $\delta > 0$ *such that*

$$\Pi \bar{B}_\delta(0) \subseteq \bigcap_{u_2 \in U_2} \left[\Pi e^{As}(B_1 U_1 - B_2 u_2) \right]. \tag{5.10}$$

Then system (5.6) *is globally capturable.*

Proof. We define $\alpha_1[\,\cdot\,] \in \mathcal{A}_1[0, \infty)$ such that

$$\begin{cases} \Pi e^{As}\left(B_1 \alpha_1[u_2(\cdot)](s) - B_2 u_2(s) \right) = \Pi e^{As} B_1 v_1(s), \\ v_1(s) = -B_1^T e^{A^T(t-s)} \Pi \mathbf{W}_1(t)^\dagger \Pi \left(e^{At} x - x_0 \right). \end{cases}$$

Under (5.9), we can find a $t > 0$ such that

$$\sup_{0 \leqslant s \leqslant t} |v_1(s)| = \sup_{0 \leqslant s \leqslant t} \left| B_1^T e^{A^T(t-s)} \Pi \mathbf{W}_1(t)^\dagger \Pi \left(e^{At} x - x_0 \right) \right|$$

$$\leqslant \sup_{0 \leqslant s \leqslant t} \left| B_1^T e^{A^T(t-s)} \Pi \mathbf{W}_1(t)^\dagger \Pi e^{At} \right| |x|$$

$$+ \sup_{0 \leqslant s \leqslant t} \left| B_1^T e^{A^T(t-s)} \Pi \mathbf{W}_1(t)^\dagger \Pi \right| |x_0| \leq \delta.$$

Then

$$d_M\left(X(t; x, \alpha_1[u_2(\cdot)], u_2(\cdot)) \right) = \left| \Pi \left(e^{At} x - x_0 \right) \right.$$

$$\left. - \int_0^t \Pi e^{A(t-s)} B_1 B_1^T e^{A^T(t-s)} \Pi \mathbf{W}_1(t)^\dagger \Pi \left(e^{At} x - x_0 \right) ds \right| = 0.$$

This proves the global capturability of the game. $\qquad\qquad\square$

Let us look at an example for which the above result applies.

Example 5.1.5. Consider two objects moving in $\mathbb{R}^3$, whose coordinates are y_1 and y_2, respectively, and they satisfy the following:

$$\begin{cases} \dot{y}_i(t) = p_i(t), \\ \dot{p}_i(t) = -\mu_i p_i(t) + u_i(t), \end{cases}$$

with $\mu_i > 0$. We assume that y_1 is the pursuer and y_2 is the evader. The evader is captured at some time $t^* \geqslant 0$ if

$$y_1(t^*) = y_2(t^*).$$

Now, we set

$$X_1 = y_1 - y_2, \quad X_2 = p_1, \quad X_3 = p_2.$$

Then the state equation becomes

$$\dot{X}(t) = \begin{pmatrix} 0 & I & -I \\ 0 & -\mu_1 I & 0 \\ 0 & 0 & -\mu_2 I \end{pmatrix} X(t) + \begin{pmatrix} 0 \\ I \\ 0 \end{pmatrix} u_1(t) + \begin{pmatrix} 0 \\ 0 \\ I \end{pmatrix} u_2(t)$$

$$\equiv AX(t) + B_1 u_1(t) + B_2 u_2(t),$$

and the terminal set is given by

$$M = M_0 = \left\{ \begin{pmatrix} 0 \\ x_2 \\ x_3 \end{pmatrix} \mid x_2, x_3 \in \mathbb{R}^3 \right\}.$$

Let $\Pi : \mathbb{R}^9 \to M_0^\perp$ be the orthogonal projection. Since

$$(B_1, AB_1) = \begin{pmatrix} 0 & I \\ I & -\mu_1 I \\ 0 & 0 \end{pmatrix},$$

we have

$$\mathcal{R}((B_1, AB_1)) = \left\{ \begin{pmatrix} x_1 \\ x_2 \\ 0 \end{pmatrix} \mid x_1, x_2 \in \mathbb{R}^3 \right\}.$$

Hence,

$$\Pi\Big(\mathcal{R}((B_1, AB_1))\Big) = M_0^\perp.$$

This means that (5.8) holds. Next, we calculate

$$e^{At} = \begin{pmatrix} I & \varphi_1(t)I & \varphi_2(t)I \\ 0 & e^{-\mu_1 t}I & 0 \\ 0 & 0 & e^{-\mu_2 t}I \end{pmatrix},$$

with

$$\varphi_i(t) = \frac{e^{-\mu_i t} - 1}{\mu_i}, \qquad t \geqslant 0, \ i = 1, 2.$$

Then

$$\Pi e^{At} = \begin{pmatrix} I & \varphi_1(t)I & \varphi_2(t)I \\ 0 & 0 & 0 \\ 0 & 0 & 0 \end{pmatrix},$$

and

$$\Pi e^{At} B_1 = \varphi_1(t) \begin{pmatrix} I \\ 0 \\ 0 \end{pmatrix}, \qquad \Pi e^{At} B_2 = \varphi_2(t) \begin{pmatrix} I \\ 0 \\ 0 \end{pmatrix}.$$

We let

$$U_i = \bar{B}_{\rho_i}(0), \qquad i = 1, 2.$$

Then (5.10) holds if

$$\frac{\rho_1}{\mu_1}(1 - e^{-\mu_1 t}) > \frac{\rho_2}{\mu_2}(1 - e^{-\mu_2 t}), \qquad \forall t > 0,$$

which is guaranteed by

$$\rho_1 > \rho_2, \qquad \mu_1 \geqslant \mu_2. \tag{5.11}$$

Next, we let

$$\begin{aligned}
\mathbf{W}_1(t) &= \int_0^t \Pi e^{A(t-s)} B_1 B_1^T e^{A^T(t-s)} \Pi ds = \left[\int_0^t \left(\frac{1 - e^{-\mu_1(t-s)}}{\mu_1} \right)^2 ds \right] \Pi \\
&= \frac{1}{\mu_1^2} \left[t - \frac{2}{\mu_1}(1 - e^{-\mu_1 t}) + \frac{1}{2\mu_1}(1 - e^{-2\mu_1 t}) \right] \Pi \\
&= \frac{1}{\mu_1^2} \left[t - 2\varphi_1(t) + \frac{1}{2\mu_1} \left(1 - [1 - \mu_1 \varphi_1(t)]^2 \right) \right] \Pi \\
&= \frac{1}{\mu_1^2} \left[t - \varphi_1(t) + \frac{\mu_1}{2} \varphi_1(t)^2 \right] \Pi \equiv \theta(t)\Pi,
\end{aligned}$$

where

$$\begin{aligned}
\theta(t) &= \frac{1}{\mu_1^2} \left[t - \varphi_1(t) + \frac{\mu_1}{2}\varphi_1(t)^2 \right] \\
&= \frac{1}{\mu_1^2} \left[t - \frac{2}{\mu_1}(1 - e^{-\mu_1 t}) + \frac{1}{2\mu_1}(1 - e^{-2\mu_1 t}) \right].
\end{aligned}$$

One sees that

$$\theta(t) > 0, \quad \theta'(t) = \varphi_1(t)^2 > 0, \qquad \forall t > 0,$$

and

$$\lim_{t \to \infty} \theta(t) = \infty.$$

Further,

$$
\begin{cases}
B_1^T e^{A^T(t-s)} \Pi \mathbf{W}_1(t)^\dagger \Pi = \dfrac{\varphi_1(t-s)}{\theta(t)} \Pi, \\[2mm]
B_1^T e^{A^T(t-s)} \Pi \mathbf{W}_1(t)^\dagger \Pi e^{At} = \dfrac{\varphi_1(t-s)}{\theta(t)} (I, \varphi_1(t), \varphi_2(t)).
\end{cases}
$$

Thus,

$$
\begin{cases}
\displaystyle\sup_{0 \leqslant s \leqslant t} \left| B_1^T e^{A^T(t-s)} \Pi \mathbf{W}_1(t)^\dagger \Pi e^{At} \right| = \dfrac{\varphi_1(t)}{\theta(t)} \to 0, \\[3mm]
\displaystyle\sup_{0 \leqslant s \leqslant t} \left| B_1^T e^{A^T(t-s)} \Pi \mathbf{W}_1(t)^\dagger \Pi \right| = \dfrac{\varphi_1(t)\sqrt{1 + \varphi_1(t)^2 + \varphi_2(t)^2}}{\theta(t)} \to 0,
\end{cases}
$$

as $t \to \infty$. Hence, (5.9) holds. Combining the above, we see that under (5.11), the differential game is globally capturable.

We point out that under conditions of Theorem 5.1.4, we only have the global capturability, not the small time global capturability.

5.1.2 *Characterization of capturable set*

For any $x \in \mathbb{R}^n$, $\alpha_1[\,\cdot\,] \in \mathcal{A}_1[0,\infty)$ and $u_2(\cdot) \in \mathcal{U}_2[0,\infty)$, we define

$$
\begin{aligned}
& T(x; \alpha_1[u_2(\cdot)], u_2(\cdot)) \\
& = \inf \Big\{ s \geqslant 0 \mid d\big(X(s; x, \alpha_1[u_2(\cdot)], u_2(\cdot)), M \big) = 0 \Big\},
\end{aligned}
\tag{5.12}
$$

with the convention that $\inf \phi = \infty$, and define

$$
\mathcal{T}(x) = \inf_{\alpha_1[\,\cdot\,] \in \mathcal{A}_1[0,\infty)} \sup_{u_2(\cdot) \in \mathcal{U}_2[0,\infty)} T(x; \alpha_1[u_2(\cdot)], u_2(\cdot)),
\tag{5.13}
$$

which is called the *minimum terminating time* of x. The map $x \mapsto \mathcal{T}(x)$ is called the *minimum terminating time function*. Next, we let

$$
\begin{aligned}
J(x; \alpha_1[u_2(\cdot)], u_2(\cdot)) &= 1 - e^{-T(x;\alpha_1[u_2(\cdot)],u_2(\cdot))} \\
&= \int_0^{T(x;\alpha_1[u_2(\cdot)],u_2(\cdot))} e^{-s} ds,
\end{aligned}
$$

with the convention that $e^{-\infty} = 0$. Define

$$
V^-(x) = \inf_{\alpha_1[\,\cdot\,] \in \mathcal{A}_1[0,\infty)} \sup_{u_2(\cdot) \in \mathcal{U}_2[0,\infty)} J(x; \alpha_1[u_2(\cdot)], u_2(\cdot)).
$$

Since the map $t \mapsto 1 - e^{-t}$ is increasing, we see that

$$
V^-(x) = 1 - e^{-\mathcal{T}(x)}, \qquad x \in \mathbb{R}^n.
\tag{5.14}
$$

The following result is comparable with Theorem 2.6.13.

Theorem 5.1.6. *Let* (PE1) *hold. Let* $M \subseteq \mathbb{R}^n$ *be closed and non-empty with* ∂M *being compact. Let* (5.2) *be STLC to* M. *Then* $\mathcal{P}(M)$ *is open,* $V^-(\cdot)$ *is continuous on* $\mathbb{R}^n$ *and*

$$\begin{cases} V^-(x) = 0, & \forall x \in M, \\ 0 < V^-(x) < 1, & \forall x \in \mathcal{P}(M) \setminus M, \\ V^-(x) = 1, & \forall x \in \mathbb{R}^n \setminus \mathcal{P}(M). \end{cases} \tag{5.15}$$

Proof. First of all, for any $x \in \mathcal{P}(M)$, there exists an $\alpha_1[\,\cdot\,] \in \mathcal{A}_1[0,\infty)$ such that for any $u_2(\cdot) \in \mathcal{U}_2[0,\infty)$, $X(s; x, \alpha_1[u_2(\cdot)], u_2(\cdot)) \in M$, for some $s \in [0, \mathcal{T}(x, \alpha_1)]$, which yields

$$0 \leqslant V^-(x) \leqslant J(x; \alpha_1[u_2(\cdot)], u_2(\cdot)) \leqslant 1 - e^{-\mathcal{T}(x,\alpha_1)} < 1.$$

Further, since $M \subseteq \mathcal{P}(M)$, for any $x \in M$, we trivially have $\mathcal{T}(x, \alpha_1) = 0$ leading to

$$V^-(x) = 0, \qquad \forall x \in M.$$

Also, by our convention, we see that

$$V^-(x) = 1, \qquad \forall x \in \mathbb{R}^n \setminus \mathcal{P}(M).$$

Hence, (5.15) holds.

We now show that $\mathcal{P}(M)$ is open. Since M is closed and $M \subseteq \mathcal{P}(M)$, it suffices to show that $\mathcal{P}(M) \setminus M$ is open.

Fix an $x \in \mathcal{P}(M) \setminus M$. By definition of $\mathcal{T}(x)$, we see that for any $\varepsilon > 0$, there exists an $\bar{\alpha}_1^\varepsilon[\,\cdot\,] \in \mathcal{A}_1[0,\infty)$ such that

$$\mathcal{T}(x) \leqslant \sup_{u_2(\cdot) \in \mathcal{U}_2[0,\infty)} T(x; \bar{\alpha}_1^\varepsilon[u_2(\cdot)], u_2(\cdot)) \leqslant \mathcal{T}(x) + \varepsilon.$$

It follows from Proposition 3.1.1 that for any $\bar{x} \in \mathbb{R}^n$, we have

$$|X(s; x, \bar{\alpha}_1^\varepsilon[u_2(\cdot)], u_2(\cdot)) - X(s; \bar{x}, \bar{\alpha}_1^\varepsilon[u_2(\cdot)], u_2(\cdot))|$$
$$\leqslant e^{L(\mathcal{T}(x)+1)}|x - \bar{x}|, \qquad \forall s \in [0, \mathcal{T}(x) + 1], \quad u_2(\cdot) \in \mathcal{U}_2[0,\infty).$$

By the definition of $T(x; \bar{\alpha}_1^\varepsilon[u_2(\cdot)], u_2(\cdot)) \equiv \bar{T}^\varepsilon(x)$, one has

$$X(\bar{T}^\varepsilon(x); x, \bar{\alpha}_1^\varepsilon[u_2(\cdot)], u_2(\cdot)) \in \partial M.$$

Now, since our game is STLC, and ∂M is compact, one can find a $\delta > 0$ and an $\alpha_1^\varepsilon[\,\cdot\,] \in \mathcal{A}_1[0,\infty)$ such that for any $u_2(\cdot), \tilde{u}_2(\cdot) \in \mathcal{U}_2[0,\infty)$, the following holds:

$$X\big(s_\varepsilon; X(\bar{T}^\varepsilon(x); \bar{x}, \bar{\alpha}_1^\varepsilon[u_2(\cdot)], u_2(\cdot)), \alpha_1^\varepsilon[\tilde{u}_2(\cdot)], \tilde{u}_2(\cdot)\big) \in \partial M,$$

for some $s_\varepsilon \in [0, \varepsilon]$. We define $\widetilde{\alpha}_1^\varepsilon[\,\cdot\,] \in \mathcal{A}_1[0, \infty)$ as follows: For any $u_2(\cdot) \in \mathcal{U}_2[0, \infty)$,

$$\widetilde{\alpha}_1^\varepsilon[u_2(\cdot)](s) = \begin{cases} \bar{\alpha}_1^\varepsilon[u_2(\cdot)](s), & s \in [0, \bar{T}^\varepsilon(x)), \\ \alpha_1^\varepsilon[u_2(\cdot + \bar{T}^\varepsilon(x))](s - \bar{T}^\varepsilon(x)), & s \in [\bar{T}^\varepsilon(x), \infty). \end{cases}$$

Then,

$$X(\bar{T}^\varepsilon(x) + s_\varepsilon; \bar{x}; \widetilde{\alpha}_1^\varepsilon[u_2(\cdot)], u_2(\cdot)) \in \partial M.$$

Thus, $\bar{x} \in \mathcal{P}(M)$, which leads to

$$B_\delta(x) \subseteq \mathcal{P}(M),$$

proving that $\mathcal{P}(M)$ is open. Also,

$$\mathcal{T}(\bar{x}) \leqslant \bar{T}^\varepsilon(x) + s_\varepsilon \leqslant \mathcal{T}(x) + 2\varepsilon.$$

This leads to

$$\begin{aligned} V^-(\bar{x}) = 1 - e^{-\mathcal{T}(\bar{x})} &\leqslant 1 - e^{-\mathcal{T}(x) - 2\varepsilon} \\ &= V^-(x) + e^{-\mathcal{T}(x)}(1 - e^{-2\varepsilon}), \qquad \forall \bar{x} \in B_\delta(x). \end{aligned}$$

Next, for any $x \in \mathcal{P}(M) \setminus M$ and $\delta > 0$, we take any $\bar{x}_1, \bar{x}_2 \in B_\delta(x)$. From the above, we have

$$\mathcal{T}(\bar{x}_1), \mathcal{T}(\bar{x}_2) \leqslant \mathcal{T}(x) + 2\varepsilon,$$

and

$$V^-(\bar{x}_1), V^-(\bar{x}_2) \leqslant V^-(x) + e^{-\mathcal{T}(x)}(1 - e^{-2\varepsilon}).$$

Now, replace x and $\bar{x}$ by $\bar{x}_1$ and $\bar{x}_2$, respectively, using the above argument, we can obtain that

$$V^-(\bar{x}_2) \leqslant V^-(\bar{x}_1) + e^{-\mathcal{T}(\bar{x}_1)}(1 - e^{-2\varepsilon}).$$

Exchange $\bar{x}_1$ and $\bar{x}_2$, we obtain

$$V^-(\bar{x}_1) \leqslant V^-(\bar{x}_2) + e^{-\mathcal{T}(\bar{x}_2)}(1 - e^{-2\varepsilon}).$$

Hence,

$$|V^-(\bar{x}_1) - V^-(\bar{x}_2)| \leqslant e^{-\mathcal{T}(\bar{x}_1) \wedge \mathcal{T}(\bar{x}_2)}(1 - e^{-2\varepsilon}), \qquad \forall \bar{x}_1, \bar{x}_2 \in B_\delta(x).$$

This gives the continuity of $V^-(\cdot)$ on $\mathcal{P}(M) \setminus M$.

By the STLC of the game, we see that

$$\lim_{d(x, M) \to 0} V^-(x) = 0.$$

To complete the proof for the continuity of $V^-(\cdot)$ on $\mathbb{R}^n$, we need to show that for any $\bar{x} \in \partial \mathcal{P}(M)$,

$$\lim_{x \to \bar{x}} V^-(x) = 1.$$

We prove it by contradiction. Suppose the above is not the case, i.e., there exists a sequence $x_k \in \mathcal{P}(M)$ and an $\varepsilon_0 \in (0,1)$ such that

$$|x_k - \bar{x}| \leqslant \frac{1}{k}, \quad V^-(x_k) \leqslant 1 - \varepsilon_0, \qquad \forall k \geqslant 1.$$

Then by (5.14), we have

$$\mathcal{T}(x_k) = \ln \frac{1}{1 - V^-(x_k)} \leqslant \ln \frac{1}{\varepsilon_0} < \infty.$$

Therefore, there exists a sequence $\alpha_1^k \in \mathcal{A}_1[0, \infty)$ with

$$\mathcal{T}(x_k, \alpha_1^k) \leqslant T_0 \equiv \ln \frac{1}{\varepsilon_0} + 1, \qquad \forall k \geqslant 1,$$

such that for any $u_2(\cdot) \in \mathcal{U}_2[0, \infty)$,

$$X(t_k; x_k, \alpha_1^k[u_2(\cdot)], u_2(\cdot))) \in \partial M, \qquad k \geqslant 1,$$

for some $t_k \in [0, \mathcal{T}(x_k, \alpha_1^k)] \subseteq [0, T_0]$. Consequently,

$$\begin{aligned} &d(X(t_k; \bar{x}, \alpha_1^k[u_2(\cdot)], u_2(\cdot)), M) \\ &\leqslant |X(t_k; \bar{x}, \alpha_1^k[u_2(\cdot)], u_2(\cdot)) - X(t_k; x_k, \alpha_1^k[u_2(\cdot)], u_2(\cdot))| \\ &\leqslant e^{Lt_k} |\bar{x} - x_k| \leqslant \frac{e^{LT_0}}{k}, \qquad k \geqslant 1. \end{aligned}$$

Also,

$$|X(t_k; \bar{x}, \alpha_1^k[u_2(\cdot)], u_2(\cdot))| \leqslant e^{Lt_k}(1 + |\bar{x}|) \leqslant e^{LT_0}(1 + |\bar{x}|), \qquad k \geqslant 1.$$

Thus, for $\delta > 0$ small enough, when k is large enough, we will have (noting that ∂M is compact)

$$X(t_k; \bar{x}, \alpha_1^k[u_2(\cdot)], u_2(\cdot)) \in B_\delta(M) \subseteq \mathcal{P}(M).$$

Namely, $\bar{x} \in \mathcal{P}(M)$. Since $\mathcal{P}(M)$ is open, we conclude that $\bar{x} \notin \partial \mathcal{P}(M)$, a contradiction. This completes the proof. $\qquad \square$

Theorem 5.1.7. *Let* (PE1) *hold. Suppose system* (5.2) *is STLC to* M. *Then for any* $x \in \mathcal{P}(M) \setminus M$, *there exists an* $\alpha_1[\cdot] \in \mathcal{A}_1[0, \infty)$ *and some* $\bar{s} \equiv \bar{s}(x) > 0$ *such that*

$$V^-(x) = \inf_{\alpha_1 \in \mathcal{A}_1[0,\infty)} \sup_{u_2(\cdot) \in \mathcal{U}_2[0,\infty)} \left\{ 1 - e^{-t} + e^{-t} V^- \big(X(t; x, \alpha_1[u_2(\cdot)], u_2(\cdot)) \big) \right\}, \tag{5.16}$$

$$\forall t \in [0, \bar{s}].$$

Proof. First of all, under our conditions, $\mathcal{P}(M) \setminus M$ is open. We fix an $x \in \mathcal{P}(M) \setminus M$. Under (PE1), taking into account the conclusion of Proposition 2.1.1, there exists an $s_0 \equiv s_0(x) > 0$ such that

$$X(t; x, \alpha_1[u_2(\cdot)], u_2(\cdot)) \in \mathcal{P}(M) \setminus M, \tag{5.17}$$
$$\forall (\alpha_1[\,\cdot\,], u_2(\cdot)) \in \mathcal{A}_1[0, \infty) \times \mathcal{U}_2[0, \infty), \quad t \in [0, 2s_0].$$

Hence,

$$T\big(x; \alpha_1[u_2(\cdot)], u_2(\cdot)\big) > s_0, \quad \forall (\alpha_1[\,\cdot\,], u_2(\cdot)) \in \mathcal{A}_1[0, \infty) \times \mathcal{U}_2[0, \infty).$$

Now, for any $(t, \alpha_1[\,\cdot\,], u_2(\cdot)) \in (0, s_0) \times \mathcal{A}_1[0, \infty) \times \mathcal{U}_2[0, \infty)$, we define

$$u_2^t(s) = u_2(s + t), \quad s \in [0, \infty),$$

and for any $\bar{u}_2(\cdot) \in \mathcal{U}_2[0, \infty)$, we define

$$\big(u_2(\cdot) \oplus_t \bar{u}_2(\cdot)\big)(s) = \begin{cases} u_2(s), & s \in [0, t), \\ \bar{u}_2(s - t), & s \in [t, \infty). \end{cases}$$

Clearly, $u_2^t(\cdot), u_2(\cdot) \oplus_t \bar{u}_2(\cdot) \in \mathcal{U}_2[0, \infty)$. Next, for any $\bar{\alpha}_1[\,\cdot\,] \in \mathcal{A}_1[0, \infty)$, we define

$$(\alpha_1 \oplus_t \bar{\alpha}_1)[\tilde{u}_2(\cdot)](s) = \begin{cases} \alpha_1[\tilde{u}_2(\cdot)](s), & s \in [0, t), \\ \bar{\alpha}_1[\tilde{u}_2^t(\cdot)](s - t), & s \in [t, \infty), \end{cases} \quad \forall \tilde{u}_2(\cdot) \in \mathcal{U}_2[0, \infty).$$

Thus, $\alpha_1 \oplus_t \bar{\alpha}_2 \in \mathcal{A}_1[0, \infty)$, and

$$(\alpha_1 \oplus_t \bar{\alpha}_1)[u_2(\cdot) \oplus_t \bar{u}_2(\cdot)](s) = \begin{cases} \alpha_1[u_2(\cdot)](s), & s \in [0, t), \\ \bar{\alpha}_1[\bar{u}_2(\cdot)](s - t), & s \in [t, \infty). \end{cases}$$

Now, by (5.17), there exists an $\bar{\alpha}_1[\,\cdot\,] \in \mathcal{A}_1[0, \infty)$ such that for any $\bar{u}_2(\cdot) \in \mathcal{U}_2[0, \infty)$, we have some $\tau > 0$ satisfying

$$M \ni X\big(\tau; X(t; x, \alpha_1[u_2(\cdot)], u_2(\cdot)), \bar{\alpha}_1[\bar{u}_2(\cdot)], \bar{u}_2(\cdot)\big)$$
$$= X\big(t + \tau; x, (\alpha_1 \oplus_t \bar{\alpha}_1)[u_2(\cdot) \oplus_t \bar{u}_2(\cdot)], u_2(\cdot) \oplus_t \bar{u}_2(\cdot)\big).$$

Hence,

$$T\big(x; (\alpha_1 \oplus_t \bar{\alpha}_1)[u_2(\cdot) \oplus_t \bar{u}_2(\cdot)], u_2(\cdot) \oplus_t \bar{u}_2(\cdot)\big)$$
$$= t + T\big(X(t; x, \alpha_1[u_2(\cdot)], u_2(\cdot)); \bar{\alpha}_1[\bar{u}_2(\cdot)], \bar{u}_2(\cdot)\big) \leqslant t + \tau,$$

which leads to the following:

$$J\big(x; (\alpha_1 \oplus_t \bar{\alpha}_1)[u_2(\cdot) \oplus_t \bar{u}_2(\cdot)], u_2(\cdot) \oplus_t \bar{u}_2(\cdot)\big)$$
$$= 1 - e^{-T(x;(\alpha_1 \oplus_t \bar{\alpha}_1)[u_2(\cdot) \oplus_t \bar{u}_2(\cdot)], u_2(\cdot) \oplus_t \bar{u}_2(\cdot))}$$
$$= 1 - e^{-t - T(X(t;x,\alpha_1[u_2(\cdot)],u_2(\cdot));\bar{\alpha}_1[\bar{u}_2(\cdot)],\bar{u}_2(\cdot))}$$
$$= 1 - e^t + e^{-t}\left[1 - e^{-T(X(t;x,\alpha_1[u_2(\cdot)],u_2(\cdot));\bar{\alpha}_1[\bar{u}_2(\cdot)],\bar{u}_2(\cdot))}\right]$$
$$= 1 - e^{-t} + e^{-t} J\big(X(t; x, \alpha_1[u_2(\cdot)], u_2(\cdot)); \bar{\alpha}_1[\bar{u}_2(\cdot)], \bar{u}_2(\cdot)\big).$$

Having the above, using a similar argument used in the proof of Theorem 3.3.5, we are able to obtain (5.16). □

Next result is a natural consequence of the above result.

Theorem 5.1.8. *Let* (PE1) *hold and the system* (5.2) *is STLC to* M. *Then* $V^-(\cdot)$ *is the unique viscosity solution to the following HJI equation:*

$$\begin{cases} V^-(x) - H^-(x, V_x^-(x)) = 0, & x \in \mathbb{R}^n \setminus M, \\ V^-\big|_{\partial M} = 0, \end{cases} \tag{5.18}$$

where

$$H^-(x, p) = 1 + \sup_{u_2 \in U_2} \inf_{u_1 \in U_1} \langle p, f(x, u_1, u_2) \rangle, \qquad \forall x, p \in \mathbb{R}^n.$$

Note that HJ equation (5.18) is a special case of (2.60) (with $\lambda = 1$). Therefore, the uniqueness of the viscosity solution follows from Theorem 2.6.18. Also, from Theorem 5.1.7, we can show that $V^-(\cdot)$ is a viscosity solution to (5.18). We leave the details to the readers.

5.2 Differential Evasion Games

Now, we look at the evasion situation. Under condition (PE1), for any $x \in \mathbb{R}^n$, $(u_1(\cdot), \alpha_2[\,\cdot\,]) \in \mathcal{U}_1[0, \infty) \times \mathcal{A}_2[0, \infty)$, the following

$$\begin{cases} \dot{X}(s) = f(X(s), u_1(s), \alpha_2[u_1(\cdot)](s)), & s \in [0, \infty), \\ X(0) = x, \end{cases}$$

admits a unique solution $X(\cdot) \equiv X(\cdot\,; x, u_1(\cdot), \alpha_2[u_1(\cdot)])$. We formulate the following problem.

Problem (E). For given $x \in \mathbb{R}^n \setminus M$, find an $\alpha_2[\,\cdot\,] \in \mathcal{A}_2[0, \infty)$ such that for any $u_1(\cdot) \in \mathcal{U}_1[0, \infty)$,

$$X(s; x, u_1(\cdot), \alpha_2[u_1(\cdot)]) \notin M, \qquad \forall s \geqslant 0. \tag{5.19}$$

The above is called a *differential evasion game*, and $\alpha_2[\,\cdot\,] \in \mathcal{A}_2[0, \infty)$ satisfying (5.19) is called an *evasion strategy* for the initial state x. In the above, M is called a *terminating set* and $\Omega \setminus M$ is called a *survival set*.

5.2.1 *Evadability*

We introduce the following definition.

Definition 5.2.1. (i) The game is said to be *evadable from M* at $x \in \mathbb{R}^n \setminus M$ if there exists an $\alpha_2[\cdot] \in \mathcal{A}_2[0, \infty)$ such that (5.19) holds for any $u_1(\cdot) \in \mathcal{U}_1[0, \infty)$. The game is said to be *evadable from M* if it is evadable from M at any $x \in \mathbb{R}^n \setminus M$.

(ii) The game is said to be *uniformly evadable from M* at $x \in \mathbb{R}^n \setminus M$ if there exists a $\delta > 0$ and a strategy $\alpha_2[\cdot] \in \mathcal{A}_2[0, \infty)$ such that

$$d\big(X(s; x, u_1(\cdot), \alpha_2[u_1(\cdot)]), M\big) \geqslant \delta, \quad \forall s \geqslant 0, \ u_1(\cdot) \in \mathcal{U}_1[0, \infty).$$

The game is said to be *uniformly evadable from M* if it is uniformly evadable from M at any $x \in \mathbb{R}^n \setminus M$.

We define

$$\mathcal{E}(M) = \Big\{ x \in \mathbb{R}^n \setminus M \ \big| \ \exists \alpha_2[\cdot] \in \mathcal{A}_2[0, \infty), \ \text{such that}$$
$$d\big(X(s; x, u_1(\cdot), \alpha_2[u_1(\cdot)]), M\big) > 0,$$
$$\forall u_1(\cdot) \in \mathcal{U}_1[0, \infty), \ s \geqslant 0 \Big\},$$

and for any $\delta > 0$, define

$$\mathcal{E}_\delta(M) = \Big\{ x \in \mathbb{R}^n \setminus B_\delta(M) \ \big| \ \exists \alpha_2[\cdot] \in \mathcal{A}_2[0, \infty), \ \text{such that}$$
$$d\big(X(s; x, u_1(\cdot), \alpha_2[u_1(\cdot)]), M\big) \geqslant \delta,$$
$$\forall u_1(\cdot) \in \mathcal{U}_1[0, \infty), \ s \geqslant 0 \Big\}.$$

We call $\mathcal{E}(M)$ and $\mathcal{E}_\delta(M)$ the *evadable set* and a *δ-uniformly evadable set* of Problem (E), respectively. By definition, we have

$$\mathcal{E}_\delta(M) \subseteq \mathcal{E}(M), \qquad \forall \delta > 0.$$

The following gives a necessary condition and a sufficient condition for the evadability of Problem (E).

Theorem 5.2.2. *Let (PE1) hold. Let $M \subseteq \mathbb{R}^n$ be the closure of a C^1 domain, with $\nu : \partial M \to \partial B_1(0)$ being its outward normal map.*

(i) *Suppose the game is evadable from M. Then*

$$\inf_{u_1 \in U_1} \sup_{u_2 \in U_2} \langle \nu(x), f(x, u_1, u_2) \rangle \geqslant 0, \qquad \forall x \in \partial M. \tag{5.20}$$

(ii) *Suppose*

$$\inf_{u_1 \in U_1} \sup_{u_2 \in U_2} \langle \nu(x), f(x, u_1, u_2) \rangle \geqslant \mu, \qquad \forall x \in \partial M, \tag{5.21}$$

for some $\mu > 0$. Then the game is uniformly evadable from M.

Proof. (i) Suppose (5.20) fails. Then there exist an $x_0 \in \partial M$ and a $\bar{u}_1 \in U_1$ such that

$$\sup_{u_2 \in U_2} \langle \nu(y), f(y, \bar{u}_1, u_2) \rangle \leqslant -\varepsilon < 0, \qquad \forall y \in B_\delta(x_0) \cap \partial M,$$

for some $\varepsilon, \delta > 0$. Then mimicking the proof of Theorem 2.6.6, we can show that there exists a $\delta' \in (0, \delta)$ such that for any $x \in B_{\delta'}(x_0) \cap M^c$, and any $\alpha_2[\,\cdot\,] \in \mathcal{A}_2[0, \infty)$, there exists a $\tau > 0$ such that

$$X(\tau; x, \bar{u}_1(\cdot), \alpha_2[\bar{u}_1(\cdot)]) \in M,$$

where $\bar{u}_1(\cdot) \equiv \bar{u}_1$ is a constant control. This contradicts the evadability of the game. Hence, (5.20) must be true.

(ii) Suppose (5.21) holds. Then making use of the fact that ∂M is C^1, there exists a $\delta > 0$ such that the map $x \mapsto d_M(x)$ is differentiable in $B_\delta(M) \setminus M$, and

$$\lim_{x \to \bar{x}, x \notin M} \nabla d_M(x) = \nu(\bar{x}), \qquad \forall \bar{x} \in \partial M.$$

Thus, by the continuity of $x \mapsto f(x, u_1, u_2)$, we have (if necessary, we might shrink $\delta > 0$)

$$\inf_{u_1 \in U_1} \sup_{u_2 \in U_2} \langle \nabla d_M(x), f(x, u_1, u_2) \rangle \geqslant \frac{\mu}{2}, \qquad \forall x \in B_\delta(x) \setminus M.$$

Then, for any $x \in B_\delta(M) \setminus M$, and any $(u_1(\cdot), \alpha_2[\,\cdot\,]) \in \mathcal{U}_1[0, \infty) \times \mathcal{A}_2[0, \infty)$,

$$\frac{d}{ds} d_M(X(s))^2 = 2 d_M(X(s)) \langle \nabla d_M(X(r)), f(X(r), u_1(r), \alpha_2[u_1(\cdot)](r)) \rangle$$

$$\geqslant \mu d_M(X(s)), \qquad \text{as long as } X(s) \in B_\delta(M) \setminus M.$$

Hence, we can construct an $\alpha_2[\,\cdot\,] \in \mathcal{A}_2[0, \infty)$, depending on the initial state x, such that for any $u_1(\cdot) \in \mathcal{U}_1[0, \infty)$, one always has

$$d_M\big(X(s; x, u_1(\cdot), \alpha_2[u_1(\cdot)])\big) \geqslant \varepsilon > 0, \qquad \forall s \in [0, \infty),$$

for some $\varepsilon > 0$. Therefore, the game is uniformly evadable. $\qquad \square$

Note that in the above result, the target set M has a non-empty interior. We now consider the case that the target set M is a linear manifold:

$$M = M_0 + x_0, \tag{5.22}$$

with $x_0 \in \mathbb{R}^n$ and M_0 being a subspace of $\mathbb{R}^n$ with

$$\dim M_0 \leqslant n - 2. \tag{5.23}$$

Clearly, in this case, M has an empty interior. Unlike the pursuit game, we have assumed the dimension of M_0 is no more than $(n - 2)$, instead of

$(n-1)$. In what follows, we let $\Pi \equiv \Pi_{M_0^\perp} : \mathbb{R}^n \to M_0^\perp$ be the orthogonal projection onto $M_0^\perp$. Thus,

$$d_M(X(t)) = |\Pi[X(t) - x_0]|$$
$$= |\Pi(x - x_0) + \int_0^t \Pi f(X(s), u_1(s), u_2(s))ds|, \quad t \geqslant 0.$$

It is possible that

$$f^1(x, u_1, u_2) \equiv \Pi f(x, u_1, u_2) = \varphi^1(x),$$
$$(x, u_1, u_2) \in \mathbb{R}^n \times U_1 \times U_2,$$

is independent of (u_1, u_2). If this happens and $x \mapsto f(x, u_1, u_2)$ is differentiable, we have the differentiability of $\varphi^1(\cdot)$ and

$$\varphi^1(X(t)) = \varphi^1(x) + \int_0^t \varphi_x^1(X(s))f(X(s), u_1(s), u_2(s))ds, \quad t \geqslant 0.$$

If we further have that

$$f^2(x, u_1, u_2) \equiv \varphi_x^1(x)f(x, u_1, u_2) = \varphi^2(x),$$
$$(x, u_1, u_2) \in \mathbb{R}^n \times U_1 \times U_2,$$

is independent of (u_1, u_2), by assuming the second order differentiability of the map $x \mapsto f(x, u_1, u_2)$, one has the following:

$$\varphi^2(X(t)) = \varphi^2(x) + \int_0^t \varphi_x^2(X(s))f(X(s), u_1(s), u_2(s))ds.$$

Inductively, assuming $x \mapsto f(x, u_1, u_2)$ to be smooth enough, we may have that

$$\begin{cases} \varphi^0(x) = \Pi(x - x_0), & x \in \mathbb{R}^n, \\ \varphi^i(x) = \varphi_x^{i-1}(x)f(x, u_1, u_2), & 1 \leqslant i \leqslant k-1, \end{cases} \tag{5.24}$$

are all independent of (u_1, u_2), and

$$f^k(x, u_1, u_2) = \varphi_x^{k-1}(x)f(x, u_1, u_2), \tag{5.25}$$
$$(x, u_1, u_2) \in \mathbb{R}^n \times U_1 \times U_2,$$

does depend on u_2. Thus, $k > 0$ is the smallest integer for which the map $f^k(x, u_1, u_2)$ depends on u_2. We now introduce the following assumption.

(PE2) Let $M \subseteq \mathbb{R}^n$ be a linear manifold of form (5.22) for some $x_0 \in \mathbb{R}^n$ and some subspace M_0 of $\mathbb{R}^n$ satisfying (5.23). Let $f : \mathbb{R}^n \times U_1 \times U_2 \to \mathbb{R}^n$ such that $x \mapsto f(x, u_1, u_2)$ is k-time continuously differentiable with all the partial derivatives bounded. Let $\varphi^0(\cdot), \cdots, \varphi^{k-1}(\cdot)$ defined by (5.24) be independent of (u_1, u_2), and $f^k(x, u_1, u_2)$ defined by (5.25) depend on u_2. Moreover, for any $x \in \mathbb{R}^n$, $(u_1(\cdot), u_2(\cdot)) \in \mathcal{U}_1[0, \infty) \times \mathcal{U}_2[0, \infty)$, with $X(\cdot) \equiv X(\cdot; x, u_1(\cdot), u_2(\cdot))$, the following holds:

$$
\begin{cases}
|\Pi f(X(t), u_1, u_2)| \le F(|x|), \\
|f^k(X(t+\tau), u_1, u_2) - f^k(X(t), u_1, u_2)| \le F(|x|)\tau, \\
\qquad\qquad t, \tau \ge 0, \ (u_1, u_2) \in U_1 \times U_2,
\end{cases}
\tag{5.26}
$$

for some continuous and increasing function $F : [0, \infty) \to [0, \infty)$.

Note that in the case that

$$
\langle f(x, u_1, u_2), x \rangle \le 0, \qquad \forall |x| \ge R, \ (u_1, u_2) \in U_1 \times U_2,
$$

for some $R > 0$, for any $(x, u_1(\cdot), u_2(\cdot)) \in \mathbb{R}^n \times U_1 \times U_2$, the unique state trajectory $X(\cdot; x, u_1(\cdot), u_2(\cdot))$ is bounded with the bound only depends on $|x|$. In such a case, (5.26) holds.

We now make an observation. Under (PE2), for any $(x, u_1(\cdot), \alpha_2[\cdot]) \in \mathbb{R}^n \times \mathcal{U}_1[0, \infty) \times \mathcal{A}_2[0, \infty)$, let $X(\cdot) \equiv X(\cdot; x, u_1(\cdot), \alpha_2[u_1(\cdot)])$. Then

$$
\begin{aligned}
\Pi\big[X(t) - x_0\big] &= \Pi(x - x_0) + \int_0^t \varphi^1(X(s))ds \\
&= \varphi^0(x) + \int_0^t \Big[\varphi^1(x) + \int_0^s \varphi^2(X(\tau))d\tau\Big]ds \\
&= \varphi^0(x) + \varphi^1(x)t + \int_0^t (t-s)\varphi^2(X(s))ds \\
&= \varphi^0(x) + \varphi^1(x)t + \frac{t^2}{2}\varphi^2(x) + \int_0^t \frac{(t-s)^2}{2}\varphi^3(X(s))ds = \cdots \\
&= \sum_{j=0}^{k-1} \frac{t^j}{j!}\varphi^j(x) + \int_0^t \frac{(t-s)^{k-1}}{(k-1)!} f^k(X(s), u_1(s), u_2(s))ds \\
&= \sum_{j=0}^{k} \frac{t^j}{j!}\varphi^j(x) + \int_0^t \frac{(t-s)^{k-1}}{(k-1)!}\Big(f^k(x, u_1(s), u_2(s)) - \varphi^k(x)\Big)ds \\
&\quad + \int_0^t \frac{(t-s)^{k-1}}{(k-1)!}\Big(f^k(X(s), u_1(s), u_2(s)) - f^k(x, u_1(s), u_2(s))\Big)ds.
\end{aligned}
$$

Consequently,

$$
d_M(X(t)) \geqslant \left| \sum_{j=0}^{k} \frac{t^j}{j!} \varphi^j(x) + \int_0^t \frac{(t-s)^{k-1}}{(k-1)!} \Big(f^k(x, u_1(s), u_2(s)) - \varphi^k(x) \Big) ds \right|
$$

$$
- F(|x|) \int_0^t \frac{(t-s)^{k-1}}{(k-1)!} s\, ds
$$

$$
= \left| \sum_{j=0}^{k-1} \frac{t^j}{j!} \varphi^j(x) + \int_0^t \frac{(t-s)^{k-1}}{(k-1)!} f^k(x, u_1(s), u_2(s)) - \varphi^k(x) \Big) ds \right|
$$

$$
- F(|x|) \frac{t^{k+1}}{(k+1)!}.
$$

We now prove the following Pontryagin's Lemma.

Lemma 5.2.3. *Let $\alpha < \beta \leqslant \infty$ and $m \geqslant 2$, $k > 0$. Let $\varepsilon > 0$ be given. Then for any polynomial $p(\cdot)$ of form*

$$
p(t) = p_0 + p_1(t-\alpha) + \cdots + p_k(t-\alpha)^k, \qquad p_i \in \mathbb{R}^m, \quad p_k \neq 0,
$$

there exists an $a \in B_\varepsilon(0)$ such that

$$
|p(t) + a(t-\alpha)^k| \geqslant \frac{\varepsilon}{\gamma(m,k)}(t-\alpha)^k, \qquad t \in [\alpha, \beta),
$$

where

$$
\gamma(m,k) = \min\{\ell \mid \ell \in \mathbb{N}, \ (2\ell)^m - (2\ell)mk > mk + 1\}.
$$

Proof. First of all, the closed ball $\bar{B}_\varepsilon(0) \subseteq \mathbb{R}^m$ is inscribed in a cubic type set in $\mathbb{R}^m$ with side length $\frac{2\varepsilon}{\sqrt{m}}$ and with center at 0. Divide $[-\frac{\varepsilon}{\sqrt{m}}, \frac{\varepsilon}{\sqrt{m}}]^m \subseteq \mathbb{R}^m$ into $(2\ell)^m$ small cubic type subsets by coordinate hyperplanes $z_i = b_{\pm j}$ with

$$
b_{\pm 0} = 0, \qquad b_{\pm j} = \pm \frac{j\varepsilon}{\ell\sqrt{m}}, \qquad 1 \leqslant j \leqslant \ell.
$$

Here, $z = (z_1, \cdots, z_m)$ represents the coordinate of $\mathbb{R}^m$, and $\ell > 1$ is undetermined. Now, suppose the curve $t \mapsto \frac{p(t)}{(t-\alpha)^k}$ (for $t \in (\alpha, \beta]$) has visited all the $(2\ell)^m$ cubic type sets obtained above. Then the curve must enter each such a set through its boundary which is a hyperplane of form $z_i = b$. Thus, without loss of generality, we may assume that the curve cross $z_1 = b$ type hyper-plans at least $\frac{(2\ell)^m - 1}{m}$ times. Since there are $2\ell + 1$ such hyperplanes, without loss of generality, we may assume that the curve cross $z_1 = 0$ at least $\frac{(2\ell)^m - 1}{m(2\ell+1)}$ times, which means that $\frac{\langle p(t), e_1 \rangle}{(t-\alpha)^k}$ has at least $\frac{(2\ell)^m-1}{m(2\ell+1)}$ roots, where $e_1 = (1, 0, \cdots, 0)^T \in \mathbb{R}^m$. This will lead to a contradiction if we choose ℓ so that

$$\frac{(2\ell)^m - 1}{m(2\ell + 1)} > k. \tag{5.27}$$

In other words, with the choice (5.27), the curve $t \mapsto \frac{p(t)}{(t-\alpha)^k}$ will not visit all these $(2\ell)^m$ cubic type sets. Hence, by choosing a cubic set among them that is not visited by the curve, and let $a = (a_1, \cdots, a_m)$ be the center of the set, then one has

$$\left| \langle \frac{p(t)}{(t-\alpha)^k}, e_i \rangle - a_i \right| \geq \frac{\varepsilon}{\ell\sqrt{m}}, \qquad 1 \leq i \leq m,$$

which is equivalent to

$$\left| \langle p(t), e_i \rangle - a_i(t-\alpha)^k \right| \geq \frac{\varepsilon}{\ell\sqrt{m}}(t-\alpha)^k, \qquad t \in [\alpha, \beta], \ 1 \leq i \leq m.$$

Consequently,

$$|p(t) - a(t-\alpha)^k| = \left(\sum_{i=1}^m |\langle p(t), e_i \rangle - a_i(t-\alpha)^k|^2 \right)^{\frac{1}{2}} \geq \frac{\varepsilon}{\ell}(t-\alpha)^k.$$

Now, (5.27) is equivalent to the following:

$$(2\ell)^m - (2\ell)mk > mk + 1.$$

Thus, our conclusion follows. $\qquad\qquad\square$

We now present the following result.

Theorem 5.2.4. *Let* (PE2) *hold, and for some* $\varepsilon > 0$,

$$\Pi\bar{B}_\varepsilon(0) \subseteq \bigcap_{u_1 \in U_1} \left[f^k(x, u_1, U_2) - \varphi^k(x) \right], \qquad \forall x \in \mathbb{R}^n.$$

Then Problem (E) *is uniformly evadable.*

Proof. Let $x \notin M = M_0 + x_0$ which is equivalent to $x - x_0 \notin M_0$. Then

$$|\varphi^0(x)| \equiv |\Pi(x - x_0)| > 0.$$

For any $(u_1(\cdot), u_2(\cdot)) \in \mathcal{U}_1[0, \infty) \times \mathcal{U}_2[0, \infty)$, let $X(\cdot) = X(\cdot; 0, x, u_1(\cdot), u_2(\cdot))$. Then

$$d_M(X(t)) = |\Pi[X(t) - x_0]|$$

$$= \left| \Pi(x - x_0) + \int_0^t \Pi f(X(s), u_1(s), u_2(s))ds \right|$$

$$\geq d_M(x) - F(|x|)t \geq \frac{d_M(x)}{2}, \qquad \forall t \in [0, \frac{d_M(x)}{2F(|x|)}].$$

Next, for the given x, the following polynomial

$$p(x;t) = \sum_{i=0}^{k} \varphi^i(x)t^i, \qquad t \in \mathbb{R}$$

is completely determined. Thus, by Lemma 5.2.3, there exists an $a_0 \in B_\varepsilon(0)$ such that

$$|p(x;t) - a_0 t^k| \geqslant \frac{\varepsilon}{\gamma(m,k)}t^k, \qquad t \geq 0.$$

Then by our assumption, we can find an $\alpha_2[\,\cdot\,] \in \mathcal{A}_1[0,\infty)$ such that

$$f^k(x, u_1(s), \alpha_2[u_1(\cdot)](s)) - \varphi^k(x) = -a_0, \qquad s \in [0,\infty).$$

Consequently,

$$d_M\big(X(t;0,x,u_1(\cdot),\alpha_2[u_1(\cdot)])\big)$$

$$\geqslant \left| \sum_{j=0}^{k} \frac{t^j}{j!}\varphi^j(x) + \int_0^t \frac{(t-s)^{k-1}}{(k-1)!}\Big(f^k(x,u_1(s),\alpha_2[u_1(\cdot)](s)) - \varphi^k(x)\Big)ds \right|$$

$$\qquad - F(|x|)\frac{t^{k+1}}{(k+1)!}$$

$$= |p(x;t) - a_0 t^k| - F(|x|)\frac{t^{k+1}}{(k+1)!}$$

$$\geqslant \frac{\varepsilon}{\gamma(m,k)}t^k - F(|x|)\frac{t^{k+1}}{(k+1)!} = \frac{F(|x|)t^k}{(k+1)!}\left[\frac{\varepsilon(k+1)!}{F(|x|)\gamma(m,k)} - t\right]$$

$$\geqslant \frac{\varepsilon}{2\gamma(m,k)}t^k, \qquad \forall t \in [0,\eta],$$

with

$$\eta = \frac{\varepsilon(k+1)!}{2F(|x|)\gamma(m,k)}.$$

Hence,

$$d_M\big(X(t;0,u_1(\cdot),\alpha_2[u_1(\cdot)])\big) \geqslant \frac{\varepsilon t^k}{2\gamma(m,k)} \vee \big[d_M(x) - F(|x|)t\big]$$

$$\geqslant \min_{t\in[0,\eta]}\left(\frac{\varepsilon t^k}{2\gamma(m,k)} \vee \big[d_M(x) - F(|x|)t\big]\right) \equiv \delta > 0, \qquad t \in [0,\eta].$$

Note that $\bar{\delta} > 0$ only depends on $|x|$. In particular,

$$d_M\big(X(\eta;0,x,u_1(\cdot),\alpha_2[u_1(\cdot)])\big)$$

$$\geqslant \frac{\varepsilon \eta^k}{2\gamma(m,k)} = \frac{\varepsilon^{k+1}[(k+1)!]^k}{[2\gamma(m,k)]^{k+1}F(|x|)^k}. \tag{5.28}$$

Now, on $[\eta, \infty)$, let

$$p(X(\eta); t) = \sum_{j=0}^{k} \frac{\varphi^j(X(\eta))}{j!} (t - \eta)^j.$$

By Lemma 5.2.3, there exists an $a_1 \in B_\varepsilon(0)$ such that

$$|p(X(\eta); t) - a_1 (t - \eta)^k| \geqslant \frac{\varepsilon}{\gamma(m,k)} (t - \eta)^k, \qquad t \geqslant \eta.$$

Then by our assumption, we can find an $\alpha_2[\cdot] \in \mathcal{A}_1[\eta, \infty)$ such that

$$f^k(X(\eta), u_1(s), \alpha_2[u_1(\cdot)](s)) - \varphi^k(X(\eta)) = -a_1, \qquad s \in [\eta, \infty).$$

Consequently,

$$d_M \big(X(t; \eta, X(\eta), u_1(\cdot), \alpha_2[u_1(\cdot)]) \big)$$

$$\geqslant \Bigg| \sum_{j=0}^{k} \frac{\varphi^j(X(\eta))}{j!} (t - \eta)^j$$

$$+ \int_{\eta}^{t} \frac{(t-s)^{k-1}}{(k-1)!} \Big(f^k(x, u_1(s), \alpha_2[u_1(\cdot)](s)) - \varphi^k(x) \Big) ds \Bigg|$$

$$- F(|x|) \frac{(t-\eta)^{k+1}}{(k+1)!}$$

$$= |p(X(\eta); t) - a_1(t-\eta)^k| - F(|x|) \frac{(t-\eta)^{k+1}}{(k+1)!}$$

$$\geqslant \frac{\varepsilon}{\gamma(m,k)} (t-\eta)^k - F(|x|) \frac{(t-\eta)^{k+1}}{(k+1)!}$$

$$= \frac{F(|x|)(t-\eta)^k}{(k+1)!} \Bigg[\frac{\varepsilon(k+1)!}{F(|x|)\gamma(m,k)} - (t-\eta) \Bigg]$$

$$\geqslant \frac{\varepsilon}{2\gamma(m,k)} (t-\eta)^k, \qquad \forall t \in [\eta, 2\eta].$$

Also,

$$d_M \big(X(t; \eta, X(\eta), u_1(\cdot), \alpha_2[u_1(\cdot)]) \big)$$

$$\geqslant d_M \big(X(\eta) \big) - F(|x|)(t-\eta)$$

$$\geqslant \frac{\varepsilon \eta^k}{2\gamma(m,k)} - F(|x|)(t-\eta), \qquad t \in [\eta, \infty).$$

Hence,

$$d_M \big(X(t; \eta, X(\eta), u_1(\cdot), \alpha_2[u_1(\cdot)]) \big)$$

$$\geqslant \frac{\varepsilon(t-\eta)^k}{2\gamma(m,k)} \vee \big[d_M(X(\eta)) - F(|x|)(t-\eta) \big]$$

$$\geqslant \min_{t \in [\eta, 2\eta]} \bigg(\frac{\varepsilon(t-\eta)^k}{2\gamma(m,k)} \vee \Big[\frac{\varepsilon^{k+1}[(k+1)!]^k}{[2\gamma(m,k)]^{k+1} F(|x|)^k} - F(|x|)(t-\eta) \Big] \bigg) \equiv \bar{\delta} > 0,$$

$$t \in [\eta, 2\eta].$$

In particular,

$$d_M\big(X(2\eta;\eta,X(\eta),u_1(\cdot),\alpha_2[u_1(\cdot)])\big) \geqslant \frac{\varepsilon\eta^k}{2\gamma(m,k)}$$

$$= \frac{\varepsilon^{k+1}[(k+1)!]^k}{[2\gamma(m,k)]^{k+1}F(|x|)^k},$$

which has the same form as (5.28). Then we can use induction to obtain an $\alpha_2[\cdot]\in\mathcal{A}_2[0,\infty)$ such that

$$d_M\big(X(t;0,x,u_1(\cdot),\alpha_2[u_1(\cdot)])\big) \geqslant \bar{\delta}, \qquad \forall t \geqslant 0,$$

proving the uniform evadability of the game. $\qquad\qquad\qquad\square$

We now present an example, which is an evasion way of viewing Example 5.1.5.

Example 5.2.5. Consider two objects moving in $\mathbb{R}^3$, whose coordinates are y_1 and y_2, respectively, and they satisfy the following:

$$\begin{cases} \dot{y}_i(t) = p_i(t), \\ \dot{p}_i(t) = -\mu_i p_i(t) + u_i(t), \end{cases}$$

with $\mu_i > 0$. The same as in Example 5.1.5, we assume that y_1 is the pursuer and y_2 is the evader. The evader is captured at some time t^* if

$$y_1(t^*) = y_2(t^*).$$

Now, we set

$$X_1 = y_1 - y_2, \quad X_2 = p_1, \quad X_3 = p_2.$$

Then the state equation becomes

$$\dot{X}(t) = \begin{pmatrix} 0 & I & -I \\ 0 & -\mu_1 I & 0 \\ 0 & 0 & -\mu_2 I \end{pmatrix} X(t) + \begin{pmatrix} 0 \\ I \\ 0 \end{pmatrix} u_1(t) + \begin{pmatrix} 0 \\ 0 \\ I \end{pmatrix} u_2(t)$$

$$\equiv AX(t) + B_1 u_1(t) + B_2 u_2(t),$$

and the terminal set is given by

$$M = M_0 = \left\{ \begin{pmatrix} 0 \\ x_2 \\ x_3 \end{pmatrix} \mid x_2, x_3 \in \mathbb{R}^3 \right\}.$$

Let $\Pi : \mathbb{R}^9 \to M^\perp$ be the orthogonal projection. Then

$$\varphi^0(x) = \Pi x = \begin{pmatrix} x_1 \\ 0 \\ 0 \end{pmatrix},$$

$$\varphi^1(x) = \Pi\Big(Ax + B_1 u_1 + B_2 u_2\Big) = \begin{pmatrix} x_2 - x_3 \\ 0 \\ 0 \end{pmatrix},$$

$$f^2(x, u_1, u_2) = \begin{pmatrix} -\mu_1 x_2 + \mu_2 x_3 + u_1 - u_2 \\ 0 \\ 0 \end{pmatrix}.$$

Thus, in our case, $k = 2$. Next, we can calculate that

$$e^{At} = \begin{pmatrix} I & \left(\int_0^t e^{-\mu_1 \tau} dt\right) I & -\left(\int_0^t e^{-\mu_2 \tau} d\tau\right) I \\ 0 & e^{-\mu_1 t} I & 0 \\ 0 & 0 & e^{-\mu_2 t} I \end{pmatrix}.$$

Hence,

$$\Pi A e^{At} = \begin{pmatrix} 0 & e^{-\mu_1 t} I & -e^{-\mu_2 t} I \\ 0 & 0 & 0 \\ 0 & 0 & 0 \end{pmatrix}.$$

Now, if the following holds:

$$\bar{B}_\varepsilon(0) \subseteq \bigcap_{u_1 \in U_1} [u_1 - U_2],$$

for some $\varepsilon > 0$, where both sides in the above are in $\mathbb{R}^3$, then the game is uniformly evadable.

5.2.2 Characterization of evadable set

Now, for any $x \in \mathbb{R}^n$, $u_1(\cdot) \in \mathcal{U}_1[0, \infty)$ and $\alpha_2[\cdot] \in \mathcal{A}_2[0, \infty)$, we let

$$T(x; u_1(\cdot), \alpha_2[u_1(\cdot)]) = \inf \{t > 0 \mid X(t; x, u_1(\cdot), \alpha_2[u_1(\cdot)]) \in M\},$$

with the convention that $\inf \phi = \infty$. Then we define

$$J(x; u_1(\cdot), \alpha_2[u_1(\cdot)]) = 1 - e^{-T(x;u_1(\cdot),\alpha_2[u_1(\cdot)])}$$

$$= \int_0^{T(x;u_1(\cdot)\alpha_2[u_1(\cdot)])} e^{-t} dt,$$

$$x \in \mathbb{R}^n, u_1(\cdot) \in \mathcal{U}_1[0, \infty), \alpha_2[\cdot] \in \mathcal{A}_2[0, \infty).$$

Clearly, we have

$$J\big(x; u_1(\cdot), \alpha_2[u_1(\cdot)]\big) = 0,$$

$$\forall x \in M, u_1(\cdot) \in \mathcal{U}_1[0,\infty), \alpha_2[\,\cdot\,] \in \mathcal{A}_2[0,\infty).$$

Define

$$V^+(x) = \sup_{\alpha_2[\,\cdot\,] \in \mathcal{A}_2[0,\infty)} \inf_{u_1(\cdot) \in \mathcal{U}_1[0,\infty)} J\big(x; u_1(\cdot), \alpha_2[u_1(\cdot)]\big), \quad \forall x \in \mathbb{R}^n.$$

The following result is comparable with Theorem 5.1.6, whose proof is obvious.

Proposition 5.2.6. *The following hold:*

$$\begin{cases} V^+(x) = 0, & x \in M, \\ 0 < V^+(x) < 1, & x \in \mathcal{E}(M)^c \setminus M, \\ V^+(x) = 1, & x \in \mathcal{E}(M). \end{cases}$$

From the above, we see that the game is evadable from M if and only if

$$V^+(x) = 1, \quad \forall x \in \mathbb{R}^n \setminus M,$$

which means

$$V^+(x) = I_{M^c}(x), \quad \forall x \in \mathbb{R}^n.$$

In this case, $V^+(\cdot)$ is discontinuous along ∂M. Therefore, the standard approach using viscosity solution does not apply directly. At the moment, we do not have a satisfactory theory for the characterization of $V^+(\cdot\,,\cdot)$.

5.3   Brief Historic Remarks

Differential games of evasion and pursuit were initiated independently by Isaacs ([57]) in the early 1950s and by Pontryagin and his colleagues ([77], [92]) in the 1950–1960s. Contributions were made by many authors. Here is a very small partial list: [56], [100], [98], [99], [93], [66], [76], [101], [104], [63], [83], [95], [84], [96], [53].

The material presented in this chapter is mainly based on the work of the author in the late 1980s ([118], [119], [120], [121], see also [122] and [123]). More precisely, Theorems 5.1.2 and 5.2.2 are modification of the relevant results found in [120] and [119], respectively (see also [118]). The idea of such an approach can be found in an earlier work of Lagunov ([66]). Theorems 5.1.3 and 5.1.4 seem to be new, which were inspired by a relevant result for controllability of linear control system to a linear manifold target set (see Theorem 2.6.10). Some idea of Theorem 5.1.4 can also be found in the Pontryagin's work [92] and [95]. The characterization of the capturable set is inspired by that of controllable set. The main idea comes from the work of Peng–Yong ([85]). Theorem 5.2.4 is a modification of a result found in [121]. Such an approach was earlier introduced by Pshenichnyi ([100]) for linear problems and later was generalized to some nonlinear cases by Pshenichnyi ([101]), Káskosz ([63]), and Ostapenko ([83], [84]). Lemma 5.2.3 is mainly due to Pontryagin ([93], [94]). We expect to have a characterization of evadable set (symmetric to that for capturable set). However, due to the discontinuity of the upper value function, the presented result in Section 5.2.2 is not yet satisfactory. Also, we failed to obtain a similar result for the evasion game similar to that for viability problem.

Chapter 6

Linear-Quadratic Differential Games

In this chapter, we concentrate on the case that the state equation is a linear ordinary differential equation (ODE, for short) and the performance criterion is a quadratic functional.

6.1 Linear-Quadratic Optimal Control Problems

In this section, we consider the following controlled linear ODE:

$$\begin{cases} \dot{X}(s) = A(s)X(s) + B(s)u(s) + f(s), & s \in [t, T], \\ X(t) = x, \end{cases} \tag{6.1}$$

with the cost functional

$$
\begin{aligned}
J(t, x; u(\cdot)) &= \frac{1}{2} \Big[\int_t^T \Big(\langle\, Q(s)X(s), X(s) \,\rangle + 2 \langle\, S(s)X(s), u(s) \,\rangle \\
&\qquad + \langle\, R(s)u(s), u(s) \,\rangle + 2 \langle\, q(s), X(s) \,\rangle + 2 \langle\, \rho(s), u(s) \,\rangle \Big) ds \\
&\qquad + \langle\, GX(T), X(T) \,\rangle + 2 \langle\, g, X(T) \,\rangle \Big] \\
&\equiv \frac{1}{2} \Big\{ \int_t^T \Big[\Big\langle \begin{pmatrix} Q(s) & S(s)^T \\ S(s) & R(s) \end{pmatrix} \begin{pmatrix} X(s) \\ u(s) \end{pmatrix}, \begin{pmatrix} X(s) \\ u(s) \end{pmatrix} \Big\rangle \\
&\qquad + 2 \Big\langle \begin{pmatrix} q(s) \\ \rho(s) \end{pmatrix}, \begin{pmatrix} X(s) \\ u(s) \end{pmatrix} \Big\rangle \Big] ds \\
&\qquad + \langle\, GX(T), X(T) \,\rangle + 2 \langle\, g, X(T) \,\rangle \Big\}.
\end{aligned}
$$

Let us first introduce the following basic assumptions. Recall that $\mathbb{S}^n$ is the set of all $(n \times n)$ symmetric matrices.

(LQ1) The coefficients of the state equation satisfy the following conditions:

$$A(\cdot) \in L^1(0,T;\mathbb{R}^{n\times n}), \quad B(\cdot) \in L^2(0,T;\mathbb{R}^{n\times m}), \quad f(\cdot) \in L^1(0,T;\mathbb{R}^n).$$

(LQ2) The weighting coefficients in the cost functional satisfy the following conditions:

$$Q(\cdot) \in L^1(0,T;\mathbb{S}^n), \quad S(\cdot) \in L^2(0,T;\mathbb{R}^{m\times n}), \quad R(\cdot) \in L^\infty(0,T;\mathbb{S}^m),$$
$$q(\cdot) \in L^1(0,T;\mathbb{R}^n), \quad \rho(\cdot) \in L^2(0,T;\mathbb{R}^m), \quad G \in \mathbb{S}^n, \quad g \in \mathbb{R}^n.$$

(LQ3) The following holds:

$$R(t) > 0, \qquad \text{a.e. } t \in [0,T]. \tag{6.2}$$

(LQ3)$'$ There exists a $\delta > 0$ such that the following holds:

$$R(t) \geqslant \delta I, \qquad \text{a.e. } t \in [0,T]. \tag{6.3}$$

Note that in (LQ2), $Q(\cdot)$, $R(\cdot)$ and G are only assumed to be symmetric, and not necessarily positive semi-definite. Also, we note that (6.2) does not necessarily imply (6.3) unless $R(\cdot)$ is assumed to be continuous at the same time. Next, we let

$$\mathcal{U}[t,T] = L^2(t,T;\mathbb{R}^m).$$

Clearly, under (LQ1), for any initial pair $(t,x) \in [0,T) \times \mathbb{R}^n$, and $u(\cdot) \in \mathcal{U}[t,T]$, there exists a unique solution $X(\cdot) \equiv X(\cdot\,;t,x,u(\cdot)) \in C([0,T];\mathbb{R}^n)$ such that

$$\|X(\cdot)\|_\infty \leqslant C\big(1 + |x| + \|u(\cdot)\|_2 + \|f(\cdot)\|_1\big),$$

with the constant $C > 0$ depending on $\|A(\cdot)\|_1$ and $\|B(\cdot)\|_2$. Then, under (LQ1)–(LQ2), cost functional $J(t,x;u(\cdot))$ is well-defined and we can state the following problem.

Problem (LQ). For given initial pair $(t,x) \in [0,T) \times \mathbb{R}^n$, find a $\bar{u}(\cdot) \in \mathcal{U}[t,T]$ such that

$$V(t,x) \stackrel{\Delta}{=} J(t,x;\bar{u}(\cdot)) = \inf_{u(\cdot)\in\mathcal{U}[t,T]} J(t,x;u(\cdot)) \leqslant J(t,x;u(\cdot)),$$
$$\forall u(\cdot) \in \mathcal{U}[t,T]. \tag{6.4}$$

6.1.1 Open-loop controls

We introduce some basic notions.

Definition 6.1.1. (i) Problem (LQ) is said to be *finite* at (t, x) if

$$\inf_{u(\cdot) \in \mathcal{U}[t,T]} J(t, x; u(\cdot)) = V(t, x) > -\infty.$$

When the above holds for every initial pair $(t, x) \in [0, T] \times \mathbb{R}^n$, we say that Problem (LQ) itself is *finite*.

(ii) A control $\bar{u}(\cdot) \in \mathcal{U}[t, T]$ is called an *open-loop optimal control* of Problem (LQ) for the initial pair $(t, x) \in [0, T) \times \mathbb{R}^n$ if (6.4) holds. The corresponding $\bar{X}(\cdot) \equiv X(\cdot; t, x, \bar{u}(\cdot))$ is called an *open-loop optimal state trajectory* and $(\bar{X}(\cdot), \bar{u}(\cdot))$ is called an *open-loop optimal pair*.

Let us present the following simple example to illustrate the above notions.

Example 6.1.2. (i) Consider the following one-dimensional controlled system:

$$\begin{cases} \dot{X}(s) = u(s), & s \in [t, T], \\ X(t) = x, \end{cases} \tag{6.5}$$

with cost functional

$$J(t, x; u(\cdot)) = -\frac{1}{2} \int_t^T |X(s)|^2 ds.$$

Then for any $(t, x) \in [0, T) \times \mathbb{R}$, by letting $u_\lambda(\cdot) = \lambda$, one has

$$J(t, x; u_\lambda(\cdot)) = -\frac{1}{2} \int_t^T |x + \lambda(s-t)|^2 ds \leqslant -\frac{1}{2} \int_t^T \left(\frac{\lambda^2}{2} (s-t)^2 - |x|^2 \right) ds$$

$$= -\frac{1}{2} \left[\frac{\lambda^2}{6} (T-t)^3 - |x|^2 (T-t) \right] \to -\infty, \quad \lambda \to \infty.$$

Thus,

$$\inf_{u(\cdot) \in \mathcal{U}[t,T]} J(t, x; u(\cdot)) = -\infty,$$

and the corresponding LQ problem is not finite at any initial pair $(t, x) \in [0, T) \times \mathbb{R}$.

(ii) Consider the same state equation (6.5) with cost functional

$$J(t, x; u(\cdot)) = \frac{1}{2} \int_t^T |X(s)|^2 ds.$$

Clearly, the corresponding LQ problem is finite since the cost functional is nonnegative. However, for any $(t, x) \in [0, T) \times (\mathbb{R} \setminus \{0\})$, by letting

$$u_\lambda(s) = -\lambda x I_{[t, t + \frac{1}{\lambda}]}(s), \qquad s \in [t, T], \quad \lambda \geqslant \frac{1}{T - t},$$

we have

$$J(t, x; u_\lambda(\cdot)) = \frac{1}{2} \int_t^{t + \frac{1}{\lambda}} |x|^2 [1 - \lambda(s - t)]^2 ds = \frac{|x|^2}{6\lambda} \to 0, \quad \lambda \to \infty.$$

Hence,

$$\inf_{u(\cdot) \in \mathcal{U}[t, T]} J(t, x; u(\cdot)) = 0.$$

But, for any $u(\cdot) \in \mathcal{U}[t, T]$,

$$J(t, x; u(\cdot)) > 0, \qquad \forall (t, x) \in [0, T) \times (\mathbb{R} \setminus \{0\}).$$

Therefore, the corresponding LQ problem does not admit an open-loop optimal control for any initial pair $(t, x) \in [0, T) \times (\mathbb{R} \setminus \{0\})$.

(iii) Consider the same state equation (6.5) with cost functional

$$J(t, x; u(\cdot)) = \frac{1}{2} |X(T)|^2.$$

It is straightforward that for any initial pair $(t, x) \in [0, T) \times \mathbb{R}^n$, one can find infinitely many open-loop optimal controls. One such family is the following:

$$\bar{u}_\lambda(s) = -\lambda x I_{[t, t + \frac{1}{\lambda}]}(s), \qquad s \in [t, T], \quad \lambda \geqslant \frac{1}{T - t},$$

with which, one has

$$J(t, x; \bar{u}_\lambda(\cdot)) = 0 = V(t, x), \qquad (t, x) \in [0, T) \times \mathbb{R}.$$

We see that $\bar{u}_\lambda(\cdot)$ explicitly depends on (t, x). By the way, it is clear that

$$V(t, x) = \begin{cases} 0, & (t, x) \in [0, T) \times \mathbb{R}, \\ \dfrac{x^2}{2}, & x \in \mathbb{R}. \end{cases}$$

This is an example that the value function $V(\cdot, \cdot)$ is discontinuous.

From the above example, we see that besides the existence issue, open-loop optimal controls, if exist, might not be unique and further, they depend on the initial pair $(t, x) \in [0, T) \times \mathbb{R}^n$, in general.

6.1.2 *A functional analysis approach**

Let us take a closer look at the map $u(\cdot) \mapsto J(t,x;u(\cdot))$ from functional analysis point of view. To this end, let $\Phi(\cdot\,,\cdot)$ be the fundamental matrix of $A(\cdot)$, i.e.,

$$\begin{cases} \Phi_s(s,t) = A(s)\Phi(s,t), & 0 \leqslant t \leqslant s \leqslant T, \\ \Phi(t,t) = I, & t \in [0,T]. \end{cases}$$

Define the following linear bounded operators

$$\begin{cases} \Gamma : \mathcal{U}[t,T] \to L^2(t,T;\mathbb{R}^n), & (\Gamma u)(\cdot) = \displaystyle\int_t^\cdot \Phi(\cdot\,,r)B(r)u(r)dr, \\ \Gamma_T : \mathcal{U}[t,T] \to \mathbb{R}^n, & \Gamma_T u = \displaystyle\int_t^T \Phi(T,r)B(r)u(r)dr. \end{cases}$$

It is straightforward that their adjoint operators are given by

$$\begin{cases} \Gamma^* : L^2(t,T;\mathbb{R}^n) \to \mathcal{U}[t,T], & (\Gamma^*\varphi)(\cdot) = \displaystyle\int_\cdot^T B(\cdot)^T \Phi(\tau,\cdot)^T \varphi(\tau)d\tau, \\ \Gamma_T^* : \mathbb{R}^n \to \mathcal{U}[t,T], & \Gamma_T^* x = B(\cdot)^T \Phi(T,\cdot)^T x. \end{cases}$$

With the above operators, by the variation of constants formula, we have

$$X(\cdot) = (\Gamma u)(\cdot) + \xi(\cdot),$$

with

$$\xi(\cdot) = \Phi(\cdot\,,t)x + \int_t^\cdot \Phi(\cdot\,,r)f(r)dr$$

and

$$X(T) = \Gamma_T u + \xi(T).$$

Then one has

$$\begin{aligned} 2J(t,x;u(\cdot)) &= \langle QX, X \rangle + 2\langle SX, u \rangle + \langle Ru, u \rangle + 2\langle q, X \rangle + 2\langle \rho, u \rangle \\ &\quad + \langle GX(T), X(T) \rangle + 2\langle g, X(T) \rangle \\ &= \langle Q(\Gamma u + \xi), \Gamma u + \xi \rangle + 2\langle S(\Gamma u + \xi), u \rangle + \langle Ru, u \rangle + 2\langle q, \Gamma u + \xi \rangle \\ &\quad + 2\langle \rho, u \rangle + \langle G(\Gamma_T u + \xi(T)), \Gamma_T u + \xi(T) \rangle + 2\langle g, \Gamma_T u + \xi(T) \rangle \\ &= \langle (\Gamma^* Q\Gamma + S\Gamma + \Gamma^* S^* + \Gamma_T^* G\Gamma_T + R)u, u \rangle \\ &\quad + 2\langle \Gamma^* Q^*\xi + S\xi + \Gamma^* q + \rho + \Gamma_T^* G\xi(T) + \Gamma_T^* g, u \rangle \\ &\quad + \langle Q\xi, \xi \rangle + 2\langle q, \xi \rangle + \langle G\xi(T), \xi(T) \rangle + 2\langle g, \xi(T) \rangle \\ &\equiv \langle \Psi u, u \rangle + 2\langle \psi, u \rangle + \psi_0, \end{aligned}$$

where

$$\begin{cases} \Psi = \Gamma^*Q\Gamma + \Gamma_T^*G\Gamma_T + S\Gamma + \Gamma^*S^* + R : \mathcal{U}[t,T] \to \mathcal{U}[t,T], \\ \psi = (\Gamma^*Q + S)\xi + \Gamma_T^*G\xi(T) + \Gamma^*q + \Gamma_T^*g + \rho \in \mathcal{U}[t,T], \\ \psi_0 = \langle Q\xi, \xi \rangle + \langle G\xi(T), \xi(T) \rangle + 2\langle q, \xi \rangle + 2\langle g, \xi(T) \rangle \in \mathbb{R}. \end{cases}$$

We see that Ψ is self-adjoint and bounded, which only depends on $t, A(\cdot)$, $B(\cdot), Q(\cdot), S(\cdot), R(\cdot)$, and G; and it is independent of $x, f(\cdot), q(\cdot), \rho(\cdot)$ and g. On the other hand, we have

$$(\Psi u)(s) = R(s)u(s) + \int_s^T B(s)^T \Phi(\tau, s)^T Q(\tau) \int_t^\tau \Phi(\tau, r) B(r) u(r) dr d\tau$$

$$+ B(s)^T \Phi(T, s)^T G \int_t^T \Phi(T, r) B(r) u(r) dr + S(s) \int_t^s \Phi(s, r) B(r) u(r) dr$$

$$+ \int_s^T B(s)^T \Phi(\tau, s)^T S(\tau)^T u(\tau) d\tau$$

$$= R(s)u(s) + \int_t^T B(s)^T \left(\int_{s \vee r}^T \Phi(\tau, s)^T Q(\tau) \Phi(\tau, r) d\tau \right) B(r) u(r) dr$$

$$+ \int_t^T B(s)^T \Phi(T, s)^T G \Phi(T, r) B(r) u(r) dr$$

$$+ \int_t^T \left(S(s) \Phi(s, r) B(r) I_{[t,s]}(r) + B(s)^T \Phi(\tau, s)^T S(\tau)^T I_{[s,T]}(r) \right) u(r) dr$$

$$\equiv R(s)u(s) + \int_t^T K(s, r) u(r) dr, \qquad s \in [t, T],$$

with

$$K(s, r) = B(s)^T \left(\int_{s \vee r}^T \Phi(\tau, s)^T Q(\tau) \Phi(\tau, r) d\tau + \Phi(T, s)^T G \Phi(T, r) \right) B(r)$$

$$+ S(s) \Phi(s, r) B(r) I_{[t,s]}(r) + B(s)^T \Phi(r, s)^T S(r) I_{[s,T]}(r).$$

Under (LQ1)–(LQ2), we have that

$$\int_t^T \int_t^T |K(s, r)|^2 ds dr < \infty.$$

Hence, the operator

$$u(\cdot) \mapsto \int_t^T K(\cdot, r) u(r) dr \equiv \mathcal{K}u(\cdot), \qquad u \in \mathcal{U}[t, T]$$

is compact. If in addition, (LQ3)$'$ holds, then the operator

$$u(\cdot) \mapsto \int_t^T R(\cdot)^{-1} K(\cdot, r) u(r) dr \equiv R^{-1}\mathcal{K}u(\cdot), \qquad u(\cdot) \in \mathcal{U}[t, T]$$

is also compact, which implies that

$$R^{-1}\Psi = I + R^{-1}\mathcal{K}$$

is a Fredholm operator.

We now look at the problem of minimizing

$$u(\cdot) \mapsto J(t, x; u(\cdot)) = \frac{1}{2}\Big\{ \langle \Psi u, u \rangle + 2 \langle \psi, u \rangle + \psi_0 \Big\},$$

over the Hilbert space $\mathcal{H} \equiv \mathcal{U}[t, T]$.

First of all, since Ψ is self-adjoint (and bounded), $\mathcal{H}_0 \equiv \mathcal{N}(\Psi)$ is a closed subspace of $\mathcal{H}$ and

$$\mathcal{H}_1 \equiv \mathcal{H}_0^\perp = \mathcal{N}(\Psi)^\perp = \overline{\mathcal{R}(\Psi)}.$$

Hence,

$$\mathcal{H} = \mathcal{H}_0 \oplus \mathcal{H}_1,$$

and under such a decomposition, the following representation for Ψ holds:

$$\Psi = \begin{pmatrix} 0 & 0 \\ 0 & \Psi_1 \end{pmatrix},$$

where $\Psi_1 : \mathcal{H}_1 \to \mathcal{R}(\Psi) \subseteq \mathcal{H}_1$ is self-adjoint, bounded, and injective. But, unless $\mathcal{R}(\Psi)$ is closed, Ψ_1 is not onto $\mathcal{H}_1$. Thus, $\Psi_1^{-1} : \mathcal{R}(\Psi) \to \mathcal{H}_1$ exists as a *closed self-adjoint operator* with domain $\mathcal{D}(\Psi_1) = \mathcal{R}(\Psi)$ which is dense in $\mathcal{H}$. Now, the *pseudo-inverse* $\Psi^\dagger$ of Ψ has the domain

$$\mathcal{D}(\Psi^\dagger) = \mathcal{N}(\Psi) + \mathcal{R}(\Psi) \equiv \{u^0 + u^1 \mid u^0 \in \mathcal{N}(\Psi), u^1 \in \mathcal{R}(\Psi)\}, \qquad (6.6)$$

and for any $u^0 + u^1 \in \mathcal{N}(\Psi) + \mathcal{R}(\Psi)$ with $u^0 \in \mathcal{N}(\Psi)$ and $u^1 \in \mathcal{R}(\Psi)$, one has

$$\Psi^\dagger u = \Psi_1^{-1} u^1,$$

or formally,

$$\Psi^\dagger = \begin{pmatrix} 0 & 0 \\ 0 & \Psi_1^{-1} \end{pmatrix}. \qquad (6.7)$$

From the above, we can easily see the following facts:

(i) $\Psi^\dagger$ is (closed, densely defined, and) self-adjoint, with the domain $\mathcal{D}(\Psi^\dagger)$ given by (6.6) and with the range $\mathcal{R}(\Psi^\dagger)$ given by the following:

$$\mathcal{R}(\Psi^\dagger) = \mathcal{H}_1 = \overline{\mathcal{R}(\Psi)}.$$

Moreover, $\mathcal{R}(\Psi)$ is closed if and only if $\Psi^\dagger$ is bounded.

(ii) By the definition of $\Psi^\dagger$ (see (6.7)), together with (6.6), one has that

$$\Psi\Psi^\dagger\Psi = \Psi, \quad \Psi^\dagger\Psi\Psi^\dagger = \Psi^\dagger, \quad (\Psi^\dagger)^\dagger \subseteq \Psi,$$

where the last relation means that Ψ is an extension of $(\Psi^\dagger)^\dagger$ whose domain is smaller than $\mathcal{H}$.

(iii) It is seen that $\Psi^\dagger\Psi \equiv P_{\mathcal{H}_1} : \mathcal{H} \to \mathcal{H}_1$ is an orthogonal projection onto $\mathcal{H}_1 = \overline{\mathcal{R}(\Psi)}$. On the other hand, $\Psi\Psi^\dagger : \mathcal{D}(\Psi^\dagger) \to \mathcal{H}$ is an orthogonal projection onto $\mathcal{R}(\Psi)$. Thus, we may naturally extend it to $\overline{\mathcal{D}(\Psi^\dagger)} = \mathcal{H}$, denoted by $\overline{\Psi\Psi^\dagger} : \mathcal{H} \to \overline{\mathcal{R}(\Psi)} \equiv \mathcal{H}_1 \subseteq \mathcal{H}$. Hence, $\overline{\Psi\Psi^\dagger} \equiv P_{\mathcal{H}_1}$ is the orthogonal projection onto $\mathcal{H}_1$. Therefore, we have

$$\overline{\Psi\Psi^\dagger} = \Psi^\dagger\Psi \equiv P_{\mathcal{H}_1} \equiv \text{orthogonal projection onto } \mathcal{H}_1.$$

Now, let us consider a quadratic functional on $\mathcal{H}$:

$$J(u) = \langle \Psi u, u \rangle + 2\langle \psi, u \rangle, \qquad u \in \mathcal{H},$$

where $\Psi : \mathcal{H} \to \mathcal{H}$ is a self-adjoint linear operator and $\psi \in \mathcal{H}$ is fixed. The following result is concerned with the completing square and critical point(s) of the functional $J(\cdot)$. Note here that we do not assume positive (negative) semi-definite condition on Ψ.

Lemma 6.1.3. *For any given $\hat{u} \in \mathcal{H}$,*

$$J(u) \equiv \langle \Psi u, u \rangle + 2\langle \psi, u \rangle$$
$$= \langle \Psi(u - \hat{u}), u - \hat{u} \rangle + 2\langle \Psi\hat{u} + \psi, u \rangle - \langle \Psi\hat{u}, \hat{u} \rangle, \quad \forall u \in \mathcal{H}. \tag{6.8}$$

Consequently, the following are equivalent:

(i) There exists a $\hat{u} \in \mathcal{H}$ such that

$$J(u) = \langle \Psi(u - \hat{u}), u - \hat{u} \rangle - \langle \Psi\hat{u}, \hat{u} \rangle, \qquad \forall u \in \mathcal{H}. \tag{6.9}$$

(ii) The following equation

$$\Psi\hat{u} + \psi = 0, \tag{6.10}$$

admits a solution $\hat{u}$.

(iii) The following holds:

$$\psi \in \mathcal{R}(\Psi)\Big(\subseteq \mathcal{D}(\Psi^\dagger)\Big). \tag{6.11}$$

In the above case, it is necessary that

$$J(u) = \langle \Psi(u - \hat{u}), u - \hat{u} \rangle - \langle \Psi^\dagger\psi, \psi \rangle, \qquad \forall u \in \mathcal{H}, \tag{6.12}$$

and the solution $\hat{u}$ of (6.10) admits the following representation:

$$\hat{u} = -\Psi^\dagger\psi + (I - \Psi^\dagger\Psi)v,$$

for some $v \in \mathcal{H}$. Moreover, such a $\hat{u}$ is unique if and only if $\mathcal{N}(\Psi) = \{0\}$.

Proof. For any $\hat{u} \in \mathcal{H}$, one has

$$
\begin{aligned}
J(u) &\equiv \langle \Psi u, u \rangle + 2 \langle \psi, u \rangle \\
&= \langle \Psi(u - \hat{u} + \hat{u}), u - \hat{u} + \hat{u} \rangle + 2 \langle \psi, u \rangle \\
&= \langle \Psi(u - \hat{u}), u - \hat{u} \rangle + 2 \langle \Psi \hat{u} + \psi, u \rangle - \langle \Psi \hat{u}, \hat{u} \rangle, \quad \forall u \in \mathcal{H}.
\end{aligned}
\tag{6.13}
$$

This gives (6.8). We now prove the three equivalent statements.

(i) $\Rightarrow$ (ii). From (6.8), we see that (6.9) holds for some $\hat{u} \in \mathcal{H}$, which implies that (6.10) holds for the same $\hat{u}$.

(ii) $\Rightarrow$ (iii) is trivial.

(iii) $\Rightarrow$ (i). Suppose (6.11) holds, then there exists a $\hat{u} \in \mathcal{H}$ such that (6.10) holds. Consequently,

$$
\begin{aligned}
&\langle \Psi(u - \hat{u}), u - \hat{u} \rangle - \langle \Psi \hat{u}, \hat{u} \rangle \\
&= \langle \Psi u, u \rangle - 2 \langle \Psi \hat{u}, u \rangle + \langle \Psi \hat{u}, \hat{u} \rangle - \langle \Psi \hat{u}, \hat{u} \rangle \\
&= \langle \Psi u, u \rangle + 2 \langle \psi, u \rangle = J(u),
\end{aligned}
$$

proving (6.9).

The rest of the conclusions are clear. $\qquad\square$

Note that (6.10) is equivalent to the following:

$$
0 = \Psi \hat{u} + \psi \equiv \frac{1}{2} \nabla J(\hat{u}).
$$

Thus, $\hat{u}$ is actually a *critical point* of functional $J(\cdot)$. Equations (6.9) and (6.12) are completion of square for the functional $J(\cdot)$ (with Ψ being only assumed to be self-adjoint).

Next, for any bounded self-adjoint operator Ψ, we have the following spectrum decomposition

$$
\Psi = \int_{\sigma(\Psi)} \lambda dP_\lambda,
\tag{6.14}
$$

where $\sigma(\Psi) \subseteq \mathbb{R}$ is the spectrum of Ψ, which is a compact set, and $\{P_\lambda \mid \lambda \in \sigma(\Psi)\}$ is a family of projection measures. For any $u \in \mathcal{H}$,

$$
\Psi u = \int_{\sigma(\Psi)} \lambda dP_\lambda u,
$$

with $\lambda \mapsto P_\lambda u$ being a vector-valued function. Further,

$$
\Psi^\dagger = \int_{\sigma(\Psi) \setminus \{0\}} \lambda^{-1} dP_\lambda.
$$

In the case that

$$\Psi \geqslant 0, \tag{6.15}$$

one has from (6.14) that $\sigma(\Psi) \subseteq [0, \infty)$, and

$$\begin{cases} \Psi^\alpha = \displaystyle\int_{\sigma(\Psi)} \lambda^\alpha dP_\lambda, & \forall \alpha \geqslant 0, \\ (\Psi^\dagger)^\alpha = (\Psi^\alpha)^\dagger = \displaystyle\int_{\sigma(\Psi)\backslash\{0\}} \lambda^{-\alpha} dP_\lambda, & \forall \alpha > 0. \end{cases} \tag{6.16}$$

The following result is concerned with the minimization problem for functional $J(\cdot)$.

Theorem 6.1.4. *Let* $\Psi : \mathcal{H} \to \mathcal{H}$ *be bounded and self-adjoint and* $\psi \in \mathcal{H}$.

(i) *The following holds:*

$$\inf_{u \in \mathcal{H}} J(u) > -\infty, \tag{6.17}$$

if and only if (6.15) *holds and*

$$\psi \in \mathcal{R}(\Psi^{\frac{1}{2}}). \tag{6.18}$$

In this case,

$$\inf_{u \in \mathcal{H}} J(u) = -|(\Psi^\dagger)^{\frac{1}{2}}\psi|^2. \tag{6.19}$$

(ii) *There exists a* $\hat{u} \in \mathcal{H}$ *such that*

$$J(\hat{u}) = \inf_{u \in \mathcal{H}} J(u),$$

if and only if (6.15) *and* (6.11) *hold.*

(iii) *If* $\mathcal{R}(\Psi)$ *is closed, then* (6.17) *holds if and only if* $J(\cdot)$ *admits a minimum.*

Proof. (i) First, let (6.17) hold. It is straightforward that one must have (6.15). Next, we prove (6.18) by contradiction. Suppose (6.18) does not hold, i.e.,

$$\int_{\sigma(\Psi)\backslash\{0\}} \lambda^{-1} d|P_\lambda \psi|^2 = \lim_{k\to\infty} \int_{\sigma(\Psi)\cap[\frac{1}{k},k]} \lambda^{-1} d|P_\lambda \psi|^2 = \infty.$$

For any $k \geqslant 1$, let

$$\psi_k = \int_{\sigma(\Psi)\cap[\frac{1}{k},k]} dP_\lambda \psi.$$

Then $\psi_k = \Psi u_k$ with

$$u_k = \int_{\sigma(\Psi) \cap [\frac{1}{k}, k]} \lambda^{-1} dP_\lambda \psi,$$

and

$$|u_k|^2 = \int_{\sigma(\Psi) \cap [\frac{1}{k}, k]} \lambda^{-2} d|P_\lambda \psi|^2 \leqslant k^2 |\psi|^2 < \infty.$$

Thus, $\psi_k \in \mathcal{R}(\Psi)$ and $u_k = \Psi^\dagger \psi_k$. Further,

$$\langle \psi, u_k \rangle = \langle \psi, \Psi^\dagger \psi_k \rangle = \int_{\sigma(\Psi) \cap [\frac{1}{k}, k]} \lambda^{-1} d|P_\lambda \psi|^2$$

$$= |(\Psi^\dagger)^{\frac{1}{2}} \psi_k|^2 \to \infty, \qquad k \to \infty.$$

Consequently,

$$J(-u_k) = \langle \Psi u_k, u_k \rangle - 2 \langle \psi, u_k \rangle = -|(\Psi^\dagger)^{\frac{1}{2}} \psi_k|^2 \to -\infty, \quad k \to \infty,$$

contradicting (6.17).

Conversely, if (6.15) and (6.18) hold, then for any $u \in \mathcal{H}$, one has

$$\begin{aligned}
J(u) &= |\Psi^{\frac{1}{2}} u|^2 + 2 \langle (\Psi^\dagger)^{\frac{1}{2}} \psi, \Psi^{\frac{1}{2}} u \rangle \\
&= |\Psi^{\frac{1}{2}} u + (\Psi^\dagger)^{\frac{1}{2}} \psi|^2 - |(\Psi^\dagger)^{\frac{1}{2}} \psi|^2 \geqslant -|(\Psi^\dagger)^{\frac{1}{2}} \psi|^2 > -\infty.
\end{aligned} \tag{6.20}$$

Hence, sufficiency follows.

Finally, from the fact that

$$\mathcal{R}((\Psi^\dagger)^{\frac{1}{2}}) \subseteq \overline{\mathcal{R}(\Psi^{\frac{1}{2}})} = \overline{\mathcal{R}(\Psi)},$$

we can always find a sequence $u_k \in \mathcal{H}$ so that (note (6.20))

$$J(u_k) = |\Psi^{\frac{1}{2}} u_k + (\Psi^\dagger)^{\frac{1}{2}} \psi|^2 - |(\Psi^\dagger)^{\frac{1}{2}} \psi|^2 \to -|(\Psi^\dagger)^{\frac{1}{2}} \psi|^2, \quad k \to \infty.$$

Thus, (6.19) follows.

(ii) By Lemma 6.1.3, we know that (6.11) holds if and only if (6.9) holds for some $\hat{u} \in \mathcal{H}$. Then (6.15) and (6.11) hold if and only if $\hat{u}$ is a minimum.

(iii) We need only to show that (6.17) implies that $J(\cdot)$ admits a minimum. Since $\mathcal{R}(\Psi)$ is closed, we have

$$\mathcal{H}_1 = \mathcal{R}(\Psi).$$

We claim that in the current case, $\psi \in \mathcal{R}(\Psi)$. If this is not the case, then

$$\psi \notin \mathcal{R}(\Psi) = \mathcal{N}(\Psi)^\perp.$$

Hence, there exists a $u \in \mathcal{N}(\Psi)$ such that $\langle \psi, u \rangle < 0$. Consequently,

$$J(\lambda u) = 2\lambda \langle \psi, u \rangle \to -\infty, \qquad \lambda \to \infty,$$

contradicting to (6.17). Then (ii) applies. $\qquad\blacksquare$

The above result tells us that, in general, the existence of minimum is strictly stronger than the finiteness of the infimum of the functional $J(\cdot)$, which have been described by conditions (6.11) and (6.18), respectively. Note here that $\mathcal{R}(\Psi) \subseteq \mathcal{R}(\Psi^{\frac{1}{2}})$ when (6.15) holds. Whereas, in the case $\mathcal{R}(\Psi)$ is closed, the finiteness of $J(\cdot)$ and the existence of a minimum of $J(\cdot)$ are equivalent.

The following example shows the necessity of condition (6.18) in a concrete way.

Example 6.1.5. Let $\{\varphi_i, i \geqslant 1\}$ be an orthonomal basis of $\mathcal{H}$. For any $u \in \mathcal{H}$, define Ψu by the following:

$$\Psi u = \sum_{i=1}^{\infty} \beta^{i-1} \langle u, \varphi_i \rangle \varphi_i,$$

where $\beta \in (0,1)$. Then $\Psi : \mathcal{H} \to \mathcal{H}$ is bounded, self-adjoint, and positive definite (but not uniformly). Clearly, for any $\alpha \in \mathbb{R}$,

$$\Psi^{\alpha} u = \sum_{i=1}^{\infty} \beta^{\alpha(i-1)} \langle u, \varphi_i \rangle \varphi_i, \qquad \forall u \in \mathcal{H}.$$

Let

$$\psi = \sum_{i=1}^{\infty} i^{-1} \varphi_i, \qquad u_k = \sum_{i=1}^{k} \frac{1}{i\beta^{i-1}} \varphi_i.$$

Then

$$\Psi u_k = \sum_{i=1}^{k} \frac{1}{i} \varphi_i \to \psi, \qquad \text{in } \mathcal{H}, \text{ as } k \to \infty,$$

which means that $\psi \in \overline{\mathcal{R}(\Psi)}$. On the other hand, we claim that $\psi \notin \mathcal{R}(\Psi^{\frac{1}{2}})$. In fact, if for some $u \in \mathcal{H}$,

$$\sum_{i=1}^{\infty} i^{-1} \varphi_i = \psi = \Psi^{\frac{1}{2}} u = \sum_{i=1}^{\infty} \beta^{\frac{i-1}{2}} \langle u, \varphi_i \rangle \varphi_i,$$

then it is necessary that

$$\langle u, \varphi_i \rangle = \frac{1}{i\beta^{\frac{i-1}{2}}}, \qquad i \geqslant 1,$$

which implies (noting $\beta \in (0,1)$)

$$|u|^2 = \sum_{i=1}^{\infty} |\langle u, \varphi_i \rangle|^2 = \sum_{i=1}^{\infty} \frac{1}{i^2 \beta^{i-1}} = \infty,$$

a contradiction.

Now, consider the quadratic functional

$$J(u) = \langle \Psi u, u \rangle + 2 \langle \psi, u \rangle = \sum_{i=1}^{\infty} \left(\beta^{i-1} \langle u, \varphi_i \rangle^2 + \frac{2 \langle u, \varphi_i \rangle}{i} \right).$$

Then by letting u_k as above, we see that

$$J(-u_k) = \langle \Psi u_k, u_k \rangle - 2 \langle \psi, u_k \rangle$$

$$= \sum_{i=1}^{k} \left[\beta^{i-1} \frac{1}{i^2 \beta^{2(i-1)}} - \frac{2}{i^2 \beta^{i-1}} \right]$$

$$= - \sum_{i=1}^{k} \frac{1}{i^2 \beta^{(i-1)}} \rightarrow -\infty, \quad \text{as} \quad k \rightarrow \infty.$$

This means that

$$\inf_{u \in \mathcal{H}} J(u) = -\infty.$$

An interesting point here is that positive semi-definiteness of Ψ is not enough to ensure the finiteness of the infimum of $J(\cdot)$.

The following corollary is very interesting.

Corollary 6.1.6. *Let* (LQ1)–(LQ2) *and* (LQ3)′ *hold. Suppose Problem* (LQ) *is finite at* $(t,x) \in [0,T) \times \mathbb{R}^n$. *Then Problem* (LQ) *admits an open-loop optimal control* $\bar{u}(\cdot)$ *which is given by*

$$\bar{u}(\cdot) = \Psi^\dagger \psi + (I - \Psi^\dagger \Psi) v(\cdot), \qquad v(\cdot) \in \mathcal{U}[t,T], \tag{6.21}$$

and the minimum value of the cost functional is given by

$$J(t,x;\bar{u}(\cdot)) = \frac{1}{2} \left\{ \psi_0 - \| \Psi^{\frac{1}{2}} \bar{u}(\cdot) \|^2 \right\}. \tag{6.22}$$

Proof. Recall that $\Psi : \mathcal{U}[t,T] \rightarrow \mathcal{U}[t,T]$ is defined by the following:

$$[\Psi u(\cdot)](s) = R(r)u(s) + \int_t^T K(s,\tau)u(\tau)d\tau, \qquad s \in [t,T],$$

and

$$J(t,x;u(\cdot)) = \frac{1}{2} \left\{ \langle \Psi u(\cdot), u(\cdot) \rangle + 2 \langle \psi, u(\cdot) \rangle + \psi_0 \right\},$$

with t suppressed in $\psi(t)$ and $\psi_0(t)$ since t is fixed. Note that Ψ is self-adjoint. Also, under our conditions (especially (LQ3)$'$), $R^{-1}\Psi$ is a Fredholm operator on the Hilbert space $\mathcal{U}[t, T]$. Therefore, the range $\mathcal{R}(R^{-1}\Psi)$ is closed, so is $\mathcal{R}(\Psi)$. Then part (iii) of Theorem 6.1.4 applies. Note that open-loop optimal control $\bar{u}(\cdot) \in \mathcal{U}[t, T]$ is characterized by

$$\psi = \Psi \bar{u}(\cdot),$$

which leads to the representation (6.21) of open-loop optimal control(s). Further, one can easily obtain (6.22). $\qquad\square$

We note that (LQ3)$'$ plays an essential role in the above. From Example 6.1.2 (ii), we realize that when $R(\cdot)$ is degenerate, the above conclusion might fail in general.

6.1.3   *A two-point boundary value problem*

We now take a different approach for Problem (LQ). The following result is a maximum principle for open-loop optimal control of Problem (LQ).

Theorem 6.1.7. *Let* (LQ1)–(LQ2) *hold. Then for given initial pair* $(t, x) \in [0, T) \times \mathbb{R}^n$, *Problem* (LQ) *admits an open-loop optimal pair* $(\bar{X}(\cdot), \bar{u}(\cdot))$ *if and only if the following two-point boundary value problem*

$$\begin{cases} \dot{\bar{X}}(s) = A(s)\bar{X}(s) + B(s)\bar{u}(s) + f(s), & s \in [t, T], \\ \dot{\bar{Y}}(s) = -A(s)^T \bar{Y}(s) - Q(s)\bar{X}(s) - S(s)^T\bar{u}(s) - q(s), & s \in [t, T], \quad (6.23) \\ \bar{X}(t) = x, \qquad \bar{Y}(T) = G\bar{X}(T) + g, \end{cases}$$

admits a solution $(\bar{X}(\cdot), \bar{Y}(\cdot))$ *satisfying the following stationarity condition:*

$$B(s)^T \bar{Y}(s) + S(s)\bar{X}(s) + R(s)\bar{u}(s) + \rho(s) = 0, \qquad s \in [t, T], \quad (6.24)$$

and the following convexity condition holds:

$$u(\cdot) \mapsto J(t, 0; u(\cdot)) \text{ is convex on } \mathcal{U}[t, T],$$

or equivalently,

$$\int_t^T \left\langle \begin{pmatrix} Q(s) & S(s)^T \\ S(s) & R(s) \end{pmatrix} \begin{pmatrix} X_0(s) \\ u(s) \end{pmatrix}, \begin{pmatrix} X_0(s) \\ u(s) \end{pmatrix} \right\rangle ds + \langle GX_0(T), X_0(T) \rangle \geqslant 0, \quad (6.25)$$

where $X_0(\cdot)$ *is the solution to the following:*

$$\begin{cases} \dot{X}_0(s) = A(s)X_0(s) + B(s)u(s), & s \in [t, T], \\ X_0(t) = 0. \end{cases} \quad (6.26)$$

Further, Problem (LQ) admits an open-loop optimal control for some initial pair $(t, x) \in [0, T) \times \mathbb{R}^n$ implies that

$$R(s) \geqslant 0, \qquad s \in [t, T]. \tag{6.27}$$

Proof. Suppose $(\bar{X}(\cdot), \bar{u}(\cdot))$ is a state-control pair for initial pair $(t, x) \in [0, T) \times \mathbb{R}^n$. For any $u(\cdot) \in \mathcal{U}[t, T]$, let $X^\varepsilon(\cdot) = X(\cdot\,; t, x, \bar{u}(\cdot) + \varepsilon u(\cdot))$. Then

$$\begin{cases} \dot{X}^\varepsilon(s) = A(s)X^\varepsilon(s) + B(s)\big[\bar{u}(s) + \varepsilon u(s)\big] + f(s), \qquad s \in [t, T], \\ X^\varepsilon(t) = x. \end{cases}$$

Thus, $X_0(\cdot) \equiv \frac{X^\varepsilon(\cdot) - \bar{X}(\cdot)}{\varepsilon}$ is independent of ε and satisfies (6.26). Further, by the Taylor expansion of the map $\varepsilon \mapsto J(t, x; \bar{u}(\cdot) + \varepsilon u(\cdot))$,

$$\begin{aligned} &J(t, x; \bar{u}(\cdot) + \varepsilon u(\cdot)) - J(t, x; \bar{u}(\cdot)) \\ &= \varepsilon\Big\{ \int_t^T \Big[\langle Q\bar{X} + S^T\bar{u} + q, X_0 \rangle + \langle S\bar{X} + R\bar{u} + \rho, u \rangle \Big] ds \\ &\qquad + \langle G\bar{X}(T) + g, X_0(T) \rangle \Big\} \\ &\quad + \frac{\varepsilon^2}{2}\Big\{ \int_t^T \Big[\langle Q(s)X_0(s), X_0(s) \rangle + 2 \langle S(s)X_0(s), u(s) \rangle \\ &\qquad + \langle R(s)u(s), u(s) \rangle \Big] ds + \langle GX_0(T), X_0(T) \rangle \Big\}. \end{aligned}$$

Let $\bar{Y}(\cdot)$ be the solution to the equation for $\bar{Y}(\cdot)$ in (6.23). Then one obtains

$$\begin{aligned} &\int_t^T \Big[\langle Q\bar{X} + S^T\bar{u} + q, X_0 \rangle + \langle S\bar{X} + R\bar{u} + \rho, u \rangle \Big] ds \\ &\qquad + \langle G\bar{X}(T) + g, X_0(T) \rangle \\ &= \int_t^T \Big[\langle Q\bar{X} + S^T\bar{u} + q, X_0 \rangle + \langle S\bar{X} + R\bar{u} + \rho, u \rangle \\ &\qquad \langle -A^T\bar{Y} - Q\bar{X} - S^T\bar{u} - q, X_0 \rangle + \langle \bar{Y}, AX_0 + Bu \rangle \Big] ds \\ &= \int_t^T \langle B^T\bar{Y} + S\bar{X} + R\bar{u} + \rho, u \rangle \, ds. \end{aligned}$$

Hence,

$$\begin{aligned} &J(t, x; \bar{u}(\cdot) + \varepsilon u(\cdot)) - J(t, x; \bar{u}(\cdot)) \\ &= \varepsilon\Big\{ \int_t^T \langle B^T\bar{Y} + S\bar{X} + R\bar{u} + \rho, u \rangle \, ds \Big\} \\ &\quad + \frac{\varepsilon^2}{2}\Big\{ \int_t^T \Big[\langle Q(s)X_0(s), X_0(s) \rangle + 2 \langle S(s)X_0(s), u(s) \rangle \\ &\qquad + \langle R(s)u(s), u(s) \rangle \Big] ds + \langle GX_0(T), X_0(T) \rangle \Big\}. \end{aligned} \tag{6.28}$$

From the above, we see that $(\bar{X}(\cdot), \bar{u}(\cdot))$ is an open-loop optimal pair if and only if (6.24) and (6.25) hold.

Finally, we claim that (6.25) implies (6.27). In fact, if (6.27) fails, then we can take a Lesbegue point $\bar{t} \in [t, T]$ such that for some $u_0 \in \mathbb{R}^m$,

$$\langle R(\bar{t})u_0, u_0 \rangle < 0.$$

Now, for any $\varepsilon > 0$, let

$$u^\varepsilon(s) = u_0 I_{[\bar{t}, \bar{t}+\varepsilon)}(s), \qquad s \in [t, T],$$

and denote the corresponding solution to (6.26) by $X_0^\varepsilon(\cdot)$. Then

$$X_0^\varepsilon(s) = \int_t^s \Phi(s, r)B(r)u^\varepsilon(r)dr = \int_{\bar{t}}^{(\bar{t}+\varepsilon)\wedge s} \Phi(s, r)B(r)u_0 dr,$$

where $\Phi(\cdot, \cdot)$ is the fundamental matrix of the map $A(\cdot)$. Then

$$|X_0^\varepsilon(s)| \leqslant \int_{\bar{t}}^{\bar{t}+\varepsilon} |\Phi(s, r)B(r)u_0| dr \leqslant \varepsilon K, \qquad \forall s \in [t, T],$$

for some constant $K > 0$. Consequently,

$$\int_t^T \langle \begin{pmatrix} Q(s) & S(s)^T \\ S(s) & R(s) \end{pmatrix} \begin{pmatrix} X_0^\varepsilon(s) \\ u^\varepsilon(s) \end{pmatrix}, \begin{pmatrix} X_0^\varepsilon(s) \\ u^\varepsilon(s) \end{pmatrix} \rangle ds + \langle GX_0^\varepsilon(T), X_0^\varepsilon(T) \rangle$$

$$= \int_{\bar{t}}^{\bar{t}+\varepsilon} \langle R(s)u_0, u_0 \rangle ds + 2 \int_{\bar{t}}^{\bar{t}+\varepsilon} \langle S(s)X_0^\varepsilon(s), u_0 \rangle ds$$

$$+ \int_{\bar{t}}^T \langle Q(s)X_0^\varepsilon(s), X_0^\varepsilon(s) \rangle ds + \langle GX_0^\varepsilon(T), X_0^\varepsilon(T) \rangle$$

$$\leqslant \varepsilon \Big(\langle Ru_0, u_0 \rangle + o(1) \Big) + \varepsilon^2 K < 0,$$

provided $\varepsilon > 0$ is small enough. This contradicts (6.25). Hence, (6.27) holds. $\qquad\square$

In the above, the equation for $\bar{Y}(\cdot)$ in (6.23) is the *adjoint equation*, and the stationarity condition (6.24) is a consequence of form of the maximum condition in the Pontryagin maximum principle.

The following result is concerned with the uniqueness of open-loop optimal controls.

Theorem 6.1.8. *Let* (LQ1)–(LQ2) *hold and let* $(t, x) \in [0, T] \times \mathbb{R}^n$ *be given. Then the following are equivalent:*

(i) *Problem* (LQ) *admits a unique open-loop optimal control* $\bar{u}(\cdot)$ *for* (t, x);

(ii) *The map* $u(\cdot) \mapsto J(t, 0; u(\cdot))$ *is strictly convex, and the two-point boundary value problem* (6.23)–(6.24) *admits a unique solution* $(\bar{X}(\cdot), \bar{Y}(\cdot), \bar{u}(\cdot))$.

Proof. (i) $\Rightarrow$ (ii). Let $\bar{u}(\cdot) \in \mathcal{U}[t, T]$ be the unique open-loop optimal control of Problem (LQ) for the given initial pair (t, x). Then by Theorem 6.1.7, the two-point boundary value problem (6.23)–(6.24) admits a solution $(\bar{X}(\cdot), \bar{Y}(\cdot), \bar{u}(\cdot))$. By the uniqueness of $\bar{u}(\cdot)$, for any $u(\cdot) \in \mathcal{U}[t, T] \setminus \{0\}$, the function $\varepsilon \mapsto J(t, x; \bar{u}(\cdot) + \varepsilon u(\cdot))$ must be strictly convex. Thus, by (6.28), we see that a strict inequality must hold in (6.25), which is equivalent to the strict convexity of $u(\cdot) \mapsto J(t, 0; u(\cdot))$. We further claim that the solution to (6.23)–(6.24) is unique. In fact, if $(\bar{X}_i(\cdot), \bar{Y}_i(\cdot), \bar{u}_i(\cdot))$ are two different solutions to (6.23)–(6.24), $i = 1, 2$, then by (6.28), both $\bar{u}_1(\cdot)$ and $\bar{u}_2(\cdot)$ are open-loop optimal controls. Now, we let $u(\cdot) = \bar{u}_1(\cdot) - \bar{u}_2(\cdot) \neq 0$, and let $X_0(\cdot)$ be the solution to (6.26) corresponding to this $u(\cdot)$. Then similar to the derivation of (6.28), we have

$$
\begin{aligned}
0 &= J(t, x; \bar{u}_1(\cdot)) - J(t, x; \bar{u}_2(\cdot)) \\
&= \frac{1}{2} \Big\{ \int_t^T \big[\langle Q(s)X_0(s), X_0(s) \rangle + 2 \langle S(s)X_0(s), u(s) \rangle \\
&\qquad + \langle R(s)u(s), u(s) \rangle \big] ds + \langle G X_0(T), X_0(T) \rangle \Big\} > 0,
\end{aligned}
$$

which is a contradiction.

(ii) $\Rightarrow$ (i). Let $(\bar{X}(\cdot), \bar{Y}(\cdot), \bar{u}(\cdot))$ be the unique solution to (6.23)–(6.24) and a strict inequality holds in (6.25). Then by Theorem 6.1.7, $(\bar{X}(\cdot), \bar{u}(\cdot))$ is an open-loop optimal control of Problem (LQ). If $(\widetilde{X}(\cdot), \widetilde{u}(\cdot))$ is another open-loop optimal pair of Problem (LQ), then for some $\widetilde{Y}(\cdot)$, $(\widetilde{X}(\cdot), \widetilde{Y}(\cdot), \widetilde{u}(\cdot))$ is a solution to (6.23)–(6.24). By the uniqueness, it is necessary that

$$
(\widetilde{X}(\cdot), \widetilde{u}(\cdot)) = (\bar{X}(\cdot), \bar{u}(\cdot)).
$$

This completes the proof. $\qquad\qquad\qquad\qquad\qquad\qquad\qquad\qquad\square$

Note that in the above, (LQ3) is not assumed. As a matter of fact, from Example 6.1.2 (iii), we see that (LQ3) is not necessary for the existence of open-loop optimal controls.

6.1.4   Closed-loop strategies

Next, for any $(t, x) \in [0, T) \times \mathbb{R}^n$, $\Theta(\cdot) \in L^2(t, T; \mathbb{R}^{m \times n}) \equiv \mathcal{Q}[t, T]$, and $v(\cdot) \in \mathcal{U}[t, T]$, we consider the following equation:

$$\begin{cases} \dot{X}(s) = [A(s) + B(s)\Theta(s)]X(s) + B(s)v(s) + f(s), & s \in [t, T], \\ X(t) = x, \end{cases}$$

which admits a unique solution $X(\cdot) \equiv X(\cdot\,; t, x, \Theta(\cdot), v(\cdot))$, depending on $\Theta(\cdot)$ and $v(\cdot)$, besides the initial pair (t, x). The above is called a *closed-loop system* of the original state equation (6.1) under *closed-loop strategy* $(\Theta(\cdot), v(\cdot))$. Note that $(\Theta(\cdot), v(\cdot))$ is independent of the initial state x. We now introduce the following definition.

Definition 6.1.9.   A pair $(\bar{\Theta}(\cdot), \bar{v}(\cdot)) \in \mathcal{Q}[t, T] \times \mathcal{U}[t, T]$ is called a *closed-loop optimal strategy* of Problem (LQ) on $[t, T]$ if

$$J(t, x; \bar{\Theta}(\cdot)\bar{X}(\cdot) + \bar{v}(\cdot)) \leqslant J(t, x; \Theta(\cdot)X(\cdot) + v(\cdot)), \tag{6.29}$$
$$\forall x \in \mathbb{R}^n, \ (\Theta(\cdot), v(\cdot)) \in \mathcal{Q}[t, T] \times \mathcal{U}[t, T],$$

where $\bar{X}(\cdot) = X(\cdot\,; t, x, \bar{\Theta}(\cdot), \bar{v}(\cdot))$, and $X(\cdot) = X(\cdot\,; t, x, \Theta(\cdot), v(\cdot))$.

We point out that in the above, both $\bar{\Theta}(\cdot)$ and $\bar{v}(\cdot)$ are required to be independent of the initial state $x \in \mathbb{R}^n$.

The following result is referred to as the *Berkovitz's Equivalence Lemma*.

Lemma 6.1.10.   *The following are equivalent:*

(i) $(\bar{\Theta}(\cdot), \bar{v}(\cdot)) \in \mathcal{Q}[t, T] \times \mathcal{U}[t, T]$ *is a closed-loop optimal strategy of Problem* (LQ) *over* $[t, T]$.

(ii) *For any* $v(\cdot) \in \mathcal{U}[t, T]$,

$$J(t, x; \bar{\Theta}(\cdot)\bar{X}(\cdot) + \bar{v}(\cdot)) \leqslant J(t, x; \bar{\Theta}(\cdot)X(\cdot) + v(\cdot)), \tag{6.30}$$

where $\bar{X}(\cdot) \equiv X(\cdot\,; t, x, \bar{\Theta}(\cdot), \bar{v}(\cdot))$ *and* $X(\cdot) \equiv X(\cdot\,; t, x, \bar{\Theta}(\cdot), v(\cdot))$.

(iii) *For any* $u(\cdot) \in \mathcal{U}[t, T]$,

$$J(t, x; \bar{\Theta}(\cdot)\bar{X}(\cdot) + \bar{v}(\cdot)) \leqslant J(t, x; u(\cdot)), \tag{6.31}$$

where $\bar{X}(\cdot) \equiv X(\cdot\,; t, x, \bar{\Theta}(\cdot), \bar{v}(\cdot))$.

Proof. (i) $\Rightarrow$ (ii). It is trivial by taking $\Theta(\cdot) = \bar{\Theta}(\cdot)$.

(ii) $\Rightarrow$ (iii). For any $(t, x) \in [0, T) \times \mathbb{R}^n$ and $u(\cdot) \in \mathcal{U}[t, T]$, let $X(\cdot) = X(\cdot\,; t, x, u(\cdot))$. Take

$$v(\cdot) = u(\cdot) - \bar{\Theta}(\cdot)X(\cdot).$$

Then

$$X(\cdot\,;t,x,u(\cdot)) = X(\cdot\,;t,x,\bar{\Theta}(\cdot)X(\cdot),v(\cdot)), \quad u(\cdot) = \bar{\Theta}(\cdot)X(\cdot) + v(\cdot).$$

Therefore, (6.31) follows from (6.30).

(iii) $\Rightarrow$ (i). For any $(\Theta(\cdot),v(\cdot)) \in \mathcal{Q}[t,T] \times \mathcal{U}[t,T]$, let

$$X(\cdot) = X(\cdot\,;t,x,\Theta(\cdot),v(\cdot)), \qquad u(\cdot) = \Theta(\cdot)X(\cdot) + v(\cdot).$$

Then by the uniqueness of the solutions to (6.1), we see that

$$X(\cdot\,;t,x,u(\cdot)) = X(\cdot\,;t,x,\Theta(\cdot)X(\cdot) + v(\cdot)).$$

Hence, (6.29) follows from (6.31). $\qquad\qquad\qquad\qquad\qquad\qquad\square$

For any fixed initial pair $(t,x) \in [0,T) \times \mathbb{R}^n$, (6.31) implies that the outcome

$$\bar{u}(\cdot) \equiv \bar{\Theta}(\cdot)\bar{X}(\cdot) + \bar{v}(\cdot)$$

of the closed-loop optimal strategy $(\bar{\Theta}(\cdot),\bar{v}(\cdot))$ is an open-loop optimal control of Problem (LQ) for (t,x). Thus, for Problem (LQ), the existence of a closed-loop optimal strategy on $[t,T]$ implies the existence of open-loop optimal control for any (t,x) with $x \in \mathbb{R}^n$.

The following gives a characterization of closed-loop optimal strategy for Problem (LQ).

Theorem 6.1.11. *Let* (LQ1)–(LQ2) *and* (6.27) *hold. Then Problem* (LQ) *admits a closed-loop optimal strategy over* $[t,T]$ *if and only if the following Riccati equation admits a solution* $P(\cdot)$:

$$\begin{cases} \dot{P}(s) + P(s)A(s) + A(s)^T P(s) + Q(s) \\ \qquad - [P(s)B(s) + S(s)^T]R(s)^\dagger[B(s)^T P(s) + S(s)] = 0, \\ \qquad\qquad\qquad\qquad\qquad s \in [t,T], \\ \mathcal{R}(B(s)^T P(s) + S(s)) \subseteq \mathcal{R}(R(s)), \qquad s \in [t,T], \\ P(T) = G, \end{cases} \tag{6.32}$$

such that

$$R(\cdot)^\dagger[B(\cdot)^T P(\cdot) + S(\cdot)] \in L^2(t,T;\mathbb{R}^{n \times n}). \tag{6.33}$$

In this case, any closed-loop optimal strategy $(\bar{\Theta}(\cdot),\bar{v}(\cdot))$ *on* $[t,T]$ *admits the following representation:*

$$\begin{cases} \bar{\Theta}(s) = -R(s)^\dagger[B(s)^T P(s) + S(s)] + [I - R(s)^\dagger R(s)]\theta(s), \\ \bar{v}(s) = -R(s)^\dagger[B(s)^T \eta(s) + \rho(s)] + [I - R(s)^\dagger R(s)]\nu(s), \tag{6.34} \\ \qquad\qquad\qquad\qquad\qquad s \in [t,T], \end{cases}$$

for some $\theta(\cdot)$ and $\nu(\cdot)$, with $\eta(\cdot)$ being the solution to the following terminal value problem:

$$
\begin{cases}
\dot{\eta}(s) = \big\{ -A(s)^T + \big[P(s)B(s) + S(s)^T\big]R(s)^\dagger B(s)^T\big\}\eta(s) \\
\qquad\qquad -P(s)f(s) - q(s) + \big[P(s)B(s) + S(s)^T\big]R(s)^\dagger \rho(s), \\
\qquad\qquad\qquad\qquad\qquad\qquad s \in [t,T], \\
B(s)^T\eta(s) + \rho(s) \in \mathcal{R}\big(R(s)\big), \qquad s \in [t,T], \\
\eta(T) = g,
\end{cases}
\tag{6.35}
$$

and

$$
\begin{aligned}
V(t,x) &\equiv \inf_{u(\cdot)\in\mathcal{U}[t,T]} J(t,x;u(\cdot)) \\
&= \frac{1}{2}\bigg\{ \langle P(t)x, x \rangle + \langle \eta(t), x \rangle + \int_t^T \big\{ 2\langle \eta(s), f(s) \rangle \\
&\qquad - \langle R(s)^\dagger\big[B(s)^T\eta(s) + \rho(s)\big], B(s)^T\eta(s) + \rho(s) \rangle \big\}ds \bigg\}.
\end{aligned}
\tag{6.36}
$$

Moreover, in the above case, the closed-loop optimal strategy is unique if and only if (LQ3) holds. When this is the case, the unique optimal closed-loop strategy is represented by

$$
\begin{cases}
\bar{\Theta}(s) = -R(s)^\dagger\big[B(s)^T P(s) + S(s)\big], \\
\bar{v}(s) = -R(s)^\dagger\big[B(s)^T\eta(s) + \rho(s)\big],
\end{cases}
\qquad s \in [t,T].
\tag{6.37}
$$

Proof. Necessity. Let $(\bar{\Theta}(\cdot), \bar{v}(\cdot)) \in \mathcal{Q}[t,T] \times \mathcal{U}[t,T]$ be a closed-loop optimal strategy of Problem (LQ) on $[t,T]$. Then, as we have remarked earlier, for any $x \in \mathbb{R}^n$, the outcome $\bar{\Theta}(\cdot)\bar{X}(\cdot) + \bar{v}(\cdot)$ of the closed-loop strategy $(\bar{\Theta}(\cdot), \bar{v}(\cdot))$ is an open-loop optimal control of Problem (LQ) for the initial pair (t,x). Hence, by Theorem 6.1.7, one has

$$
\begin{cases}
\dot{\bar{X}}(s) = \big[A(s) + B(s)\bar{\Theta}(s)\big]\bar{X}(s) + B(s)\bar{v}(s) + f(s), \quad s \in [t,T], \\
\dot{\bar{Y}}(s) = -A(s)^T\bar{Y}(s) - \big[Q(s) + S(s)^T\bar{\Theta}(s)\big]\bar{X}(s) \\
\qquad\qquad -S(s)^T\bar{v}(s) - q(s), \quad s \in [t,T], \\
\bar{X}(t) = x, \qquad \bar{Y}(T) = G\bar{X}(T) + g, \\
B(s)^T\bar{Y}(s) + \big[S(s) + R(s)\bar{\Theta}(s)\big]\bar{X}(s) + R(s)\bar{v}(s) + \rho(s) = 0, \\
\qquad\qquad\qquad\qquad\qquad s \in [t,T].
\end{cases}
\tag{6.38}
$$

Since the above admits a solution for each $x \in \mathbb{R}^n$, and $(\bar{\Theta}(\cdot), \bar{v}(\cdot))$ is independent of x, by subtracting solutions corresponding to x and 0, the latter from the former, we see that for any $x \in \mathbb{R}^n$, as long as $(X(\cdot), Y(\cdot))$ is the solution to the following decoupled two-point boundary value problem:

$$\begin{cases} \dot{X}(s) = [A(s) + B(s)\bar{\Theta}(s)]X(s), & s \in [t, T], \\ \dot{Y}(s) = -A(s)^T Y(s) - [Q(s) + S(s)^T\bar{\Theta}(s)]X(s), & s \in [t, T], \\ X(t) = x, \quad Y(T) = GX(T), \end{cases} \tag{6.39}$$

one must have the following stationarity condition:

$$B(s)^T Y(s) + [S(s) + R(s)\bar{\Theta}(s)]X(s) = 0, \quad s \in [t, T]. \tag{6.40}$$

Now, we let $(\mathbb{X}(\cdot), \mathbb{Y}(\cdot))$ solve the following decoupled two-point boundary value problem:

$$\begin{cases} \dot{\mathbb{X}}(s) = [A(s) + B(s)\bar{\Theta}(s)]\mathbb{X}(s), & s \in [t, T], \\ \dot{\mathbb{Y}}(s) = -A(s)^T\mathbb{Y}(s) - [Q(s) + S(s)^T\bar{\Theta}(s)]\mathbb{X}(s), & s \in [t, T], \\ \mathbb{X}(t) = I, \quad \mathbb{Y}(T) = G\mathbb{X}(T). \end{cases} \tag{6.41}$$

Then for any $(t, x) \in [0, T) \times \mathbb{R}^n$, the unique solution $(X(\cdot), Y(\cdot))$ of (6.39) can be represented by the following linear equation:

$$X(s) = \mathbb{X}(s)x, \quad Y(s) = \mathbb{Y}(s)x, \quad s \in [t, T].$$

Consequently, (6.40) implies

$$B(s)^T\mathbb{Y}(s) + [S(s) + R(s)\bar{\Theta}(s)]\mathbb{X}(s) = 0. \tag{6.42}$$

Clearly, $\mathbb{X}(\cdot)^{-1}$ exists and satisfies the following:

$$\begin{cases} \dfrac{d}{ds}\mathbb{X}(s)^{-1} = -\mathbb{X}(s)^{-1}[A(s) + B(s)\bar{\Theta}(s)], & s \in [t, T], \\ \mathbb{X}(t)^{-1} = I. \end{cases}$$

We now define

$$P(\cdot) = \mathbb{Y}(\cdot)\mathbb{X}(\cdot)^{-1}.$$

Then, suppressing s, we have

$$\begin{aligned} \dot{P} &= -A^T P - (Q + S^T\bar{\Theta}) - P(A + B\bar{\Theta}) \\ &= -PA - A^T P - Q - (PB + S^T)\bar{\Theta}, \quad s \in [t, T]. \end{aligned} \tag{6.43}$$

On the other hand, by (6.40), we have

$$B^T P + S + R\bar{\Theta} = 0, \quad s \in [t, T]. \tag{6.44}$$

Thus, the inclusion condition in (6.32) holds. To show (6.33), we applying $R^\dagger$ to (6.44) to achieve the following:

$$R^\dagger(B^T P + S) = -R^\dagger R\bar{\Theta}.$$

Since $R^\dagger R$ is an orthogonal projection, one has
$$\|R^\dagger(B^T P + S)\| = \|R^\dagger R\bar\Theta\| \le \|\bar\Theta\| \in L^2(t, T),$$
which gives (6.33). Further, (6.44) yields
$$\bar\Theta(s) = -R(s)^\dagger[B(s)^T P(s) + S(s)] + [I - R(s)^\dagger R(s)]\theta(s),$$
for some $\theta(\cdot)$, which is the first relation in (6.34). Then, making use of (6.44), one has
$$\begin{aligned}(PB + S^T)\bar\Theta &= -(PB + S^T)R^\dagger(B^T P + S) - \bar\Theta^T R(I - R^\dagger R)\theta \\ &= -(PB + S^T)R^\dagger(B^T P + S).\end{aligned} \tag{6.45}$$
Plug the above into (6.43), we see that $P(\cdot)$ satisfies Riccati equation (6.32). To determine $\bar v(\cdot)$, we define
$$\eta(s) = \bar Y(s) - P(s)\bar X(s), \qquad s \in [t, T].$$
According to the last relation in (6.38) and (6.44), we have
$$\begin{aligned}0 &= B(s)^T \bar Y(s) + [S(s) + R(s)\bar\Theta(s)]\bar X(s) + R(s)\bar v(s) + \rho(s) \\ &= B(s)^T[P(s)\bar X(s) + \eta(s)] + [S(s) + R(s)\bar\Theta(s)]\bar X(s) + R(s)\bar v(s) + \rho(s) \\ &= [B(s)^T P(s) + S(s) + R(s)\bar\Theta(s)]\bar X(s) + B(s)^T\eta(s) + R(s)\bar v(s) + \rho(s) \\ &= B(s)^T\eta(s) + R(s)\bar v(s) + \rho(s).\end{aligned}$$
Hence,
$$B(s)^T\eta(s) + \rho(s) = -R(s)\bar v(s) \in \mathcal{R}(R(s)), \qquad s \in [t, T].$$
Then,
$$\bar v(s) = -R(s)^\dagger[B(s)^T\eta(s) + \rho(s)] + [I - R(s)^\dagger R(s)]\nu(s), \quad s \in [t, T],$$
for some $\nu(\cdot)$, which is the second relation in (6.34). Consequently, making use of (6.44) again, we obtain
$$\begin{aligned}(PB + S^T)\bar v &= -(PB + S^T)R^\dagger(B^T\eta + \rho) + \bar\Theta^T R(I - R^\dagger R)\nu \\ &= -(PB + S^T)R^\dagger(B^T\eta + \rho).\end{aligned}$$
Now, we calculate (note (6.45))
$$\begin{aligned}\dot\eta &= \dot{\bar Y} - \dot P\bar X - P\dot{\bar X} \\ &= -A^T(P\bar X + \eta) - (Q + S^T\bar\Theta)\bar X - S^T\bar v - q \\ &\quad + [PA + A^T P + Q - (PB + S^T)R^\dagger(B^T P + S)]\bar X \\ &\quad - P(A\bar X + B\bar\Theta\bar X + B\bar v + f) \\ &= -A^T\eta - (PB + S^T)\bar v - q \\ &\quad - [(PB + S^T)R^\dagger(B^T P + S) + (PB + S^T)\bar\Theta]\bar X - Pf \\ &= -A^T\eta + (PB + S^T)R^\dagger(B^T\eta + \rho) - q - Pf \\ &= -[A^T - (PB + S^T)R^\dagger B^T]\eta - Pf - q + (PB + S^T)R^\dagger\rho.\end{aligned}$$

Therefore, $\eta(\cdot)$ satisfies (6.35). This completely determines $\bar{v}(\cdot)$.

Sufficiency. Let $(\bar{\Theta}(\cdot), \bar{v}(\cdot))$ be defined by (6.34) for some $\theta(\cdot)$ and $\nu(\cdot)$, with $P(\cdot)$ and $\eta(\cdot)$ being the solutions to Riccati equation (6.32) and terminal value problem (6.35), respectively. We look at the following:

$$
\begin{aligned}
J(t, x; u(\cdot)) &= \frac{1}{2}\Big\{ \langle GX(T), X(T) \rangle + 2\langle g, X(T) \rangle + \int_t^T [\langle QX, X \rangle \\
&\quad + 2\langle SX, u \rangle + \langle Ru, u \rangle + 2\langle q, X \rangle + 2\langle \rho, u \rangle] ds \Big\} \\
&= \frac{1}{2}\langle P(t)x, x \rangle + \langle \eta(t), x \rangle + \frac{1}{2}\int_t^T [\langle \dot{P}X, X \rangle + \langle P\dot{X}, X \rangle + \langle PX, \dot{X} \rangle \\
&\quad + 2\langle \dot{\eta}, X \rangle + 2\langle \eta, \dot{X} \rangle + \langle QX, X \rangle \\
&\quad + 2\langle SX, u \rangle + \langle Ru, u \rangle + 2\langle q, X \rangle + 2\langle \rho, u \rangle] ds \\
&= \frac{1}{2}\langle P(t)x, x \rangle + \langle \eta(t), x \rangle \\
&\quad + \frac{1}{2}\int_t^T \big\{ \langle [-PA - A^T P - Q + (PB + S^T)R^{\dagger}(B^T P + S)]X, X \rangle \\
&\quad + \langle P(AX + Bu + f), X \rangle + \langle PX, AX + Bu + f \rangle \\
&\quad + 2\langle [-A^T + (PB + S^T)R^{\dagger}B^T]\eta - Pf - q + (PB + S^T)R^{\dagger}\rho, X \rangle \\
&\quad + 2\langle \eta, AX + Bu + f \rangle + \langle QX, X \rangle + 2\langle SX, u \rangle + \langle Ru, u \rangle \\
&\quad + 2\langle q, X \rangle + 2\langle \rho, u \rangle \big\} ds \\
&= \frac{1}{2}\langle P(t)x, x \rangle + \langle \eta(t), x \rangle + \frac{1}{2}\int_t^T \big\{ \langle (PB + S^T)R^{\dagger}(B^T P + S)X, X \rangle \\
&\quad + 2\langle \eta, f \rangle + 2\langle (PB + S^T)R^{\dagger}(B^T \eta + \rho), X \rangle \\
&\quad + 2\langle (B^T P + S)X + B^T \eta + \rho, u \rangle + \langle Ru, u \rangle \big\} ds \\
&= \frac{1}{2}\langle P(t)x, x \rangle + \langle \eta(t), x \rangle + \frac{1}{2}\int_t^T \big\{ \langle R^{\dagger}(B^T P + S)X, (B^T P + S)X \rangle \\
&\quad + 2\langle \eta, f \rangle + 2\langle R^{\dagger}(B^T \eta + \rho), (B^T P + S)X \rangle \\
&\quad + 2\langle (B^T P + S)X + B^T \eta + \rho, u \rangle + \langle Ru, u \rangle \big\} ds.
\end{aligned}
$$

Since

$$
\mathcal{R}(B^T P + S) \subseteq \mathcal{R}(R), \quad B^T \eta + \rho \in \mathcal{R}(R),
$$

we have the following:

$$
\begin{aligned}
R\bar{\Theta} &= -RR^{\dagger}(B^T P + S) = -(B^T P + S), \\
R\bar{v} &= -RR^{\dagger}(B^T \eta + \rho) = -(B^T \eta + \rho).
\end{aligned}
$$

Then

$$J(t,x;u(\cdot)) = \frac{1}{2}\langle P(t)x,x\rangle + \langle \eta(t),x\rangle$$

$$+\frac{1}{2}\int_t^T \big\{\langle R^\dagger(B^TP+S)X,(B^TP+S)X\rangle$$

$$+2\langle \eta,f\rangle +2\langle R^\dagger(B^T\eta+\rho),(B^TP+S)X\rangle$$

$$+2\langle (B^TP+S)X+B^T\eta+\rho,u\rangle + \langle Ru,u\rangle\big\}ds$$

$$= \frac{1}{2}\langle P(t)x,x\rangle + \langle \eta(t),x\rangle +\frac{1}{2}\int_t^T \big\{\langle R^\dagger R\bar\Theta X,R\bar\Theta X\rangle +2\langle \eta,f\rangle$$

$$+2\langle R^\dagger R\bar v,R\bar\Theta X\rangle -2\langle R(\bar\Theta X+\bar v),u\rangle + \langle Ru,u\rangle\big\}ds$$

$$= \frac{1}{2}\langle P(t)x,x\rangle + \langle \eta(t),x\rangle +\frac{1}{2}\int_t^T \big\{\langle R\bar\Theta X,\bar\Theta X\rangle +2\langle \eta,f\rangle$$

$$+2\langle R\bar v,\bar\Theta X\rangle -2\langle R(\bar\Theta X+\bar v),u\rangle + \langle Ru,u\rangle\big\}ds$$

$$= \frac{1}{2}\langle P(t)x,x\rangle + \langle \eta(t),x\rangle +\frac{1}{2}\int_t^T \big\{\langle R(u-\bar\Theta X-\bar v),(u-\bar\Theta X-\bar v)\rangle$$

$$+2\langle \eta,f\rangle - \langle R\bar v,\bar v\rangle\big\}ds$$

$$= \frac{1}{2}\langle P(t)x,x\rangle + \langle \eta(t),x\rangle +\frac{1}{2}\int_t^T \big\{\langle R(u-\bar\Theta X-\bar v),(u-\bar\Theta X-\bar v)\rangle$$

$$+2\langle \eta,f\rangle - \langle R^\dagger(B^T\eta+\rho),(B^T\eta+\rho)\rangle\big\}ds$$

$$\geqslant \frac{1}{2}\langle P(t)x,x\rangle + \langle \eta(t),x\rangle +\frac{1}{2}\int_t^T \big\{2\langle \eta(s),f(s)\rangle$$

$$- \langle R(s)^\dagger[B(s)^T\eta(s)+\rho(s)],B(s)^T\eta(s)+\rho(s)\rangle\big\}ds$$

$$= J(t,x;\bar\Theta(\cdot)\bar X(\cdot)+\bar v(\cdot)) = V(t,x).$$

This proves the sufficiency.

Finally, from (6.34), we see that the closed-loop optimal strategy is unique if and only if (LQ3) holds and then (6.37) holds. $\qquad\square$

From the above result, we see that if Problem (LQ) admits a closed-loop optimal strategy $(\bar\Theta(\cdot),\bar v(\cdot))$ on $[t,T]$, then for any $\tau \in [t,T]$, the restriction $(\bar\Theta(\cdot),\bar v(\cdot))\big|_{[\tau,T]}$ of $(\bar\Theta(\cdot),\bar v(\cdot))$ on $[\tau,T]$ is a closed-loop optimal strategy of Problem (LQ) on $[\tau,T]$. Therefore, if Problem (LQ) admits a closed-loop optimal strategy on $[0,T]$, so does it on any $[t,T]$. Such a property is usually referred to as the *time-consistency* of Problem (LQ).

We point out that in the above result, (LQ3) is not necessary for the equivalence between the existence of a closed-loop optimal strategy and the solvability of the corresponding Riccati equation. (LQ3) is only used to ensure the uniqueness of the closed-loop optimal strategy. We now present an example for which the above theorem applies.

Example 6.1.12. Consider the following controlled system:

$$\begin{cases} \dot{X}(s) = AX(s) + Bu(s), & s \in [t,T], \\ X(t) = x, \end{cases}$$

with

$$A = \begin{pmatrix} 1 & -1 \\ 1 & -1 \end{pmatrix}, \quad B = \begin{pmatrix} 1 & 1 \\ 1 & -1 \end{pmatrix}.$$

The cost functional is given by the following:

$$J(t,x;u(\cdot)) = \frac{1}{2}\Big\{ \langle GX(T), X(T) \rangle + \int_t^T \langle Ru(s), u(s) \rangle \, ds \Big\},$$

with

$$G = \begin{pmatrix} 1 & -1 \\ -1 & 1 \end{pmatrix}, \quad R = \begin{pmatrix} 0 & 0 \\ 0 & 1 \end{pmatrix}.$$

Note that in this case, both G and R are degenerate. Also, it is clear that $R^\dagger = R$. Now, we look at the corresponding Riccati equation:

$$0 = \dot{P} + PA + A^T P + Q - PBR^\dagger B^T P$$

$$= \begin{pmatrix} \dot{P}_1 & \dot{P}_2 \\ \dot{P}_2 & \dot{P}_3 \end{pmatrix} + \begin{pmatrix} P_1 & P_2 \\ P_2 & P_3 \end{pmatrix}\begin{pmatrix} 1 & -1 \\ 1 & -1 \end{pmatrix} + \begin{pmatrix} 1 & 1 \\ -1 & -1 \end{pmatrix}\begin{pmatrix} P_1 & P_2 \\ P_2 & P_3 \end{pmatrix}$$

$$- \begin{pmatrix} P_1 & P_2 \\ P_2 & P_3 \end{pmatrix}\begin{pmatrix} 1 & 1 \\ 1 & -1 \end{pmatrix}\begin{pmatrix} 0 & 0 \\ 0 & 1 \end{pmatrix}\begin{pmatrix} 1 & 1 \\ 1 & -1 \end{pmatrix}\begin{pmatrix} P_1 & P_2 \\ P_2 & P_3 \end{pmatrix}$$

$$= \begin{pmatrix} \dot{P}_1 & \dot{P}_2 \\ \dot{P}_2 & \dot{P}_3 \end{pmatrix} + \begin{pmatrix} P_1+P_2 & -(P_1+P_2) \\ P_2+P_3 & -(P_2+P_3) \end{pmatrix} + \begin{pmatrix} P_1+P_2 & P_2+P_3 \\ -(P_1+P_2) & -(P_2+P_3) \end{pmatrix}$$

$$- \begin{pmatrix} (P_1-P_2)^2 & (P_1-P_2)(P_2-P_3) \\ (P_1-P_2)(P_2-P_3) & (P_2-P_3)^2 \end{pmatrix}$$

$$= \begin{pmatrix} \dot{P}_1+2(P_1+P_2)-(P_1-P_2)^2 & \dot{P}_2-P_1+P_3-(P_1-P_2)(P_2-P_3) \\ \dot{P}_2-P_1+P_3-(P_1-P_2)(P_2-P_3) & \dot{P}_3-2(P_2+P_3)-(P_2-P_3)^2 \end{pmatrix}.$$

Then it suffices to let

$$P_1(\cdot) = -P_2(\cdot) = P_3(\cdot),$$

with $P_3(\cdot)$ satisfying

$$\begin{cases} \dot{P}_3(s) - 4P_3(s)^2 = 0, & s \in [t,T], \\ P_3(T) = 1, \end{cases}$$

whose solution is given by

$$P_3(s) = \frac{1}{1 + 4(T - s)}, \qquad s \in [0, T].$$

Hence,

$$P(s) = \frac{1}{1 + 4(T - s)} G, \qquad s \in [t, T],$$

and

$$B^T P(s) = \frac{1}{1 + 4(T - s)} B^T G = \frac{1}{1 + 4(T - s)} \begin{pmatrix} 1 & 1 \\ 1 & -1 \end{pmatrix} \begin{pmatrix} 1 & -1 \\ -1 & 1 \end{pmatrix}$$

$$= \frac{1}{1 + 4(T - s)} \begin{pmatrix} 0 & 0 \\ 2 & -2 \end{pmatrix}.$$

Thus,

$$\mathcal{R}(B^T P(s)) = \left\{ \begin{pmatrix} 0 \\ \lambda \end{pmatrix} \mid \lambda \in \mathbb{R} \right\} = \mathcal{R}(R).$$

Hence, Theorem 6.1.11 applies and we have the following representation of closed-loop optimal strategies:

$$\begin{cases} \bar{\Theta}(s) = -R^\dagger B^T P(s) + (I - R^\dagger R)\theta(s) \\ \qquad = -\dfrac{1}{1 + 4(T - s)} \begin{pmatrix} 0 & 0 \\ 2 & -2 \end{pmatrix} + \begin{pmatrix} \theta(s) & 0 \\ 0 & 0 \end{pmatrix}, \qquad s \in [t, T], \\ \bar{v}(s) = (I - R^\dagger R)\nu(s) = \begin{pmatrix} \nu(s) & 0 \\ 0 & 0 \end{pmatrix}, \end{cases}$$

for some $\theta(\cdot), \nu(\cdot)$.

Let us now re-look at Example 6.1.2 (ii)–(iii), for which there are infinitely many open-loop optimal controls.

Example 6.1.2 (ii). (Continued) For this example, we have

$$A = f = R = G = q = \rho = g = 0, \quad B = Q = 1.$$

We already know that open-loop optimal control does not exist. The Riccati equation reads

$$\begin{cases} \dot{P}(s) + 1 = 0, & s \in [t, T], \\ \mathcal{R}(P(s)) \subseteq \mathcal{R}(0) = \{0\}, & s \in [t, T], \\ P(T) = 0. \end{cases}$$

Clearly, this Riccati equation does not have a solution. Hence, there is no closed-loop optimal strategy for this problem.

Example 6.1.2 (iii). (Continued). Recall that for this example,

$$A = f = Q = R = q = \rho = g = 0, \quad B = G = 1.$$

Thus, the two-point boundary value problem reads:

$$\begin{cases} \dot{\bar{X}}(s) = \bar{u}(s), & s \in [t, T], \\ \dot{\bar{Y}}(s) = 0, & s \in [t, T], \\ \bar{X}(t) = x, & \bar{Y}(T) = \bar{X}(T), \end{cases}$$

which admits infinitely many solutions, and the stationarity condition reads

$$\bar{Y}(s) = 0, \quad s \in [t, T].$$

The corresponding Riccati equation looks like the following:

$$\begin{cases} \dot{P}(s) = 0, & s \in [t, T], \\ \mathcal{R}(P(s)) \subseteq \mathcal{R}(0) = \{0\}, & s \in [t, T], \\ P(T) = 1, \end{cases}$$

which does not have a solution. Thus, there exists no closed-loop optimal strategy for the corresponding Problem (LQ). Consequently, although for any $(t, x) \in [0, T) \times (\mathbb{R} \setminus \{0\})$, the problem has infinitely many open-loop optimal controls, none of them admits a closed-loop representation.

The following result is concerned with the existence of closed-loop optimal strategy when open-loop optimal controls exist. In this result, we need to assume (LQ3)'.

Theorem 6.1.13. *Let (LQ1)–(LQ2) and (LQ3)' hold. Suppose for any $(t, x) \in [0, T) \times \mathbb{R}^n$, Problem (LQ) admits a unique open-loop optimal control. Then Problem (LQ) admits a unique closed-loop optimal strategy $(\bar{\Theta}(\cdot), \bar{v}(\cdot))$ over $[0, T]$, and for any initial pair $(t, x) \in [0, T) \times \mathbb{R}^n$, the corresponding open-loop optimal control $\bar{u}(\cdot)$ is an outcome of $(\bar{\Theta}(\cdot), \bar{v}(\cdot))$:*

$$\begin{aligned} \bar{u}(s) &= \bar{\Theta}(s)\bar{X}(s) + \bar{v}(s) \\ &= -R(s)^{-1}[B(s)^T P(s) + S(s)]\bar{X}(s) \\ &\quad -R(s)^{-1}[B(s)^T \eta(s) + \rho(s)], \quad s \in [t, T], \end{aligned}$$

where $P(\cdot)$ is the solution the following Riccati equation:

$$\begin{cases} \dot{P}(s) + P(s)A(s) + A(s)^T P(s) + Q(s) \\ \quad -[P(s)B(s) + S(s)^T]R(s)^{-1}[B(s)^T P(s) + S(s)] = 0, \\ \hspace{5cm} s \in [t, T], \\ P(T) = G, \end{cases} \quad (6.46)$$

with $\bar{X}(\cdot)$ being the solution to the following closed-loop system:

$$\begin{cases} \dot{\bar{X}}(s) = \{A(s) - B(s)R(s)^{-1}[B(s)^T P(s) + S(s)]\}\bar{X}(s) \\ \qquad\qquad - B(s)R(s)^{-1}[B(s)^T \eta(s) + \rho(s)] + f(s), \quad s \in [t, T], \\ \bar{X}(t) = x, \end{cases}$$

and $\eta(\cdot)$ being the solution to the following:

$$\begin{cases} \dot{\eta}(s) = \{-A(s)^T + [P(s)B(s) + S(s)^T]R(s)^{-1}B(s)^T\}\eta(s) \\ \qquad\qquad - P(s)f(s) - q(s) + [P(s)B(s) + S(s)^T]R(s)^{-1}\rho(s), \\ \qquad\qquad\qquad\qquad\qquad\qquad s \in [t, T], \\ \eta(T) = g. \end{cases}$$

In this case, (6.36) holds.

Proof. By Theorem 6.1.7, we know that for any initial pair $(t, x) \in [0, T) \times \mathbb{R}^n$, the following admits a solution $(\bar{X}(\cdot\,; t, x), \bar{Y}(\cdot\,; t, x))$:

$$\begin{cases} \dot{\bar{X}}(s) = \widehat{A}(s)\bar{X}(s) - M(s)\bar{Y}(s) + \widehat{f}(s), \\ \dot{\bar{Y}}(s) = -\widehat{A}(s)^T\bar{Y}(s) - \widehat{Q}(s)\bar{X}(s) - \widehat{q}(s), \\ \bar{X}(t) = x, \qquad \bar{Y}(T) = G\bar{X}(T) + g, \end{cases}$$

where

$$\widehat{A}(s) = A(s) - B(s)R(s)^{-1}S(s), \quad M(s) = B(s)R(s)^{-1}B(s)^T,$$
$$\widehat{Q}(s) = Q(s) - S(s)^T R(s)^{-1}S(s),$$
$$\widehat{f}(s) = f(s) - B(s)R(s)^{-1}\rho(s), \quad \widehat{q}(s) = q(s) - S(s)^T R(s)^{-1}\rho(s).$$

For any $x \in \mathbb{R}^n$, let

$$X(\cdot\,; t, x) = \bar{X}(\cdot\,; t, x) - \bar{X}(\cdot\,; t, 0),$$
$$Y(\cdot\,; t, x) = \bar{Y}(\cdot\,; t, x) - \bar{Y}(\cdot\,; t, 0).$$

Then $(X(\cdot\,; t, x), Y(\cdot\,; t, x))$ is the unique solution to the following:

$$\begin{cases} \dot{X}(s) = \widehat{A}(s)X(s) - M(s)Y(s), \\ \dot{Y}(s) = -\widehat{A}(s)^T Y(s) - \widehat{Q}(s)X(s), \\ X(t) = x, \qquad Y(T) = GX(T). \end{cases} \qquad (6.47)$$

Thus, $x \mapsto (X(\cdot\,; t, x), Y(\cdot\,; t, x))$ is a well-defined linear map, and we may let

$$X(\cdot\,; t, x) = \mathbb{X}(\cdot)x, \quad Y(\cdot\,; t, x) = \mathbb{Y}(\cdot)x.$$

We claim that $\mathbb{X}(s)^{-1}$ exists for all $s \in [t,T]$. If this is not the case, then for some $\tau \in [t,T]$, one can find an $x_0 \in \mathbb{R}^n \setminus \{0\}$ such that

$$\mathbb{X}(\tau)x_0 = 0.$$

It is clear that

$$X_0(\cdot) = \mathbb{X}(\cdot)x_0, \quad Y_0(\cdot) = \mathbb{Y}(\cdot)x_0$$

is a solution to (6.47) on $[\tau, T]$ corresponding to $(\tau, 0)$. Then by our above arguments, this solution has to be zero. Hence,

$$\mathbb{X}(s)x_0 = \mathbb{Y}(s)x_0 = 0, \quad s \in [\tau, T].$$

Consequently, $X(T) = Y(T) = 0$ and $(X(\cdot), Y(\cdot))$ is the solution to the following:

$$\begin{cases} \dot{X}(s) = \widehat{A}(s)X(s) - M(s)Y(s), & s \in [t,T], \\ \dot{Y}(s) = -\widehat{A}(s)^T Y(s) - \widehat{Q}(s)X(s), & s \in [t,T], \\ X(T) = Y(T) = 0, \end{cases}$$

whose solution $(X(\cdot), Y(\cdot))$ must be identically equal to 0 over $[t,T]$. In particular, $x_0 = X(t) = 0$ which is a contradiction.

It is clear that $(\mathbb{X}(\cdot), \mathbb{Y}(\cdot))$ satisfies the following:

$$\begin{cases} \dot{\mathbb{X}}(s) = \widehat{A}(s)\mathbb{X}(s) - M(s)\mathbb{Y}(s), & s \in [t,T], \\ \dot{\mathbb{Y}}(s) = -\widehat{A}(s)^T \mathbb{Y}(s) - \widehat{Q}(s)\mathbb{X}(s), & s \in [t,T], \\ \mathbb{X}(t) = I, \quad \mathbb{Y}(T) = G\mathbb{X}(T). \end{cases} \tag{6.48}$$

Also, $\mathbb{X}(\cdot)^{-1}$ satisfies

$$\begin{cases} \dfrac{d}{ds}\mathbb{X}(s)^{-1} = -\mathbb{X}(s)^{-1}\widehat{A}(s) + \mathbb{X}(s)^{-1}M(s)\mathbb{Y}(s)\mathbb{X}(s)^{-1}, & s \in [t,T], \\ \mathbb{X}(t)^{-1} = I. \end{cases}$$

Now, we define

$$P(\cdot) = \mathbb{Y}(\cdot)\mathbb{X}(\cdot)^{-1}.$$

Then

$$\begin{cases} \dot{P}(s) + P(s)\widehat{A}(s) + \widehat{A}(s)^T P(s) + \widehat{Q}(s) - P(s)M(s)P(s) = 0, \\ \qquad\qquad\qquad\qquad\qquad\qquad\qquad\qquad s \in [t,T], \\ P(T) = G. \end{cases}$$

Since $\widehat{Q}(\cdot)$, $M(\cdot)$ and G are symmetric, $P(\cdot)^T$ satisfies the equation that $P(\cdot)$ satisfies. By the uniqueness, it is necessary that $P(\cdot)^T = P(\cdot)$. The above can also be written as (6.46). This means (6.32)–(6.33) hold. The rest conclusions then follow from Theorem 6.1.11.   □

Let us make some remarks.

(i) For closed-loop optimal strategies, we have a characterization in terms of the solvability of Riccati equation. The outcome $\bar{\Theta}(\cdot)\bar{X}(\cdot) + \bar{v}(\cdot)$ of a closed-loop optimal strategy $(\bar{\Theta}(\cdot), \bar{v}(\cdot))$ over $[t, T]$ is an open-loop optimal control for the initial pair $(t, \bar{X}(t))$. From this, we see that the existence of a closed-loop optimal strategy implies the existence of an open-loop optimal control.

(ii) For open-loop optimal controls, we show that if Problem (LQ) admits a unique open-loop optimal control for any initial pairs $(t, x) \in [0, T] \times \mathbb{R}^n$, then, under (LQ3)$'$, Problem (LQ) admits a closed-loop optimal strategy on $[0, T]$. Assumption (LQ3)$'$ ensures the invertibility of $\mathbb{X}(\cdot)$. Note that (6.48) is different from (6.41), the former is coupled, whereas the latter is decoupled. From the proof, we see that (LQ3)$'$ can be replaced by (LQ3) together with the following additional assumption:

$$\begin{cases} B(\cdot)R(\cdot)^{-1}S(\cdot) \in L^1(0, T; \mathbb{R}^{n \times n}), \\ B(\cdot)R(\cdot)^{-1}B(\cdot)^T, S(\cdot)^T R(\cdot)^{-1}S(\cdot) \in L^1(0, T; \mathbb{S}^n), \\ B(\cdot)R(\cdot)^{-1}\rho(\cdot), S(\cdot)^T R(\cdot)^{-1}\rho(\cdot) \in L^1(0, T; \mathbb{R}^n). \end{cases}$$

To conclude this subsection, we present several more examples.

Example 6.1.14. Consider the following controlled ODE:

$$\begin{cases} \dot{X}(s) = u(s), & s \in [t, T], \\ X(t) = x, \end{cases}$$

with cost functional

$$J(t, x; u(\cdot)) = \frac{1}{2} \int_t^T \left(-|X(s)|^2 + |u(s)|^2 \right) ds.$$

For the corresponding LQ problem, the Riccati equation reads

$$\begin{cases} \dot{P}(t) - P(t)^2 - 1 = 0, & t \in [0, T], \\ P(T) = 0, \end{cases}$$

whose solution is only defined on $(T - \frac{\pi}{2}, T]$, given by

$$P(t) = \tan(t - T), \qquad t \in (T - \frac{\pi}{2}, T].$$

Hence, according to Theorem 6.1.11, we see that the closed-loop optimal strategy is given by

$$\bar{\Theta}(s) = -P(s) = -\tan(s - T) = \tan(T - s), \qquad s \in (T - \frac{\pi}{2}, T],$$

and the value function is given by

$$V(t,x) = \frac{1}{2}P(t)x^2 = \frac{x^2\tan(t-T)}{2}, \qquad (t,x) \in (T - \frac{\pi}{2}, T] \times \mathbb{R}.$$

Note that if $T \geqslant \frac{\pi}{2}$, then for any $t \in [0, T - \frac{\pi}{2}]$, any small $\varepsilon > 0$, and $u(\cdot) \in \mathcal{U}[T - \frac{\pi}{2} + \varepsilon, T]$, we denote

$$u_\varepsilon(s) = \begin{cases} 0, & s \in [t, T - \frac{\pi}{2} + \varepsilon), \\ u(s), & s \in [T - \frac{\pi}{2} + \varepsilon, T]. \end{cases}$$

Then $u_\varepsilon(\cdot) \in \mathcal{U}[t, T]$, and it follows from optimality principle that

$$\begin{aligned} V(t,x) &\leqslant \inf_{u(\cdot) \in \mathcal{U}[T - \frac{\pi}{2} + \varepsilon, T]} J(t, x; u_\varepsilon(\cdot)) \\ &= -\frac{x^2(T - \frac{\pi}{2} + \varepsilon - t)}{2} + V(T - \frac{\pi}{2} + \varepsilon, x) \\ &= -\frac{x^2(T - \frac{\pi}{2} + \varepsilon - t)}{2} + \frac{x^2}{2}\tan(-\frac{\pi}{2} + \varepsilon). \end{aligned}$$

Sending $\varepsilon \downarrow 0$, we see that

$$V(t,x) = -\infty, \qquad \forall (t,x) \in \left[0, T - \frac{\pi}{2}\right] \times \mathbb{R}.$$

In other words, the corresponding LQ problem has a unique closed-loop optimal strategy on any $[t, T]$ $(t \in (T - \frac{\pi}{2}, T))$, and the corresponding LQ problem is not finite at any $(t, x) \in [0, T - \frac{\pi}{2}] \times (\mathbb{R} \setminus \{0\})$, if $T \geqslant \frac{\pi}{2}$.

Example 6.1.15. Consider the following state equation:

$$\begin{cases} \dot{X}(s) = u(s) + f(s), & s \in [t, T], \\ X(t) = x, \end{cases}$$

and performance functional

$$J(t, x; u(\cdot)) = \int_t^T |u(s)|^2 ds + g|X(T)|^2,$$

with $g \in \mathbb{R}$, which could be positive or negative. For the current case, the Riccati equation reads

$$\begin{cases} \dot{P}(s) - P(s)^2 = 0, & s \in [0, T], \\ P(T) = g, \end{cases}$$

whose solution is given by

$$P(s) = \frac{g}{1 + g(T-s)},$$

provided $1 + g(T - s) > 0$. The corresponding equation for $\eta(\cdot)$ is given by

$$\begin{cases} \dot{\eta}(s) - P(s)\eta(s) + P(s)f(s) = 0, & s \in [t, T], \\ \eta(T) = 0. \end{cases}$$

Then

$$\eta(s) = \int_s^T e^{\int_r^s P(\tau)d\tau} P(r)f(r)dr$$

$$= \int_s^T e^{-\int_r^s \frac{d[1+g(T-\tau)]}{1+g(T-\tau)}} \frac{g}{1+g(T-r)} f(r)dr$$

$$= \int_s^T \frac{1+g(T-r)}{1+g(T-s)} \frac{g}{1+g(T-r)} f(r)dr$$

$$= \frac{g}{1+g(T-s)} \int_s^T f(r)dr \equiv P(s) \int_s^T f(r)dr.$$

The closed-loop system is

$$\begin{cases} \dot{\bar{X}}(s) = -P(s)\bar{X}(s) + f(s) - P(s)\int_s^T f(r)dr, & s \in [t, T], \\ \bar{X}(t) = x. \end{cases}$$

Thus,

$$\left[\bar{X}(s) + \int_s^T f(r)dr \right]' = -P(s)\left[\bar{X}(s) + \int_s^T f(r)dr \right].$$

Then

$$\bar{X}(s) + \int_s^T f(r)dr = e^{-\int_t^s P(\tau)d\tau}\left[x + \int_t^T f(r)dr \right]$$

$$= e^{\int_t^s \frac{d[1+g(T-\tau)]}{1+g(T-\tau)}}\left[x + \int_t^T f(r)dr \right] = \frac{1+g(T-s)}{1+g(T-t)}\left[x + \int_t^T f(r)dr \right].$$

Hence,

$$\bar{X}(s) = -\int_s^T f(r)dr + \frac{1+g(T-s)}{1+g(T-t)}\left[x + \int_t^T f(r)dr \right].$$

The optimal control $\bar{u}(\cdot)$ is given by

$$\bar{u}(s) = -P(s)\bar{X}(s) - \eta(s) = -P(s)\left[\bar{X}(s) + \int_s^T f(r)dr \right]$$

$$= -\frac{g}{1+g(T-t)}\left[x + \int_t^T f(r)dr \right].$$

Further, the optimal value of the cost functional is given by

$$V(t,x) \equiv J(t,x;\bar{u}(\cdot)) = \inf_{u(\cdot)\in\mathcal{U}[t,T]} J(t,x;u(\cdot))$$

$$= \frac{1}{2}\langle P(t)x, x\rangle + \langle \eta(t), x\rangle - \frac{1}{2}\int_t^T [|\eta(s)|^2 - 2\langle \eta(s), f(s)\rangle]ds.$$

In the case $g \geq 0$, the above $P(s)$ is defined for $s \in [0,T]$, and the corresponding LQ problem is uniquely solvable on any $[t,T]$ with $t \in [0,T]$. However, if $g < 0$, the above $P(t)$ is defined only for

$$T - \frac{1}{|g|} < t \leq T.$$

Therefore, the LQ problem is uniquely solvable on $[t,T]$ with t satisfies the above. Now, if

$$0 \leq t \leq T - \frac{1}{|g|},$$

which means $t \in [0,T]$ and

$$1 + g(T-t) \leq 0,$$

then by taking $u(\cdot) = u^\lambda(\cdot) \equiv \lambda \in \mathbb{R}$, we have

$$J(t,x;u^\lambda(\cdot)) = \lambda^2(T-t) + g\left[x + \lambda(T-t) + \int_t^T f(r)dr\right]^2$$

$$= \lambda^2(T-t)[1 + g(T-t)] + 2g\lambda(T-t)\left[x + \int_t^T f(r)dr\right]$$

$$+ g\left[x + \int_t^T f(r)dr\right]^2.$$

Hence, we obtain

$$\inf_{u(\cdot)\in[t,T]} J(t,x;u(\cdot)) = -\infty,$$

provided either

$$1 + g(T-t) < 0,$$

or

$$1 + g(T-t) = 0, \quad x + \int_t^T f(r)dr \neq 0.$$

Example 6.1.16. Consider the following controlled state equation:

$$\begin{cases} \dot{X}(t) = u(t), & t \in [0,1], \\ X(0) = x, \end{cases}$$

with the cost functional
$$J(x, u(\cdot)) = \frac{1}{2}\Big[X^2(1) + \int_0^1 t^2 u^2(t)dt\Big].$$

In this example,
$$\begin{cases} A = 0, & B = 1, & f = 0, \\ G = 1, & g = 0, & Q = 0, & S = 0, & R(t) = t^2, & q = 0, & \rho = 0. \end{cases}$$

The corresponding Riccati equation reads
$$\begin{cases} \dot{P}(t) = \dfrac{P(t)^2}{t^2}, & t \in [0,1], \\ P(1) = 1. \end{cases} \tag{6.49}$$

It is easy to see that $P(t) = t$ is the unique solution of (6.49), and
$$\begin{aligned} B(t)^T P(t) &= t, \\ R(t) &= t^2 \geqslant 0, \end{aligned} \qquad t \in [0,1].$$

Thus,
$$\mathcal{R}(B(t)^T P(t)) \subseteq \mathcal{R}(R(t)), \qquad t \in [0,1].$$

But
$$R(t)^{-1} B(t)^T P(t) = \frac{1}{t},$$
which is not in $L^2(0, 1; \mathbb{R})$. Hence, by Theorem 6.1.11, closed-loop optimal strategy does not exist.

6.2 Differential Games

We now consider the following controlled state equation:
$$\begin{cases} \dot{X}(s) = A(s)X(s) + B_1(s)u_1(s) + B_2(s)u_2(s) + f(s), & s \in [t, T], \\ X(t) = x, \end{cases}$$

and the performance functional
$$\begin{aligned} J(t, x; u_1(\cdot), u_2(\cdot)) &= J(t, x; u(\cdot)) \\ &= \frac{1}{2}\Big\{ \langle GX(T), X(T) \rangle + 2\langle g, X(T) \rangle \\ &\quad + \int_t^T \Big[\Big\langle \begin{pmatrix} Q(s) & S_1(s)^T & S_2(s)^T \\ S_1(s) & R_{11}(s) & R_{12}(s) \\ S_2(s) & R_{21}(s) & R_{22}(s) \end{pmatrix} \begin{pmatrix} X(s) \\ u_1(s) \\ u_2(s) \end{pmatrix}, \begin{pmatrix} X(s) \\ u_1(s) \\ u_2(s) \end{pmatrix} \Big\rangle \\ &\quad + 2\Big\langle \begin{pmatrix} q(s) \\ \rho_1(s) \\ \rho_2(s) \end{pmatrix}, \begin{pmatrix} X(s) \\ u_1(s) \\ u_2(s) \end{pmatrix} \Big\rangle \Big] ds \Big\}. \end{aligned}$$

In the above, $u_i(\cdot) \in \mathcal{U}_i[t, T]$ is called a control of Player i ($i = 1, 2$) on $[t, T]$, with

$$\mathcal{U}_i[t, T] = L^2(t, T; \mathbb{R}^{m_i}), \qquad i = 1, 2.$$

We formally pose the following problem.

Problem (LQG). For given $(t, x) \in [0, T) \times \mathbb{R}^n$, Player 1 wants to find a control $\bar{u}_1(\cdot) \in \mathcal{U}_1[t, T]$ minimizing $J(t, x; u_1(\cdot), u_2(\cdot))$, and Player 2 wants to find a control $\bar{u}_2(\cdot) \in \mathcal{U}_2[t, T]$ maximizing $J(t, x; u_1(\cdot), u_2(\cdot))$.

The above is referred to as a *linear-quadratic two-person zero-sum differential game*. When $f(\cdot), q(\cdot), \rho_1(\cdot), \rho_2(\cdot), g = 0$, we denote the problem by Problem (LQG)*, which is a special case of Problem (LQG).

For notational simplicity, we let $m = m_1 + m_2$ and denote

$$B(\cdot) = (B_1(\cdot), B_2(\cdot)), \quad D(\cdot) = (D_1(\cdot), D_2(\cdot)),$$

$$S(\cdot) = \begin{pmatrix} S_1(\cdot) \\ S_2(\cdot) \end{pmatrix}, \quad R(\cdot) = \begin{pmatrix} R_{11}(\cdot) & R_{12}(\cdot) \\ R_{21}(\cdot) & R_{22}(\cdot) \end{pmatrix} \equiv \begin{pmatrix} R_1(\cdot) \\ R_2(\cdot) \end{pmatrix},$$

$$\rho(\cdot) = \begin{pmatrix} \rho_1(\cdot) \\ \rho_2(\cdot) \end{pmatrix}, \quad u(\cdot) = \begin{pmatrix} u_1(\cdot) \\ u_2(\cdot) \end{pmatrix}.$$

Naturally, we identify $\mathcal{U}[t, T] = \mathcal{U}_1[t, T] \times \mathcal{U}_2[t, T]$. With such notations, the state equation becomes

$$\begin{cases} \dot{X}(s) = A(s)X(s) + B(s)u(s) + f(s), & s \in [t, T], \\ X(t) = x, \end{cases} \tag{6.50}$$

and the performance functional becomes

$$\begin{aligned} J(t, x; u_1(\cdot), u_2(\cdot)) &= J(t, x; u(\cdot)) \\ &= \frac{1}{2} \Big\{ \langle GX(T), X(T) \rangle + 2 \langle g, X(T) \rangle \\ &\quad + \int_t^T \Big[\langle \begin{pmatrix} Q(s) & S(s)^T \\ S(s) & R(s) \end{pmatrix} \begin{pmatrix} X(s) \\ u(s) \end{pmatrix}, \begin{pmatrix} X(s) \\ u(s) \end{pmatrix} \rangle \\ &\quad + 2 \langle \begin{pmatrix} q(s) \\ \rho(s) \end{pmatrix}, \begin{pmatrix} X(s) \\ u(s) \end{pmatrix} \rangle \Big] ds \Big\}. \end{aligned} \tag{6.51}$$

With the above notation, we introduce the following standard assumptions:

(LQG1) The coefficients of the state equation satisfy the following:

$$A(\cdot) \in L^1(0, T; \mathbb{R}^{n \times n}), \quad B(\cdot) \in L^2(0, T; \mathbb{R}^{n \times m}), \quad f(\cdot) \in L^1(0, T; \mathbb{R}^n).$$

(LQG2) The weighting coefficients in the cost functional satisfy the following:

$$\begin{cases} Q(\cdot) \in L^1(0,T;\mathbb{S}^n), & S(\cdot) \in L^2(0,T;\mathbb{R}^{m\times n}), & R(\cdot) \in L^\infty(0,T;\mathbb{S}^m), \\ q(\cdot) \in L^1(0,T;\mathbb{R}^n), & \rho(\cdot) \in L^2(0,T;\mathbb{R}^m), & G \in \mathbb{S}^n, & g \in \mathbb{R}^n. \end{cases}$$

Under (LQG1), for any $(t,x) \in [0,T) \times \mathbb{R}^n$, and

$$u(\cdot) \equiv (u_1(\cdot), u_2(\cdot)) \in \mathcal{U}_1[t,T] \times \mathcal{U}_2[t,T] \equiv \mathcal{U}[t,T],$$

equation (6.50) admits a unique solution

$$X(\cdot) \stackrel{\Delta}{=} X(\cdot\,;t,x,u_1(\cdot),u_2(\cdot)) \equiv X(\cdot\,;t,x,u(\cdot)) \in C([0,T];\mathbb{R}^n).$$

Moreover, the following estimate holds:

$$\sup_{t\leqslant s\leqslant T} |X(s)| \leqslant K\Big\{|x| + \int_t^T |f(s)|ds + \Big(\int_t^T |u(s)|^2 ds\Big)^{\frac{1}{2}}\Big\}.$$

Therefore, under (LQG1)–(LQG2), the quadratic performance functional $J(t,x;u(\cdot)) \equiv J(t,x;u_1(\cdot),u_2(\cdot))$ is well defined for all $(t,x) \in [0,T] \times \mathbb{R}^n$ and $(u_1(\cdot), u_2(\cdot)) \in \mathcal{U}_1[t,T] \times \mathcal{U}_2[t,T]$.

We now introduce the following definitions.

Definition 6.2.1. (i) A pair $(u_1^*(\cdot), u_2^*(\cdot)) \in \mathcal{U}_1[t,T] \times \mathcal{U}_2[t,T]$ is called an *open-loop saddle point* of Problem (LQG) for the initial pair $(t,x) \in [0,T) \times \mathbb{R}^n$ if for any $(u_1(\cdot), u_2(\cdot)) \in \mathcal{U}_1[t,T] \times \mathcal{U}_2[t,T]$,

$$J(t,x;u_1^*(\cdot), u_2(\cdot)) \leqslant J(t,x;u_1^*(\cdot), u_2^*(\cdot)) \leqslant J(x;u_1(\cdot), u_2^*(\cdot)). \tag{6.52}$$

(ii) The *open-loop upper value* $V^+(t,x)$ of Problem (LQG) at $(t,x) \in [0,T) \times \mathbb{R}^n$ and the *open-loop lower value* $V^-(t,x)$ of Problem (LQG) at $(t,x) \in [0,T) \times \mathbb{R}^n$ are defined by the following:

$$\begin{cases} V^+(t,x) = \inf_{u_1(\cdot)\in\mathcal{U}_1[t,T]} \sup_{u_2(\cdot)\in\mathcal{U}_2[t,T]} J(t,x;u_1(\cdot), u_2(\cdot)), \\ V^-(t,x) = \sup_{u_2(\cdot)\in\mathcal{U}_2[t,T]} \inf_{u_1(\cdot)\in\mathcal{U}_1[t,T]} J(t,x;u_1(\cdot), u_2(\cdot)), \end{cases}$$

which automatically satisfy the following:

$$V^-(t,x) \leqslant V^+(t,x), \qquad (t,x) \in [0,T) \times \mathbb{R}^n.$$

In the case that

$$V^-(t,x) = V^+(t,x) \equiv V(t,x), \tag{6.53}$$

we say that Problem (LQG) admits an *open-loop value* $V(t,x)$ at (t,x). The maps $(t,x) \mapsto V^\pm(t,x)$ and $(t,x) \mapsto V(t,x)$ are called *open-loop upper value function*, *open-loop lower value function*, and *open-loop value function*, respectively.

(iii) For any given initial pair $(t, x) \in [0, T) \times \mathbb{R}^n$, we say that $V^+(t, x)$ is (uniquely) *achievable* if there exists a (unique) $\mu_2 : \mathcal{U}_1^0[t, T] \to \mathcal{U}_2[t, T]$ (depending on (t, x)) such that for any $u_1(\cdot) \in \mathcal{U}_1^0[t, T]$

$$J(t, x; u_1(\cdot), \mu_2(u_1(\cdot))) = \sup_{u_2(\cdot) \in \mathcal{U}_2[t,T]} J(t, x; u_1(\cdot), u_2(\cdot)),$$

where

$$\mathcal{U}_1^0[t, T] = \left\{ u_1(\cdot) \in \mathcal{U}_1[t, T] \mid \sup_{u_2(\cdot) \in \mathcal{U}_2[t,T]} J(t, x; u_1(\cdot), u_2(\cdot)) < \infty \right\}$$

is a non-empty subset of $\mathcal{U}_1[t, T]$; and there exists a $\bar{u}_1(\cdot) \in \mathcal{U}_1^0[t, T]$ (also depending on (t, x)) such that

$$J(t, x; \bar{u}_1(\cdot), \mu_2(\bar{u}_1(\cdot))) = \inf_{u_1(\cdot) \in \mathcal{U}_1[t,T]} J(t, x; u_1(\cdot), \mu_2(u_1(\cdot)))$$

$$= \inf_{u_1(\cdot) \in \mathcal{U}_1^0[t,T]} \sup_{u_2(\cdot) \in \mathcal{U}_2[t,T]} J(t, x; u_1(\cdot), u_2(\cdot)) \equiv V^+(t, x).$$

In this case, we say that $V^+(t, x)$ is achieved by $(\bar{u}_1(\cdot), \mu_2(\cdot))$. Similarly, we can define the (unique) achievability of $V^-(t, x)$. Further, in the case (6.53) holds, we say that $V(\cdot, \cdot)$ is (uniquely) achievable if $V(\cdot, \cdot)$, regarded as $V^+(\cdot, \cdot)$ and $V^-(\cdot, \cdot)$, both are (uniquely) achievable.

From (6.52), we see that if $(u_1^*(\cdot), u_2^*(\cdot))$ is an open-loop saddle point of Problem (LQG), then $u_1^*(\cdot)$ is an open-loop optimal control for the LQ problem with the cost functional $J(t, x; u_1(\cdot), u_2^*(\cdot))$, and $u_2^*(\cdot)$ is an open-loop optimal control for the LQ problem with the cost functional $-J(t, x; u_1^*(\cdot), u_2(\cdot))$.

As a special case of Proposition 1.2.13, we have the following result.

Proposition 6.2.2. (i) *Suppose for each initial pair $(t, x) \in [0, T] \times \mathbb{R}^n$, Problem (LQG) admits an open-loop saddle point $(\bar{u}_1(\cdot\,; t, x), \bar{u}_2(\cdot\,; t, x)) \in \mathcal{U}_1[t, T] \times \mathcal{U}_2[t, T]$ at $(t, x) \in [0, T] \times \mathbb{R}^n$. Then Problem (LQG) has the open-loop value function $V(\cdot, \cdot)$. Moreover,*

$$V(t, x) = J(t, x; \bar{u}_1(\cdot\,; t, x), \bar{u}_2(\cdot\,; t, x)).$$

(ii) *Suppose Problem (LQG) admits an open-loop value $V(t, x)$ at initial pair $(t, x) \in [0, T) \times \mathbb{R}^n$ for which $V(t, x)$ is achievable. Then Problem (LQG) admits an open-loop saddle point at (t, x). More precisely, if $V(t, x)$ is achieved by $(\bar{u}_1(\cdot), \mu_2(\cdot))$ and $(\mu_1(\cdot), \bar{u}_2(\cdot))$, then $(\bar{u}_1(\cdot), \bar{u}_2(\cdot)) \in \mathcal{U}_1[t, T] \times \mathcal{U}_2[t, T]$ is an open-loop saddle point of Problem (LQG) at (t, x).*

Next, we let

$$\mathcal{Q}_i[t,T] = L^2(t,T;\mathbb{R}^{m_i \times n}), \qquad i = 1,2.$$

For any initial pair $(t,x) \in [0,T) \times \mathbb{R}^n$ and

$$\Theta(\cdot) \equiv (\Theta_1(\cdot), \Theta_2(\cdot)) \in \mathcal{Q}_1[t,T] \times \mathcal{Q}_2[t,T],$$
$$v(\cdot) \equiv (v_1(\cdot), v_2(\cdot)) \in \mathcal{U}_1[t,T] \times \mathcal{U}_2[t,T],$$

consider the following system:

$$\begin{cases} \dot{X}(s) = [A(s) + B(s)\Theta(s)]X(s) + B(s)v(s) + f(s), & s \in [t,T], \\ X(t) = x. \end{cases} \tag{6.54}$$

Clearly, under (LQG1), the above admits a unique solution $X(\cdot) \equiv X(\cdot\,;t,x,\Theta_1(\cdot),v_1(\cdot);\Theta_2(\cdot),v_2(\cdot))$. If we denote

$$u_i(\cdot) = \Theta_i(\cdot)X(\cdot) + v_i(\cdot), \qquad i = 1,2,$$

then the above (6.54) coincides with the original state equation (6.50). We refer to (6.54) as a *closed-loop system* of the original system. With the solution $X(\cdot)$ to (6.54), we denote

$$J\big(t,x;\Theta_1(\cdot)X(\cdot) + v_1(\cdot), \Theta_2(\cdot)X(\cdot) + v_2(\cdot)\big) \equiv J(t,x;\Theta(\cdot)X(\cdot) + v(\cdot))$$
$$= \frac{1}{2}\Big\{ \langle GX(T), X(T) \rangle + 2\langle g, X(T) \rangle$$
$$+ \int_t^T \Big[\Big\langle \begin{pmatrix} Q(s) & S(s)^T \\ S(s) & R(s) \end{pmatrix} \begin{pmatrix} X(s) \\ \Theta(s)X(s) + v(s) \end{pmatrix}, \begin{pmatrix} X(s) \\ \Theta(s)X(s) + v(s) \end{pmatrix} \Big\rangle$$
$$+ 2\Big\langle \begin{pmatrix} q(s) \\ \rho(s) \end{pmatrix}, \begin{pmatrix} X(s) \\ \Theta(s)X(s) + v(s) \end{pmatrix} \Big\rangle \Big] ds \Big\}$$
$$= \frac{1}{2}\Big\{ \langle GX(T), X(T) \rangle + 2\langle g, X(T) \rangle$$
$$+ \int_t^T \Big[\Big\langle \begin{pmatrix} Q+\Theta^T S+S^T\Theta+\Theta^T R\Theta & S^T+\Theta^T R \\ S+R\Theta & R \end{pmatrix} \begin{pmatrix} X \\ v \end{pmatrix}, \begin{pmatrix} X \\ v \end{pmatrix} \Big\rangle$$
$$+ 2\Big\langle \begin{pmatrix} q+\Theta^T \rho \\ \rho \end{pmatrix}, \begin{pmatrix} X \\ v \end{pmatrix} \Big\rangle \Big] ds \Big\}.$$

One can define $J(t,x;\Theta_1(\cdot)X(\cdot) + v_1(\cdot), u_2(\cdot))$ and $J(t,x;u_1(\cdot), \Theta_2(\cdot)X(\cdot) + v_2(\cdot))$ similarly. We now introduce the following definition.

Definition 6.2.3. (i) A 4-tuple $(\Theta_1^*(\cdot), v_1^*(\cdot); \Theta_2^*(\cdot), v_2^*(\cdot)) \in \mathcal{Q}_1[t,T] \times \mathcal{U}_1[t,T] \times \mathcal{Q}_2[t,T] \times \mathcal{U}_2[t,T]$ is called a *closed-loop saddle point* of Problem

(LQG) on $[t, T]$ if for any $x \in \mathbb{R}^n$ and $(\Theta_1(\cdot), v_1(\cdot); \Theta_2(\cdot), v_2(\cdot)) \in \mathcal{Q}_1[t, T] \times \mathcal{U}_1[t, T] \times \mathcal{Q}_2[t, T] \times \mathcal{U}_2[t, T]$, the following holds:

$$
\begin{aligned}
& J(t, x; \Theta_1^*(\cdot)X(\cdot) + v_1^*(\cdot), \Theta_2(\cdot)X(\cdot) + v_2(\cdot)) \\
& \leqslant J(t, x; \Theta_1^*(\cdot)X^*(\cdot) + v_1^*(\cdot), \Theta_2^*(\cdot)X^*(\cdot) + v_2^*(\cdot)) \qquad (6.55) \\
& \leqslant J(t, x; \Theta_1(\cdot)X(\cdot) + v_1(\cdot), \Theta_2^*(\cdot)X(\cdot) + v_2^*(\cdot)).
\end{aligned}
$$

(ii) The *closed-loop upper value function* $(t, x) \mapsto \bar{V}^+(t, x)$ of Problem (LQG) on $[0, T) \times \mathbb{R}^n$ and the *closed-loop lower value function* $(t, x) \mapsto \bar{V}^-(t, x)$ of Problem (LQG) on $[0, T) \times \mathbb{R}^n$ are defined by the following:

$$
\begin{cases}
\bar{V}^+(t, x) = \displaystyle\inf_{\substack{\Theta_1(\cdot) \in \mathcal{Q}_1[t,T] \\ v_1(\cdot) \in \mathcal{U}_1[t,T]}} \sup_{\substack{\Theta_2(\cdot) \in \mathcal{Q}_2[t,T] \\ v_2(\cdot) \in \mathcal{U}_2[t,T]}} J(t, x; \Theta_1(\cdot)X(\cdot) + v_1(\cdot), \Theta_2(\cdot)X(\cdot) + v_2(\cdot)), \\[4mm]
\bar{V}^-(t, x) = \displaystyle\sup_{\substack{\Theta_2(\cdot) \in \mathcal{Q}_2[t,T] \\ v_2(\cdot) \in \mathcal{U}_2[t,T]}} \inf_{\substack{\Theta_1(\cdot) \in \mathcal{Q}_1[t,T] \\ v_1(\cdot) \in \mathcal{U}_1[t,T]}} J(t, x; \Theta_1(\cdot)X(\cdot) + v_1(\cdot), \Theta_2(\cdot)X(\cdot) + v_2(\cdot)).
\end{cases}
$$

One can show that

$$
\bar{V}^-(t, x) \leqslant \bar{V}^+(t, x), \qquad (t, x) \in [0, T) \times \mathbb{R}^n.
$$

In the case that

$$
\bar{V}^-(t, x) = \bar{V}^+(t, x) \equiv \bar{V}(t, x), \qquad (t, x) \in [0, T) \times \mathbb{R}^n,
$$

we say that Problem (LQG) admits a *closed-loop value function* $\bar{V}(\cdot, \cdot)$.

There are some remarks in order.

(i) An open-loop saddle point $(u_1^*(\cdot), u_2^*(\cdot))$ usually depends on the initial state x, whereas, a closed-loop saddle point $(\Theta_1^*(\cdot), v_1^*(\cdot); \Theta_2^*(\cdot), v_2^*(\cdot))$ is required to be independent of the initial state x.

(ii) In (6.55), the state process $X(\cdot)$ appearing in

$$
J(t, x; \Theta_1^*(\cdot)X(\cdot) + v_1^*(\cdot), \Theta_2(\cdot)X(\cdot) + v_2(\cdot))
$$

is different from that in

$$
J(t, x; \Theta_1(\cdot)X(\cdot) + v_1(\cdot), \Theta_2^*(\cdot)X(\cdot) + v_2^*(\cdot));
$$

and both are different from $X^*(\cdot) \equiv X(\cdot; t, x, \Theta_1^*(\cdot), v_1^*(\cdot); \Theta_2^*(\cdot), v_2^*(\cdot))$ which is the solution of (6.54) corresponding to

$$
(\Theta_1(\cdot), v_1(\cdot); \Theta_2(\cdot), v_2(\cdot)) = (\Theta_1^*(\cdot), v_1^*(\cdot); \Theta_2^*(\cdot), v_2^*(\cdot)).
$$

On the other hand, we have the *Berkovitz's Equivalence Lemma*, similar to that for Problem (LQ).

Lemma 6.2.4. *Let* (LQG1)–(LQG2) *hold. For* $(\Theta_i^*(\cdot), v_i^*(\cdot)) \in \mathcal{Q}_i[t, T]$ $\times \mathcal{U}_i[t, T]$, *the following statements are equivalent:*

(i) $(\Theta_1^*(\cdot), v_1^*(\cdot); \Theta_2^*(\cdot), v_2^*(\cdot))$ *is a closed-loop saddle point of Problem* (LQG) *on* $[t, T]$.

(ii) *For any* $x \in \mathbb{R}^n$ *and* $(v_1(\cdot), v_2(\cdot)) \in \mathcal{U}_1[t, T] \times \mathcal{U}_2[t, T]$,

$$J(t, x; \Theta_1^*(\cdot)X(\cdot) + v_1^*(\cdot), \Theta_2^*(\cdot)X(\cdot) + v_2(\cdot))$$
$$\leqslant J(t, x; \Theta_1^*(\cdot)X^*(\cdot) + v_1^*(\cdot), \Theta_2^*(\cdot)X^*(\cdot) + v_2^*(\cdot))$$
$$\leqslant J(t, x; \Theta_1^*(\cdot)X(\cdot) + v_1(\cdot), \Theta_2^*(\cdot)X(\cdot) + v_2^*(\cdot)).$$

(iii) *For any* $x \in \mathbb{R}^n$, *and* $(u_1(\cdot), u_2(\cdot)) \in \mathcal{U}_1[t, T] \times \mathcal{U}_2[t, T]$,

$$J(t, x; \Theta_1^*(\cdot)X(\cdot) + v_1^*(\cdot), u_2(\cdot))$$
$$\leqslant J(t, x; \Theta_1^*(\cdot)X^*(\cdot) + v_1^*(\cdot), \Theta_2^*(\cdot)X^*(\cdot) + v_2^*(\cdot)) \qquad (6.56)$$
$$\leqslant J(t, x; u_1(\cdot), \Theta_2^*(\cdot)X(\cdot) + v_2^*(\cdot)).$$

Proof. (i) $\Rightarrow$ (ii) is trivial, by taking $\Theta_i(\cdot) = \Theta_i^*(\cdot)$, $i = 1, 2$.

(ii) $\Rightarrow$ (iii). For any $x \in \mathbb{R}^n$, and any $u_1(\cdot) \in \mathcal{U}_1[t, T]$, let $X(\cdot)$ be the solution of the following ODE:

$$\begin{cases} \dot{X}(s) = \left[A(s) + B_2(s)\Theta_2^*(s)\right]X(s) \\ \qquad\quad + B_1(s)u_1(s) + B_2(s)v_2^*(s) + f(s), \qquad s \in [t, T], \\ X(t) = x. \end{cases}$$

Set

$$v_1(\cdot) = u_1(\cdot) - \Theta_1^*(\cdot)X(\cdot) \in \mathcal{U}_1[t, T],$$

then $X(\cdot)$ is also the solution to the following ODE:

$$\begin{cases} \dot{X}(s) = \left[A(s) + B_1(s)\Theta_1^*(s) + B_2(s)\Theta_2^*(s)\right]X(s) \\ \qquad\quad + B_1(s)v_1(s) + B_2(s)v_2^*(s) + f(s), \qquad s \in [t, T], \\ X(t) = x. \end{cases}$$

Therefore,

$$J(t, x; \Theta_1^*(\cdot)X^*(\cdot) + v_1^*(\cdot), \Theta_2^*(\cdot)X^*(\cdot) + v_2^*(\cdot))$$
$$\leqslant J(t, x; \Theta_1^*(\cdot)X(\cdot) + v_1(\cdot), \Theta_2^*(\cdot)X^*(\cdot) + v_2^*(\cdot))$$
$$= J(t, x; u_1(\cdot), \Theta_2^*(\cdot)X^*(\cdot) + v_2^*(\cdot)).$$

Similarly, for any $u_2(\cdot) \in \mathcal{U}_2[t, T]$, we can show that

$$J(t, x; \Theta_1^*(\cdot)X(\cdot) + v_1^*(\cdot), u_2(\cdot))$$
$$\leqslant J(t, x; \Theta_1^*(\cdot)X^*(\cdot) + v_1^*(\cdot), \Theta_2^*(\cdot)X^*(\cdot) + v_2^*(\cdot)).$$

Thus, (iii) holds.

(iii) $\Rightarrow$ (i). For any $\Theta_i(\cdot) \in \mathcal{Q}_i[t,T]$ and $v_i(\cdot) \in \mathcal{U}_i[t,T]$, $i = 1,2$, let $X(\cdot)$ be the solution to the following ODE:

$$\begin{cases} \dot{X}(s) = \big[A(s) + B_1(s)\Theta_1(s) + B_2(s)\Theta_2^*(s)\big]X(s) \\ \qquad\quad +B_1(s)v_1(s) + B_2(s)v_2^*(s) + f(s), \qquad s \in [t,T], \\ X(t) = x. \end{cases}$$

Set

$$u_1(\cdot) \overset{\Delta}{=} \Theta_1(\cdot)X(\cdot) + v_1(\cdot) \in \mathcal{U}_1[t,T].$$

By uniqueness, $X(\cdot)$ also solves the following ODE:

$$\begin{cases} \dot{X}(s) = \big[A(s) + B_2(s)\Theta_2^*(s)\big]X(s) \\ \qquad\quad +B_1(s)u_1(s) + B_2(s)v_2^*(s) + f(s), \quad s \in [t,T], \\ X(t) = x. \end{cases}$$

Therefore,

$$\begin{aligned} &J(t,x;\Theta_1^*(\cdot)X^*(\cdot) + v_1^*(\cdot), \Theta_2^*(\cdot)X^*(\cdot) + v_2^*(\cdot)) \\ &\leqslant J(t,x;u_1(\cdot), \Theta_2^*(\cdot)X(\cdot) + v_2^*(\cdot)) \\ &= J(t,x;\Theta_1(\cdot)X(\cdot) + v_1(\cdot), \Theta_2^*(\cdot)X(\cdot) + v_2^*(\cdot)). \end{aligned}$$

Similarly, we have

$$\begin{aligned} &J(t,x;\Theta_1^*(\cdot)X(\cdot) + v_1^*(\cdot), \Theta_2(\cdot)X(\cdot) + v_2(\cdot)) \\ &\leqslant J(t,x;\Theta_1^*(\cdot)X^*(\cdot) + v_1^*(\cdot), \Theta_2^*(\cdot)X^*(\cdot) + v_2^*(\cdot)). \end{aligned}$$

This completes the proof. $\qquad\qquad\qquad\qquad\qquad\qquad\qquad\qquad\square$

We note that (ii) of Lemma 6.2.4 tells us that if we consider the following state equation

$$\begin{cases} \dot{X} = (A + B\Theta^*)X + B_1 v_1 + B_2 v_2^* + f, \\ X(t) = x, \end{cases} \tag{6.57}$$

with the cost functional

$$J_1(t,x;v_1(\cdot)) = J(t,x;\Theta_1^*(\cdot)X(\cdot) + v_1(\cdot), \Theta_2^*(\cdot)X(\cdot) + v_2^*(\cdot)), \tag{6.58}$$

then $v_1^*(\cdot)$ is an open-loop optimal control of the corresponding Problem (LQ). Likewise, if we consider the following state equation

$$\begin{cases} \dot{X} = (A + B\Theta^*)X + B_2 v_2 + B_1 v_1^* + f, \\ X(t) = x, \end{cases} \tag{6.59}$$

with the cost functional

$$J_2(t, x; v_2(\cdot)) = -J(t, x; \Theta_1^*(\cdot)X(\cdot) + v_1^*(\cdot), \Theta_2^*(\cdot)X(\cdot) + v_2(\cdot)), \qquad (6.60)$$

then $v_2^*(\cdot)$ is an open-loop optimal control of the corresponding Problem (LQ).

On the other hand, comparing with (6.52), we see that (6.56) does not imply that $(\Theta_1^*(\cdot)X^*(\cdot) + v_1^*(\cdot), \Theta_2^*(\cdot)X^*(\cdot) + v_2^*(\cdot))$ is an open-loop saddle point of Problem (LQG), for the initial pair $(t, X^*(t))$. This is different from Problem (LQG) for which the outcome $\bar{\Theta}(\cdot)\bar{X}(\cdot) + \bar{v}(\cdot)$ of a closed-loop optimal strategy $(\bar{\Theta}(\cdot), \bar{v}(\cdot))$ is an open-loop optimal control for the initial pair $(t, \bar{X}(t))$.

More precisely, let us compare the following two inequalities:

$$J(t, x; u_1^*(\cdot), u_2^*(\cdot)) \leqslant J(t, x; u_1(\cdot), u_2^*(\cdot)), \qquad (6.61)$$

and

$$\begin{aligned} J(t, x; \Theta_1^*(\cdot)X^*(\cdot) &+ v_1^*(\cdot), \Theta_2^*(\cdot)X^*(\cdot) + v_2^*(\cdot)) \\ &\leqslant J(t, x; u_1(\cdot), \Theta_2^*(\cdot)X(\cdot) + v_2^*(\cdot)). \end{aligned} \qquad (6.62)$$

For (6.61), we look at the following state equation:

$$\begin{cases} \dot{X}(s) = A(s)X(s) + B_1(s)u_1(s) + B_2(s)u_2^*(s) + f(s), & s \in [t, T], \\ X(t) = x, \end{cases}$$

and the following cost functional

$$\begin{aligned} J_1(t, x; u_1(\cdot)) &\equiv J(t, x; u_1(\cdot), u_2^*(\cdot)) \\ &= \frac{1}{2}\Big\{ \langle GX(T), X(T) \rangle + 2\langle g, X(T) \rangle + \int_t^T \Big[\langle QX, X \rangle + 2\langle S_1 X, u_1 \rangle \\ &\quad + \langle R_{11}u_1, u_1 \rangle + \langle R_{22}u_2^*, u_2^* \rangle + 2\langle R_{12}u_2^*, u_1 \rangle + 2\langle S_2 X, u_2^* \rangle \\ &\quad + 2\langle q, X \rangle + 2\langle \rho_1, u_1 \rangle + 2\langle \rho_2, u_2^* \rangle \Big] ds \Big\} \\ &= \frac{1}{2}\Big\{ \langle GX(T), X(T) \rangle + 2\langle g, X(T) \rangle + \int_t^T \Big[\langle QX, X \rangle + 2\langle S_1 X, u_1 \rangle \\ &\quad + \langle R_{11}u_1, u_1 \rangle + 2\langle q + S_2^T u_2^*, X \rangle + 2\langle \rho_1 + R_{12}u_2^*, u_1 \rangle \\ &\quad + \langle R_{22}u_2^*, u_2^* \rangle + 2\langle \rho_2, u_2^* \rangle \Big] ds \Big\}. \end{aligned}$$

Therefore, (6.61) holds if and only if $u_1^*(\cdot)$ is an open-loop optimal control of Problem (LQ) with the coefficients given by the following: (We use $\widetilde{A}$, $\widetilde{B}$, etc. to distinguish them from the original ones)

$$\begin{cases} \widetilde{A} = A, & \widetilde{B} = B_1, & \widetilde{f} = f + B_2 u_2^*, & \widetilde{G} = G, & \widetilde{g} = g, & \widetilde{Q} = Q, \\ \widetilde{S} = S_1, & \widetilde{R} = R_{11}, & \widetilde{q} = q + S_2^T u_2^*, & \widetilde{\rho} = \rho_1 + R_{12}u_2^*. \end{cases}$$

$$(6.63)$$

However, for (6.62), we look at the following state equation:

$$\begin{cases} \dot{X}_1(s) = \left[A(s) + B_2(s)\Theta_2^*(s) \right] X_1(s) \\ \qquad\qquad + B_1(s)u_1(s) + B_2(s)v_2^*(s) + f(s), \quad s \in [t, T], \\ X_1(t) = x, \end{cases}$$

and the following cost functional

$$\begin{aligned} \bar{J}_1(t, x; u_1(\cdot)) &= J(t, x; u_1(\cdot), \Theta_2^*(\cdot)X_1(\cdot) + v_2^*(\cdot)) \\ &= \frac{1}{2} \Big\{ \langle GX_1(T), X_1(T) \rangle + 2\langle g, X_1(T) \rangle \\ &\qquad + \int_t^T \Big[\langle QX_1, X_1 \rangle + \langle R_{11}u_1, u_1 \rangle \\ &\qquad + \langle R_{22}(\Theta_2^*X_1 + v_2^*), \Theta_2^*X_1 + v_2^* \rangle + 2\langle S_1 X_1, u_1 \rangle \\ &\qquad + 2\langle S_2 X_1, \Theta_2^*X_1 + v_2^* \rangle + 2\langle R_{21}u_1, \Theta_2^*X_1 + v_2^* \rangle \\ &\qquad + 2\langle q, X_1 \rangle + 2\langle \rho_1, u_1 \rangle + 2\langle \rho_2, \Theta_2^*X_1 + v_2^* \rangle \Big] ds \Big\} \\ &= \frac{1}{2} \Big\{ \langle GX_1(T), X_1(T) \rangle + 2\langle g, X_1(T) \rangle \\ &\qquad + \int_t^T \Big[\langle [Q + (\Theta_2^*)^T R_{22}\Theta_2^* + (\Theta_2^*)^T S_2 + S_2^T \Theta_2^*] X_1, X_1 \rangle \\ &\qquad + \langle R_{11}u_1, u_1 \rangle + 2\langle (S_1 + R_{12}\Theta_2^*)X_1, u_1 \rangle \\ &\qquad + 2\langle q + [S_2^T + (\Theta_2^*)^T R_{22}]v_2^* + (\Theta_2^*)^T \rho_2, X_1 \rangle \\ &\qquad + 2\langle \rho_1 + R_{12}v_2^*, u_1 \rangle + \langle R_{22}v_2^*, v_2^* \rangle + 2\langle \rho_2, v_2^* \rangle \Big] ds \Big\}. \end{aligned}$$

Then $(\Theta_1^*(\cdot), v_1^*(\cdot))$ is a closed-loop optimal strategy for a Problem (LQ), with

$$\begin{cases} \widetilde{A} = A + B_2\Theta_2^*, \quad \widetilde{B} = B_1, \quad \widetilde{f} = f + B_2 v_2^*, \\ \widetilde{Q} = Q + (\Theta_2^*)^T R_{22}\Theta_2^* + (\Theta_2^*)^T S_2 + S_2^T \Theta_2^*, \\ \widetilde{S} = S_1 + R_{12}\Theta_2^*, \quad \widetilde{R} = R_{11}, \\ \widetilde{q} = q + [S_2^T + (\Theta_2^*)^T R_{22}]v_2^* + (\Theta_2^*)^T \rho_2, \quad \widetilde{\rho} = \rho_1 + R_{12}v_2^*, \\ \widetilde{G} = G, \quad \widetilde{g} = g. \end{cases} \tag{6.64}$$

Comparing (6.63) and (6.64), we see that one cannot say anything whether the outcome $\Theta_1^*(\cdot)X^*(\cdot) + v_1^*(\cdot)$ of $(\Theta_1^*(\cdot), v_1^*(\cdot))$ for the initial pair (t, x) has anything to do with $u_1^*(\cdot)$.

6.3 A Quadratic Game in a Hilbert Space*

With the compact notations leading to the state equation (6.50) and performance functional (6.51) for Problem (LQG), we may apply the same technique of Section 6.1.2 to represent the performance functional as follows:

$$J(t, x; u(\cdot)) = \frac{1}{2} \Big\{ \langle \Psi u, u \rangle + 2 \langle \psi, u \rangle + \psi_0 \Big\},$$

with $\Psi : \mathcal{U}[t, T] \to \mathcal{U}[t, T]$ being bounded and self-adjoint, and $\psi \in \mathcal{U}[t, T]$, $\psi_0 \in \mathbb{R}$. Also, Ψ admits the following representation:

$$(\Psi u)(s) = R(s)u(s) + \int_t^T K(s, r)u(r)dr.$$

Inspired by Section 6.1.2, in this section, we will carefully discuss a quadratic zero-sum game in a Hilbert space, the results will be useful in studying our Problem (LQG) from the open-loop point of view.

Let $\mathcal{H} = \mathcal{H}_1 \times \mathcal{H}_2$ with $\mathcal{H}_1$ and $\mathcal{H}_2$ being two Hilbert spaces, and consider a quadratic functional on $\mathcal{H}$:

$$\begin{aligned}
J(u) \equiv J(u_1, u_2) &= \langle \Psi u, u \rangle + 2 \langle \psi, u \rangle \\
&\equiv \langle \begin{pmatrix} \Psi_{11} & \Psi_{12} \\ \Psi_{21} & \Psi_{22} \end{pmatrix} \begin{pmatrix} u_1 \\ u_2 \end{pmatrix}, \begin{pmatrix} u_1 \\ u_2 \end{pmatrix} \rangle + 2 \langle \begin{pmatrix} \psi_1 \\ \psi_2 \end{pmatrix}, \begin{pmatrix} u_1 \\ u_2 \end{pmatrix} \rangle, \\
&\quad \forall u \equiv (u_1, u_2) \in \mathcal{H}.
\end{aligned} \tag{6.65}$$

We assume that $\Psi_{ij} : \mathcal{H}_j \to \mathcal{H}_i$ are bounded, $\Psi \equiv \begin{pmatrix} \Psi_{11} & \Psi_{12} \\ \Psi_{21} & \Psi_{22} \end{pmatrix}$ is self-adjoint, and $\psi \equiv \begin{pmatrix} \psi_1 \\ \psi_2 \end{pmatrix} \in \mathcal{H}_1 \times \mathcal{H}_2$. Consider a two-person zero-sum game with the cost/payoff functional given by (6.65). In the game, Player 1 takes $u_1 \in \mathcal{H}_1$ to minimize $J(u_1, u_2)$ and Player 2 takes $u_2 \in \mathcal{H}_2$ to maximize $J(u_1, u_2)$. For such a zero-sum game, we may define saddle point and upper and lower values in an obvious way. The following result is an extension of Theorem 6.1.4.

Theorem 6.3.1. *Let us list the following statements:*

(i) *The game has a saddle point* $(\hat{u}_1, \hat{u}_2) \in \mathcal{H}_1 \times \mathcal{H}_2$, *i.e.,*

$$J(\hat{u}_1, u_2) \leqslant J(\hat{u}_1, \hat{u}_2) \leqslant J(u_1, \hat{u}_2), \quad \forall (u_1, u_2) \in \mathcal{H}_1 \times \mathcal{H}_2. \tag{6.66}$$

(ii) *The following hold:*

$$\Psi_{11} \geqslant 0, \qquad \Psi_{22} \leqslant 0, \tag{6.67}$$

and

$$\psi \in \mathcal{R}(\Psi). \tag{6.68}$$

(iii) *The game has the value, i.e., the upper value V^+ and the lower value V^- are equal:*

$$V^+ \equiv \inf_{u_1 \in \mathcal{H}_1} \sup_{u_2 \in \mathcal{H}_2} J(u_1, u_2) = \sup_{u_2 \in \mathcal{H}_2} \inf_{u_1 \in \mathcal{H}_1} J(u_1, u_2) \equiv V^-.$$

(iv) *The upper value V^+ and the lower value V^- are finite.*

(v) *Conditions in (6.67) are satisfied and the following holds:*

$$\psi \in \overline{\mathcal{R}(\Psi)}.$$

Then the following relations hold:

(i) $\iff$ (ii) $\Rightarrow$ (iii) $\Rightarrow$ (iv) $\Rightarrow$ (v).

Further, in the case that (i) *holds, each saddle point $\hat{u} = (\hat{u}_1, \hat{u}_2) \in \mathcal{H}_1 \times \mathcal{H}_2$ is a solution of the following equation:*

$$\Psi\hat{u} + \psi = 0,$$

and it admits a representation

$$\hat{u} = -\Psi^\dagger \psi + (I - \Psi^\dagger \Psi)v,$$

for some $v \in \mathcal{H}$. Moreover, $\hat{u}$ is unique if and only if $\mathcal{N}(\Psi) = \{0\}$. Finally, in the case that $\mathcal{R}(\Psi)$ is closed, all the above five statements are equivalent.

Proof. (i) $\Rightarrow$ (ii): We first show (6.67) by a contradiction argument. If $\Psi_{11} \geqslant 0$ is not true, then $\langle \Psi_{11}u_1, u_1 \rangle < 0$, for some $u_1 \in \mathcal{H}_1$. Consequently, we have that (note (6.66))

$$J(\hat{u}_1, \hat{u}_2) \leqslant \lim_{\lambda \to \infty} J(\lambda u_1, \hat{u}_2)$$

$$= \lim_{\lambda \to \infty} \left\{ \langle \begin{pmatrix} \Psi_{11} & \Psi_{12} \\ \Psi_{21} & \Psi_{22} \end{pmatrix} \begin{pmatrix} \lambda u_1 \\ \hat{u}_2 \end{pmatrix}, \begin{pmatrix} \lambda u_1 \\ \hat{u}_2 \end{pmatrix} \rangle + 2 \langle \begin{pmatrix} \psi_1 \\ \psi_2 \end{pmatrix}, \begin{pmatrix} \lambda u_1 \\ \hat{u}_2 \end{pmatrix} \rangle \right\}$$

$$= \lim_{\lambda \to \infty} \lambda^2 \left\{ \langle \Psi_{11}u_1, u_1 \rangle + \frac{2}{\lambda} \langle \Psi_{12}\hat{u}_2, u_1 \rangle + \frac{1}{\lambda^2} \langle \Psi_{22}\hat{u}_2, \hat{u}_2 \rangle \right.$$

$$\left. + \frac{2}{\lambda} \langle \psi_1, u_1 \rangle + \frac{2}{\lambda^2} \langle \psi_2, \hat{u}_2 \rangle \right\} = -\infty.$$

This is a contradiction. Hence, $\Psi_{11} \geqslant 0$ must be true. Similarly, $\Psi_{22} \leqslant 0$ must hold.

Next, by (6.66) (comparing with (6.13)), we have

$$
\begin{aligned}
0 \geqslant\ & J(\hat{u}_1, u_2) - J(\hat{u}_1, \hat{u}_2) \\
=\ & \langle\, \Psi_{11}\hat{u}_1, \hat{u}_1 \,\rangle + 2\,\langle\, \Psi_{12}u_2, \hat{u}_1 \,\rangle + \langle\, \Psi_{22}u_2, u_2 \,\rangle + 2\,\langle\, \psi_1, \hat{u}_1 \,\rangle + 2\,\langle\, \psi_2, u_2 \,\rangle \\
& - \Big[\, \langle\, \Psi_{11}\hat{u}_1, \hat{u}_1 \,\rangle + 2\,\langle\, \Psi_{12}\hat{u}_2, \hat{u}_1 \,\rangle + \langle\, \Psi_{22}\hat{u}_2, \hat{u}_2 \,\rangle + 2\,\langle\, \psi_1, \hat{u}_1 \,\rangle + 2\,\langle\, \psi_2, \hat{u}_2 \,\rangle \,\Big] \\
=\ & \langle\, \Psi_{22}(\hat{u}_2 + u_2 - \hat{u}_2), \hat{u}_2 + u_2 - \hat{u}_2 \,\rangle - \langle\, \Psi_{22}\hat{u}_2, \hat{u}_2 \,\rangle \\
& + 2\,\langle\, \Psi_{21}\hat{u}_1 + \psi_2, u_2 - \hat{u}_2 \,\rangle \\
=\ & \langle\, \Psi_{22}(u_2 - \hat{u}_2), u_2 - \hat{u}_2 \,\rangle + 2\,\langle\, \Psi_{21}\hat{u}_1 + \Psi_{22}\hat{u}_2 + \psi_2, u_2 - \hat{u}_2 \,\rangle,
\end{aligned}
$$

for all $u_2 \in \mathcal{H}_2$. Hence, it is necessary that

$$
\Psi_{21}\hat{u}_1 + \Psi_{22}\hat{u}_2 + \psi_2 = 0.
$$

Similarly,

$$
\Psi_{11}\hat{u}_1 + \Psi_{12}\hat{u}_2 + \psi_1 = 0.
$$

Thus, (6.68) follows.

(ii) $\Rightarrow$ (i): Let (6.68) hold. Then the map $(u_1, u_2) \mapsto J(u_1, u_2)$ admits a critical point $\hat{u} \equiv (\hat{u}_1, \hat{u}_2)$. By Lemma 6.1.3, we get

$$
J(u_1, u_2) = \langle\, \Psi(u - \hat{u}), u - \hat{u} \,\rangle - \langle\, \Psi^\dagger \psi, \psi \,\rangle.
$$

Thus, $J(\hat{u}_1, \hat{u}_2) = -\langle\, \Psi^\dagger \psi, \psi \,\rangle$. Since $\Psi_{11} \geqslant 0$ and $\Psi_{22} \leqslant 0$, it follows that

$$
\begin{aligned}
J(\hat{u}_1, u_2) &= J(\hat{u}_1, \hat{u}_2) + \langle\, \Psi_{22}(u_2 - \hat{u}_2), u_2 - \hat{u}_2 \,\rangle \leqslant J(\hat{u}_1, \hat{u}_2), \\
J(u_1, \hat{u}_2) &= J(\hat{u}_1, \hat{u}_2) + \langle\, \Psi_{11}(u_1 - \hat{u}_1), u_1 - \hat{u}_1 \,\rangle \geqslant J(\hat{u}_1, \hat{u}_2).
\end{aligned}
$$

Hence, (6.66) follows.

(i) $\Rightarrow$ (iii) $\Rightarrow$ (iv) are trivial.

(iv) $\Rightarrow$ (v): We first show (6.67) by a contradiction argument. If $\Psi_{11} \geqslant 0$ is not true, then $\langle\, \Psi_{11}u_1, u_1 \,\rangle < 0$, for some $u_1 \in \mathcal{H}_1$. Consequently, for any fixed $u_2 \in \mathcal{H}_2$, we have that

$$
\lim_{\lambda \to \infty} J(\lambda u_1, u_2) = \lim_{\lambda \to \infty} \lambda^2 \langle\, \Psi_{11}u_1, u_1 \,\rangle = -\infty.
$$

This contradicts the finiteness of V^-. Hence, $\Psi_{11} \geqslant 0$ must be true. Similarly, by the finiteness of V^+, $\Psi_{22} \leqslant 0$ holds.

Now we show that $\psi \in \overline{\mathcal{R}(\Psi)} = \mathcal{N}(\Psi)^\perp$. Let

$$
\Psi\hat{u} \equiv \begin{pmatrix} \Psi_{11} & \Psi_{12} \\ \Psi_{21} & \Psi_{22} \end{pmatrix} \begin{pmatrix} \hat{u}_1 \\ \hat{u}_2 \end{pmatrix} = \begin{pmatrix} \Psi_{11}\hat{u}_1 + \Psi_{12}\hat{u}_2 \\ \Psi_{21}\hat{u}_1 + \Psi_{22}u_2 \end{pmatrix} = 0. \tag{6.69}
$$

We want to show that $\langle \psi, \hat{u} \rangle = 0$. To this end, we note that $\Psi_{12} = \Psi_{21}^*$. Hence, by (6.69), one has

$$\langle \Psi_{11}\hat{u}_1, \hat{u}_1 \rangle = -\langle \Psi_{12}\hat{u}_2, \hat{u}_1 \rangle = -\langle \hat{u}_2, \Psi_{21}\hat{u}_1 \rangle = \langle \Psi_{22}\hat{u}_2, \hat{u}_2 \rangle.$$

Due to (6.67), we must have

$$\Psi_{11}\hat{u}_1 = 0, \qquad \Psi_{22}\hat{u}_2 = 0.$$

Hence, it follows from (6.69) that

$$\Psi_{12}\hat{u}_2 = 0, \qquad \Psi_{21}\hat{u}_1 = 0.$$

Consequently,

$$\begin{aligned}
J(\lambda\hat{u}_1, u_2) &= \lambda^2 \langle \Psi_{11}\hat{u}_1, \hat{u}_1 \rangle + 2\lambda \langle \Psi_{21}\hat{u}_1, u_2 \rangle + \langle \Psi_{22}u_2, u_2 \rangle \\
&\quad + 2\lambda \langle \psi_1, \hat{u}_1 \rangle + 2 \langle \psi_2, u_2 \rangle \\
&= 2\lambda \langle \psi_1, \hat{u}_1 \rangle + 2 \langle \psi_2, u_2 \rangle + \langle \Psi_{22}u_2, u_2 \rangle.
\end{aligned}$$

By the finiteness of V^-, we can find some $\bar{u}_2 \in \mathcal{H}_2$ such that

$$-\infty < \inf_{u_1 \in \mathcal{H}_1} J(u_1, \bar{u}_2) \leqslant \inf_{\lambda \in \mathbb{R}} J(\lambda\hat{u}_1, \bar{u}_2).$$

Hence, we must have $\langle \psi_1, \hat{u}_1 \rangle = 0$. Similarly, one can obtain $\langle \psi_2, \hat{u}_2 \rangle = 0$. These imply $\langle \psi, \hat{u} \rangle = 0$, proving (v).

The rest of the proof is clear. ☐

Note that condition (6.67) is equivalent to the following *convexity-concavity condition* for the performance functional:

$$\begin{cases} u_1 \mapsto J(u_1, u_2) \text{ is convex,} \\ u_2 \mapsto J(u_1, u_2) \text{ is concave.} \end{cases}$$

Hence, according to the above theorem, we see that the above convexity-concavity condition is necessary for the finiteness of the upper and lower values. Similar to Corollary 6.1.6, we have the following interesting result for Problem (LQG).

Corollary 6.3.2. *Let (LQG1)–(LQG2) hold. Let $R(\cdot)^{-1}$ exist and bounded. Then for any $x \in \mathbb{R}^n$, Problem (LQG) admits an open-loop saddle point at (t, x) if and only if both upper and lower open-loop values $V^{\pm}(t, x)$ are finite.*

Proof. It suffices to prove the sufficiency. By the representation of Ψ, we see that $R^{-1}\Psi$ is a Fredholm operator on $\mathcal{U}[t, T]$. Then $\mathcal{R}(\Psi)$ is closed. Next, by the finiteness of the upper and lower values $V^{\pm}(t, x)$, we see that (6.67) holds. Consequently, applying Theorem 6.3.1, we obtain the existence of an open-loop saddle point for Problem (LQG). □

In Proposition 6.2.2, we need the existence of the open-value and the achievability of the value in order to get the existence of an open-loop saddle point. Here, we do not assume those, thanks to the Fredholm operator theory. Note that the invertibility condition for $R(\cdot)$ plays an essential role in the above. We will see an example that if $R(\cdot)$ has some degeneracy, the above conclusion could fail.

6.4 Open-Loop Saddle Points and Two-Point Boundary Value Problems

In this section, we present a characterization of open-loop saddle points of Problem (LQG) in terms of two-point boundary value problems. The main result of this section can be stated as follows.

Theorem 6.4.1. *Let (LQG1)–(LQG2) hold and let $(t, x) \in [t, T) \times \mathbb{R}^n$ be given. Then Problem (LQG) admits an open-loop saddle point $u^*(\cdot) \equiv (u_1^*(\cdot), u_2^*(\cdot)) \in \mathcal{U}_1[t, T] \times \mathcal{U}_2[t, T]$ with $X^*(\cdot) \equiv X(\cdot\,; t, x, u^*(\cdot))$ being the corresponding state trajectory if and only if the following two-point boundary value problem admits a solution $(X^*(\cdot), Y^*(\cdot))$:*

$$\begin{cases} \dot{X}^*(s) = A(s)X^*(s) + B(s)u^*(s) + f(s), & s \in [t, T], \\ \dot{Y}^*(s) = -\left[A(s)^T Y^*(s) + Q(s)X^*(s) + S(s)^T u^*(s) + q(s)\right], & s \in [t, T], \quad (6.70) \\ X^*(t) = x, \qquad Y^*(T) = GX^*(T) + g, \end{cases}$$

such that the following stationarity condition holds:

$$B(s)^T Y^*(s) + S(s)X^*(s) + R(s)u^*(s) + \rho(s) = 0, \tag{6.71}$$
$$\text{a.e. } s \in [t, T],$$

and the following convexity-concavity conditions hold:

$$\begin{cases} u_1(\cdot) \mapsto J(t, 0; u_1(\cdot), u_2(\cdot)) \text{ is convex,} \\ u_2(\cdot) \mapsto J(t, 0; u_1(\cdot), u_2(\cdot)) \text{ is concave.} \end{cases} \tag{6.72}$$

Or equivalently, for $i = 1, 2$,

$$(-1)^{i-1} \left\{ \langle GX_i(T), X_i(T) \rangle + \int_t^T \left[\langle Q(s)X_i(s), X_i(s) \rangle \right. \right.$$
$$\left. \left. + 2\langle S_i(s)X_i(s), u_i(s) \rangle + \langle R_{ii}(s)u_i(s), u_i(s) \rangle \right] ds \right\} \geqslant 0, \tag{6.73}$$
$$\forall u_i(\cdot) \in \mathcal{U}_i[t, T],$$

where $X_i(\cdot)$ solves the following:

$$\begin{cases} \dot{X}_i(s) = A(s)X_i(s) + B_i(s)u_i(s), & s \in [t,T], \\ X_i(t) = 0. \end{cases} \tag{6.74}$$

Proof. Let $u^*(\cdot) \equiv (u_1^*(\cdot), u_2^*(\cdot)) \in \mathcal{U}_1[t,T] \times \mathcal{U}_2[t,T]$ and $X^*(\cdot)$ be the corresponding state process. Further, let $Y^*(\cdot)$ be the solution to the second equation in (6.70). For any $u_1(\cdot) \in \mathcal{U}_1[t,T]$ and $\varepsilon \in \mathbb{R}$, let $X^\varepsilon(\cdot)$ be the solution to the following perturbed state equation:

$$\begin{cases} \dot{X}^\varepsilon(s) = A(s)X^\varepsilon(s) + B_1(s)[u_1^*(s) + \varepsilon u_1(s)] \\ \qquad\qquad + B_2(s)u_2^*(s) + f(s), \quad s \in [t,T], \\ X^\varepsilon(t) = x. \end{cases}$$

Then $X_1(\cdot) = \dfrac{X^\varepsilon(\cdot) - X^*(\cdot)}{\varepsilon}$ is independent of ε satisfying (6.74), and

$$J(t,x; u_1^*(\cdot) + \varepsilon u_1(\cdot), u_2^*(\cdot)) - J(t,x; u_1^*(\cdot), u_2^*(\cdot))$$

$$= \frac{\varepsilon}{2} \Big\{ \langle G[2X^*(T) + \varepsilon X_1(T)], X_1(T) \rangle + 2\langle g, X_1(T) \rangle$$

$$+ \int_t^T \Big[\Big\langle \begin{pmatrix} Q & S_1^T & S_2^T \\ S_1 & R_{11} & R_{12} \\ S_2 & R_{21} & R_{22} \end{pmatrix} \begin{pmatrix} 2X^* + \varepsilon X_1 \\ 2u_1^* + \varepsilon u_1 \\ 2u_2^* \end{pmatrix}, \begin{pmatrix} X_1 \\ u_1 \\ 0 \end{pmatrix} \Big\rangle$$

$$+ 2\Big\langle \begin{pmatrix} q \\ \rho_1 \end{pmatrix}, \begin{pmatrix} X_1 \\ u_1 \end{pmatrix} \Big\rangle \Big] ds \Big\}$$

$$= \varepsilon \Big\{ \langle GX^*(T) + g, X_1(T) \rangle + \int_t^T \Big[\langle QX^* + S^T u^* + q, X_1 \rangle$$

$$+ \langle S_1 X^* + R_{11}u_1^* + R_{12}u_2^* + \rho_1, u_1 \rangle \Big] ds \Big\}$$

$$+ \frac{\varepsilon^2}{2} \Big\{ \langle GX_1(T), X_1(T) \rangle + \int_t^T \Big[\langle QX_1, X_1 \rangle + 2\langle S_1 X_1, u_1 \rangle$$

$$+ \langle R_{11}u_1, u_1 \rangle \Big] ds \Big\}.$$

On the other hand, we have

$$\langle GX^*(T) + g, X_1(T) \rangle + \int_t^T \Big[\langle QX^* + S^T u^* + q, X_1 \rangle$$

$$+ \langle S_1 X^* + R_{11}u_1^* + R_{12}u_2^* + \rho_1, u_1 \rangle \Big] ds$$

$$= \int_t^T \Big[\langle -(A^T Y^* + QX^* + S^T u^* + q), X_1 \rangle + \langle Y^*, AX_1 + B_1 u_1 \rangle$$

$$+ \langle QX^* + S^T u^* + q, X_1 \rangle + \langle S_1 X^* + R_{11}u_1^* + R_{12}u_2^* + \rho_1, u_1 \rangle \Big] ds$$

$$= \mathbb{E} \int_t^T \langle B_1^T Y^* + S_1 X^* + R_{11}u_1^* + R_{12}u_2^* + \rho_1, u_1 \rangle \, ds.$$

Hence,

$$
J(t, x; u_1^*(\cdot) + \varepsilon u_1(\cdot), u_2^*(\cdot)) - J(t, x; u_1^*(\cdot), u_2^*(\cdot))
$$

$$
= \varepsilon \Big\{ \int_t^T \langle B_1^T Y^* + S_1 X^* + R_{11} u_1^* + R_{12} u_2^* + \rho_1, u_1 \rangle \, ds \Big\}
$$

$$
+ \frac{\varepsilon^2}{2} \Big\{ \langle G X_1(T), X_1(T) \rangle
$$

$$
+ \int_t^T \Big[\langle Q X_1, X_1 \rangle + 2 \langle S_1 X_1, u_1 \rangle + \langle R_{11} u_1, u_1 \rangle \Big] ds \Big\}.
$$

Therefore,

$$
J(t, x; u_1^*(\cdot), u_2^*(\cdot)) \leqslant J(t, x; u_1^*(\cdot) + \varepsilon u_1(\cdot), u_2^*(\cdot)),
$$

$$
\forall u_1(\cdot) \in \mathcal{U}_1[t, T], \quad \varepsilon \in \mathbb{R},
$$

if and only if (6.73) holds for $i = 1$, and

$$
B_1^T Y^* + S_1 X^* + R_{11} u_1^* + R_{12} u_2^* + \rho_1 = 0, \quad \text{a.e. } s \in [t, T]. \tag{6.75}
$$

Similarly,

$$
J(t, x; u_1^*(\cdot), u_2^*(\cdot)) \geqslant J(t, x; u_1^*(\cdot), u_2^*(\cdot) + \varepsilon u_2(\cdot)),
$$

$$
\forall u_2(\cdot) \in \mathcal{U}_2[t, T], \quad \varepsilon \in \mathbb{R},
$$

if and only if (6.73) holds for $i = 2$, and

$$
B_2^T Y^* + S_2 X^* + R_{21} u_1^* + R_{22} u_2^* + \rho_2 = 0, \quad \text{a.e. } s \in [t, T]. \tag{6.76}
$$

Combining (6.75)–(6.76), we obtain (6.71). □

The following result is concerned with the uniqueness of open-loop saddle points.

Theorem 6.4.2. *Let (LQG1)–(LQG2) hold, and let $(t, x) \in [0, T) \times \mathbb{R}^n$ be given. Suppose Problem (LQG) admits a unique open-loop saddle point $u^*(\cdot)$ at (t, x). Then (6.70)–(6.71) admits a unique solution $(X^*(\cdot), Y^*(\cdot), u^*(\cdot))$. Conversely, if the convexity-concavity conditions (6.72) hold and (6.70)–(6.71) admit a unique adapted solution $(X^*(\cdot), Y^*(\cdot), u^*(\cdot))$, then $u^*(\cdot)$ is the unique saddle point of Problem (LQG).*

Proof. Suppose $u^*(\cdot) \in \mathcal{U}[t, T]$ is a unique open-loop saddle point of Problem (LQG). Then by Theorem 6.4.1, (6.70) admits a solution $(X^*(\cdot), Y^*(\cdot), u^*(\cdot))$, and the convex-concave conditions (6.73) hold. Now, if (6.70) admits another different adapted solution $(\bar{X}(\cdot), \bar{Y}(\cdot), \bar{u}(\cdot))$. Since the convexity-concavity conditions are satisfied, by the sufficiency part of Theorem 6.4.1, $\bar{u}(\cdot)$ is a different open-loop saddle point, a contradiction.

Conversely, if Problem (LQG) has two different open-loop saddle points, then (6.70) will have two different solutions. □

We now present an example that due to the degeneracy of $R(\cdot)$, the conclusion of Corollary 6.3.2 fails.

Example 6.4.3. Consider the following one-dimensional state equation:

$$\begin{cases} \dot{X}(s) = u_1(s) + u_2(s), & s \in [t, T], \\ X(t) = x, \end{cases}$$

with $T \in (0, \frac{\pi}{2})$ and with the performance functional:

$$J(t, x; u_1(\cdot), u_2(\cdot)) = \frac{1}{2} \int_t^T [X(s)^2 - u_2(s)^2] ds.$$

The open-loop lower value function satisfies

$$\begin{aligned} V^-(t, x) &= \sup_{u_2(\cdot) \in \mathcal{U}_2[t,T]} \inf_{u_1(\cdot) \in \mathcal{U}_1[t,T]} J(t, x; u_1(\cdot), u_2(\cdot)) \\ &\geqslant \inf_{u_1(\cdot) \in \mathcal{U}_1[t,T]} J(t, x; u_1(\cdot), 0) = 0. \end{aligned}$$

On the other hand, for any $u_2(\cdot) \in \mathcal{U}_2[t, T]$ and $u_1(\cdot) = 0$, one has

$$\begin{cases} \dot{X}(s) = u_2(s), & s \in [t, T], \\ X(t) = x. \end{cases}$$

Hence, by using the result of Example 6.1.14, one obtains

$$\begin{aligned} V^+(t, x) &= \inf_{u_1(\cdot) \in \mathcal{U}_1[t,T]} \sup_{u_2(\cdot) \in \mathcal{U}_2[t,T]} J(t, x; u_1(\cdot), u_2(\cdot)) \\ &\leqslant \sup_{u_2(\cdot) \in \mathcal{U}_2[t,T]} J(t, x; 0, u_2(\cdot)) = \frac{x^2 \tan(T - t)}{2}. \end{aligned}$$

Thus, both the open-loop lower and upper value functions are finite. Now, suppose $(u_1^*(\cdot), u_2^*(\cdot)) \in \mathcal{U}_1[t, T] \times \mathcal{U}_2[t, T]$ is an open-loop saddle point of the above problem for the initial pair $(t, x) \in [0, T) \times (\mathbb{R} \setminus \{0\})$, then by Theorem 6.4.1, we have

$$\begin{pmatrix} 1 \\ 1 \end{pmatrix} Y^*(s) + \begin{pmatrix} 0 & 0 \\ 0 & -1 \end{pmatrix} \begin{pmatrix} u_1^*(s) \\ u_2^*(s) \end{pmatrix} = 0, \qquad \text{a.e. } s \in [t, T], \qquad (6.77)$$

where $(X^*(\cdot), Y^*(\cdot))$ is the solution of the following system:

$$\begin{cases} \dot{X}^*(s) = u_1^*(s) + u_2^*(s), & s \in [t, T], \\ \dot{Y}^*(s) = -X^*(s), & s \in [t, T], \\ X^*(t) = x, & Y^*(T) = 0. \end{cases}$$

From (6.77), we have

$$Y^*(s) = 0, \qquad u_2^*(s) = 0, \qquad \text{a.e. } s \in [t, T].$$

Hence, it is necessary that

$$\begin{cases} X^*(s) = 0, & \text{a.e. } s \in [t, T], \\ u_1^*(s) = u_2^*(s) = 0, & \text{a.e. } s \in [t, T]. \end{cases}$$

This leads to a contradiction since $X^*(t) = x \neq 0$. Therefore, the corresponding differential game does not have an open-loop saddle point for $(t, x) \in [0, T) \times (\mathbb{R} \setminus \{0\})$, although both open-loop lower and upper value functions are finite.

6.5   Closed-Loop Saddle Points and Riccati Equations

We now look at closed-loop saddle points for Problem (LQG). First, we present the following result which is a consequence of Theorem 6.4.1.

Proposition 6.5.1. *Let (LQG1)–(LQG2) hold. Let $(\Theta^*(\cdot), v^*(\cdot)) \in \mathcal{Q}[t, T] \times \mathcal{U}[t, T]$ be a closed-loop saddle point of Problem (LQG). Then the following system admits a solution $(X^*(\cdot), Y^*(\cdot))$:*

$$\begin{cases} \dot{X}^* = (A + B\Theta^*)X^* + Bv^* + f, & s \in [t, T], \\ \dot{Y}^*(s) = -\{A^T Y^* + (Q + S^T\Theta^*)X^* + S^T v^* + q\}, & (6.78) \\ X^*(t) = x, \qquad Y^*(T) = GX^*(T) + g, \end{cases}$$

and the following stationarity condition holds:

$$Rv^* + B^T Y^* + (S + R\Theta^*)X^* + \rho = 0, \qquad \text{a.e.}$$

Proof. Let $(\Theta^*(\cdot), v^*(\cdot)) \in \mathcal{Q}[t, T] \times \mathcal{U}[t, T]$ be a closed-loop saddle point of Problem (LQG) with $\Theta^*(\cdot) = (\Theta_1^*(\cdot), \Theta_2^*(\cdot))$ and $v^*(\cdot) = (v_1^*(\cdot), v_2^*(\cdot)^T)$. We consider state equation (6.57) with the cost functional (6.58) for which we carry out some computation: (denoting $\tilde{v} = (v_1, v_2^*)$)

$$J_1(t, x; v_1(\cdot)) \equiv J(t, x; \Theta^* X(\cdot) + \tilde{v}(\cdot))$$

$$= \frac{1}{2} \Big\{ \langle GX(T), X(T) \rangle + 2 \langle g, X(T) \rangle + \int_t^T \Big[\langle QX, X \rangle + 2 \langle SX, \Theta^* X + \tilde{v} \rangle$$

$$+ \langle R(\Theta^* X + \tilde{v}), \Theta^* X + \tilde{v} \rangle + 2 \langle q, X \rangle + 2 \langle \rho, \Theta^* X + \tilde{v} \rangle \Big] ds \Big\}$$

$$= \frac{1}{2}\Big\{ \langle GX(T), X(T) \rangle + 2\langle g, X(T) \rangle$$

$$+ \int_t^T \Big[\langle [Q + (\Theta^*)^T S + S^T \Theta^* + (\Theta^*)^T R\Theta^*]X, X \rangle$$

$$+ 2\Big\langle \begin{pmatrix} (S_1 + R_1\Theta^*)X \\ (S_2 + R_2\Theta^*)X \end{pmatrix}, \begin{pmatrix} v_1 \\ v_2^* \end{pmatrix} \Big\rangle + \Big\langle \begin{pmatrix} R_{11} & R_{12} \\ R_{21} & R_{22} \end{pmatrix} \begin{pmatrix} v_1 \\ v_2^* \end{pmatrix}, \begin{pmatrix} v_1 \\ v_2^* \end{pmatrix} \Big\rangle$$

$$+ 2\langle q + (\Theta^*)^T \rho, X \rangle + 2\Big\langle \begin{pmatrix} \rho_1 \\ \rho_2 \end{pmatrix}, \begin{pmatrix} v_1 \\ v_2^* \end{pmatrix} \Big\rangle \Big] ds \Big\}$$

$$= \frac{1}{2}\Big\{ \langle GX(T), X(T) \rangle + 2\langle g, X(T) \rangle$$

$$+ \int_t^T \Big[\langle [Q + (\Theta^*)^T S + S^T \Theta^* + (\Theta^*)^T R\Theta^*]X, X \rangle$$

$$+ 2\langle (S_1 + R_1\Theta^*)X, v_1 \rangle + 2\langle q + (\Theta^*)^T \rho + (S_2 + R_2\Theta^*)^T v_2^*, X \rangle$$

$$+ \langle R_{11}v_1, v_1 \rangle + 2\langle \rho_1 + R_{12}v_2^*, v_1 \rangle + \langle R_{22}v_2^*, v_2^* \rangle + 2\langle \rho_2, v_2^* \rangle \Big] ds \Big\}.$$

We know that $v_1^*(\cdot)$ is an open-loop optimal control for the problem with state equation (6.57) and the above cost functional. Thus, according to Theorem 6.4.1, we have

$$0 = B_1^T Y^* + (S_1 + R_1\Theta^*)X^* + R_{11}v_1^* + \rho_1 + R_{12}v_2^*, \quad \text{a.e.} ,$$

with $Y^*(\cdot)$ being the solution to the following terminal value problem:

$$\begin{cases} \dot{Y}^* = -\big\{ (A + B\Theta^*)^T Y^* + [Q + (\Theta^*)^T S + S^T \Theta^* + (\Theta^*)^T R\Theta^*]X^* \\ \qquad + (S_1 + R_1\Theta^*)^T v_1^* + q + (\Theta^*)^T \rho + (S_2 + R_2\Theta^*)^T v_2^* \big\} \\ \quad = -\big\{ A^T Y^* + QX^* + S^T(\Theta^* X^* + v^*) + q \\ \qquad + (\Theta^*)^T [B^T Y^* + SX^* + R(\Theta^* X^* + v^*) + \rho] \big\} \\ Y^*(T) = GX^*(T) + g. \end{cases}$$

Likewise, by considering state equation (6.59) and payoff functional (6.60), we can obtain

$$0 = B_2^T Y^* + (S_2 + R_2\Theta^*)X^* + R_{21}v_1^* + \rho_2 + R_{22}v_2^*, \quad \text{a.e.}$$

with $Y^*(\cdot)$ being the solution to the same terminal value problem as above. Thus,

$$0 = B^T Y^* + (S + R\Theta^*)X^* + Rv^* + \rho, \quad \text{a.e.}$$

Then the above terminal value problem is reduced to that in (6.78). $\qquad\square$

The following result gives a characterization for closed-loop saddle points of Problem (LQG).

Theorem 6.5.2. *Let* (LQG1)–(LQG2) *hold. Then Problem* (LQG) *admits a closed-loop saddle point* $(\Theta^*(\cdot), v^*(\cdot)) \in \mathcal{Q}[t,T] \times \mathcal{U}[t,T]$ *with* $\Theta^*(\cdot) \equiv (\Theta_1^*(\cdot), \Theta_2^*(\cdot))$ *and* $v^*(\cdot) \equiv (v_1^*(\cdot), v_2^*(\cdot))$ *if and only if the following Riccati equation:*

$$\begin{cases} \dot{P}(s) + P(s)A(s) + A(s)^T P(s) + Q(s) \\ \quad - [P(s)B(s) + S(s)^T]R(s)^\dagger [B(s)^T P(s) + S(s)] = 0, \\ \qquad\qquad\qquad\qquad \text{a.e. } s \in [t,T], \\ P(T) = G, \end{cases} \quad (6.79)$$

admits a solution $P(\cdot) \in C([t,T];\mathbb{S}^n)$ *such that*

$$\begin{cases} \mathcal{R}(B(s)^T P(s) + S(s)) \subseteq \mathcal{R}(R(s)), \quad \text{a.e. } s \in [t,T], \\ R(\cdot)^\dagger [B(\cdot)^T P(\cdot) + S(\cdot)] \in L^2(t,T;\mathbb{R}^{m \times n}), \\ R_{11}(s) \geqslant 0, \quad R_{22}(s) \leqslant 0, \quad \text{a.e. } s \in [t,T], \end{cases} \quad (6.80)$$

and the solution $\eta(\cdot)$ *of the following terminal value problem:*

$$\begin{cases} \dot{\eta} = -\Big\{ [A^T - (PB + S^T)R^\dagger B^T]\eta - (PB + S^T)R^\dagger \rho + Pf + q \Big\}, \\ \eta(T) = g, \end{cases}$$

$$(6.81)$$

satisfies

$$\begin{cases} B(s)^T \eta(s) + \rho(s) \in \mathcal{R}(R(s)), \quad \text{a.e. } s \in [t,T], \\ R(\cdot)^\dagger [B(\cdot)^T \eta(\cdot) + \rho(\cdot)] \in L^2(t,T;\mathbb{R}^m). \end{cases}$$

In this case, the closed-loop saddle point $(\Theta^*(\cdot), v^*(\cdot))$ *admits the following representation:*

$$\begin{cases} \Theta^*(\cdot) = -R(\cdot)^\dagger [B(\cdot)^T P(\cdot) + S(\cdot)] + \{I - R(\cdot)^\dagger R(\cdot)\}\theta(\cdot), \\ v^*(\cdot) = -R(\cdot)^\dagger [B(\cdot)^T \eta(\cdot) + \rho(\cdot)] + \{I - R(\cdot)^\dagger R(\cdot)\}\nu(\cdot), \end{cases}$$

for some $\theta(\cdot) \in L^2(t,T;\mathbb{R}^{m \times n})$ *and* $\nu(\cdot) \in L^2(t,T;\mathbb{R}^m)$. *Further, the value function admits the following representation:*

$$V(t,x) = \frac{1}{2} \Big\{ \langle P(t)x, x \rangle + 2\langle \eta(t), x \rangle$$

$$+ \int_t^T [2\langle \eta, f \rangle - \langle R^\dagger (B^T \eta + \rho), B^T \eta + \rho \rangle] ds \Big\}.$$

Proof. Necessity. Let $(\Theta^*(\cdot), v^*(\cdot))$ be a closed-loop saddle point of Problem (LQG) over $[t,T]$, where

$$\Theta^*(\cdot) \equiv (\Theta_1^*(\cdot), \Theta_2^*(\cdot)) \in \mathcal{Q}_1[t,T] \times \mathcal{Q}_2[t,T],$$

$$v^*(\cdot) \equiv (v_1^*(\cdot), v_2^*(\cdot)) \in \mathcal{U}_1[t,T] \times \mathcal{U}_2[t,T].$$

Then, by Proposition 6.5.1, for any $x \in \mathbb{R}^n$, the following system admits a solution $(X^*(\cdot), Y^*(\cdot))$:

$$\begin{cases} \dot{X}^* = (A + B\Theta^*)X^* + Bv^* + f, & s \in [t, T], \\ \dot{Y}^*(s) = -\{A^T Y^* + (Q + S^T \Theta^*)X^* + S^T v^* + q\}, \\ X^*(t) = x, \qquad Y^*(T) = GX^*(T) + g, \end{cases}$$

and the following stationarity condition holds:

$$B^T Y^* + (S + R\Theta^*)X^* + Rv^* + \rho = 0, \quad \text{a.e.}$$

Since the above admits a solution for each $x \in \mathbb{R}^n$, and $(\Theta^*(\cdot), v^*(\cdot))$ is independent of x, by subtracting solutions corresponding x and 0, the latter from the former, we see that for any $x \in \mathbb{R}^n$, as long as $(X(\cdot), Y(\cdot))$ is the solution to the following system:

$$\begin{cases} \dot{X} = (A + B\Theta^*)X, & s \in [t, T], \\ \dot{Y} = -[A^T Y + (Q + S^T \Theta^*)X], & s \in [t, T], \\ X(t) = x, \qquad Y(T) = GX(T), \end{cases}$$

one must have the following stationarity condition:

$$B^T Y + (S + R\Theta^*)X = 0, \quad \text{a.e. } s \in [t, T].$$

Clearly, the above are the same as (6.39)–(6.40). Hence, we can copy line by line the proof of Theorem 6.1.11 to get all the necessity conclusions, except the third relation in (6.80) whose proof is contained in the proof of the sufficiency below.

Sufficiency. We take any

$$u(\cdot) = (u_1(\cdot)^T, u_2(\cdot)^T)^T \in \mathcal{U}_1[t, T] \times \mathcal{U}_2[t, T],$$

let $X(\cdot) \equiv X(\cdot; t, x, u(\cdot))$ be the corresponding state process. Then

$$J(t, x; u(\cdot)) = \frac{1}{2}\Big\{ \langle GX(T), X(T) \rangle + 2\langle g, X(T) \rangle$$
$$+ \int_t^T [\langle QX, X \rangle + 2\langle SX, u \rangle + 2\langle Ru, u \rangle + 2\langle q, X \rangle + 2\langle \rho, u \rangle] ds \Big\}$$

$$
= \frac{1}{2}\Big\{ \langle P(t)x, x \rangle + 2\langle \eta(t), x \rangle
$$

$$
+ \int_t^T \{ \langle\, [\, -PA - A^T P - Q + (PB + S^T)R^\dagger(B^T P + S)]X, X \,\rangle
$$

$$
+ \langle P(AX + Bu + f), X \rangle + \langle PX, AX + Bu + f \rangle
$$

$$
+ 2\langle\, [\, -A^T + (PB + S^T)R^\dagger B^T]\eta, X \,\rangle + 2\langle\, (PB + S^T)R^\dagger \rho - Pf - q, X \,\rangle
$$

$$
+ 2\langle \eta, AX + Bu + f \rangle + \langle QX, X \rangle + 2\langle SX, u \rangle
$$

$$
+ \langle Ru, u \rangle + 2\langle q, X \rangle + 2\langle \rho, u \rangle \} ds \Big\}
$$

$$
= \frac{1}{2}\Big\{ \langle P(t)x, x \rangle + 2\langle \eta(t), x \rangle + \int_t^T \Big[\langle (PB + S^T)R^\dagger(B^T P + S)X, X \rangle
$$

$$
+ 2\langle (B^T P + S)X + B^T \eta + \rho, u \rangle + \langle Ru, u \rangle
$$

$$
+ 2\langle (PB + S^T)R^\dagger(B^T \eta + \rho), X \rangle + 2\langle \eta, f \rangle \Big] ds \Big\}.
$$

Note that

$$
B^T P + S = -R\Theta^*, \qquad B^T \eta + \rho = -Rv^*,
$$

and

$$
\langle Rv^*, v^* \rangle = \langle RR^\dagger(B^T \eta + \rho), R^\dagger(B^T \eta + \rho) \rangle
$$

$$
= \langle R^\dagger(B^T \eta + \rho), B^T \eta + \rho \rangle.
$$

Thus,

$$
J(t, x; u(\cdot)) = \frac{1}{2}\Big\{ \langle P(t)x, x \rangle + 2\langle \eta(t), x \rangle
$$

$$
+ \int_t^T \Big[\langle (PB + S^T)R^\dagger(B^T P + S)X, X \rangle
$$

$$
+ 2\langle (B^T P + S)X + B^T \eta + \rho, u \rangle + \langle Ru, u \rangle
$$

$$
+ 2\langle (PB + S^T)R^\dagger(B^T \eta + \rho), X \rangle + 2\langle \eta, f \rangle \Big] ds \Big\}
$$

$$
= \frac{1}{2}\Big\{ \langle P(t)x, x \rangle + 2\langle \eta(t), x \rangle
$$

$$
+ \int_t^T \Big[\langle (\Theta^*)^T RR^\dagger R\Theta^* X, X \rangle - 2\langle R(\Theta^* X + v^*), u \rangle
$$

$$
+ \langle Ru, u \rangle + 2\langle (\Theta^*)^T RR^\dagger Rv^*, X \rangle + 2\langle \eta, f \rangle \Big] ds \Big\}
$$

$$
= \frac{1}{2}\Big\{ \langle P(t)x, x \rangle + 2\langle \eta(t), x \rangle + \int_t^T \Big[2\langle \eta, f \rangle + \langle R\Theta^* X, \Theta^* X \rangle
$$

$$
- 2\langle R(\Theta^* X + v^*), u \rangle + \langle Ru, u \rangle + 2\langle R\Theta^* X, v^* \rangle \Big] ds \Big\}
$$

$$= \frac{1}{2} \Big\{ \langle P(t)x, x \rangle + 2 \langle \eta(t), x \rangle$$

$$+ \int_t^T \Big[2 \langle \eta, f \rangle - \langle R^\dagger (B^T \eta + \rho), B^T \eta + \rho \rangle$$

$$+ \langle R(u - \Theta^* X - v^*), u - \Theta^* X - v^* \rangle \Big] ds \Big\}$$

$$= J\big(t, x; \Theta^*(\cdot) X^*(\cdot) + v^*(\cdot)\big)$$

$$+ \frac{1}{2} \int_t^T \langle R(u - \Theta^* X - v^*), u - \Theta^* X - v^* \rangle \, ds.$$

Consequently,

$$J(t, x; \Theta_1^*(\cdot) X(\cdot) + v_1(\cdot), \Theta_2^*(\cdot) X(\cdot) + v_2^*(\cdot))$$

$$= J(t, x; \Theta^*(\cdot) X^*(\cdot) + v^*(\cdot)) + \frac{1}{2} \int_t^T \langle R_{11}(v_1 - v_1^*), v_1 - v_1^* \rangle \, ds.$$

Hence,

$$J(t, x; \Theta^*(\cdot) X^*(\cdot) + v^*(\cdot)) \leqslant J(t, x; \Theta_1^*(\cdot) X(\cdot) + v_1(\cdot), \Theta_2^*(\cdot) X(\cdot)),$$

$$\forall v_1(\cdot) \in \mathcal{U}_1[t, T],$$

if and only if

$$R_{11} \geqslant 0, \quad \text{a.e. } s \in [t, T].$$

Similarly,

$$J(t, x; \Theta_1^*(\cdot) X(\cdot) + v_1^*(\cdot), \Theta_2^*(\cdot) X(\cdot) + v_2(\cdot))$$

$$= J(t, x; \Theta^*(\cdot) X^*(\cdot) + v^*(\cdot)) + \frac{1}{2} \int_t^T \langle R_{22}(v_2 - v_2^*), v_2 - v_2^* \rangle \, ds.$$

Hence,

$$J(t, x; \Theta^*(\cdot) X^*(\cdot) + v^*(\cdot)) \geqslant J(t, x; \Theta_1^*(\cdot) X(\cdot) + v_1^*(\cdot), \Theta_2^*(\cdot) X(\cdot) + v_2(\cdot)),$$

$$\forall v_1(\cdot) \in \mathcal{U}_1[t, T],$$

if and only if

$$R_{22} \leqslant 0, \quad \text{a.e. } s \in [t, T].$$

Thus, $(\Theta^*(\cdot), v^*(\cdot))$ is a closed-loop saddle point of Problem (LQG). $\qquad \square$

A solution $P(\cdot)$ satisfying (6.80) is called a *regular solution*. The following result shows that the regular solution of (6.79) is unique.

Corollary 6.5.3. *Let* (LQG1)–(LQG2) *hold. Then the Riccati equation* (6.79) *admits at most one solution* $P(\cdot) \in C([t, T]; \mathbb{S}^n)$ *such that* (6.80) *hold.*

Proof. Consider Problem (LQG)*. Then the solution $\eta(\cdot)$ of (6.81) is zero. Suppose that $P(\cdot)$ and $\bar{P}(\cdot)$ are two solutions of Riccati equation (6.79) satisfying (6.80). By Theorem 6.5.2, we have

$$\langle P(t)x, x \rangle = 2V(t,x) = \langle \bar{P}(t)x, x \rangle, \quad \forall x \in \mathbb{R},$$

which implies $P(t) = \bar{P}(t)$. By considering Problem (LQG)* on $[s, T], t < s < T$, we obtain

$$P(s) = \bar{P}(s), \qquad \forall s \in [t, T].$$

This completes the proof. $\qquad\qquad\qquad\qquad\qquad\qquad\qquad\qquad\qquad$ □

6.6   Solution to LQ Differential Games

In this section, we present some solutions to Problem (LQG)*, i.e., Problem (LQG) with $f(\cdot), \sigma(\cdot), q(\cdot), \rho(\cdot), g = 0$. For the simplicity of presentation, we now introduce the following assumption which is stronger than (LQG1)–(LQG2).

(LQG3) Maps $A : [0, T] \to \mathbb{R}^{n \times n}$, $B_i : [0, T] \to \mathbb{R}^{n \times m_i}$, $Q : [0, T] \to \mathbb{S}^n$, and $R = \begin{pmatrix} R_{11} & 0 \\ 0 & R_{22} \end{pmatrix} : [0, T] \to \mathbb{S}^m$ are continuous. Moreover, for some $\delta > 0$,

$$(-1)^{i-1} R_i(s) \geqslant \delta I, \qquad s \in [0, T], \quad i = 1, 2. \qquad (6.82)$$

6.6.1   *Closed-loop saddle point*

From Theorem 6.5.2, we know that Problem (LQG) admits a closed-loop saddle point provided a convexity-concavity condition holds and the corresponding Riccati equation is solvable. In this subsection, we look at some cases that the Riccati equation is solvable. Let (LQG3) hold. We consider the following Riccati equation:

$$\begin{cases} \dot{P}(s) + P(s)A(s) + A(s)^T P(s) - P(s)M(s)P(s) + Q(s) = 0, \\ \qquad\qquad\qquad\qquad\qquad\qquad\qquad\qquad s \in [t, T], \quad (6.83) \\ P(T) = G, \end{cases}$$

with

$$M(s) = B_1(s)R_{11}(s)^{-1}B_1(s)^T + B_2(s)R_{22}(s)^{-1}B_2(s)^T, \qquad (6.84)$$
$$s \in [0, T].$$

Clearly, $M(\cdot)$ is just a symmetric matrix valued function, and for each $s \in [t, T]$, $M(s)$ is indefinite in general.

Let us look at some cases that Riccati equation (6.83) is uniquely solvable.

Case 1. Let

$$M(s) = 0, \qquad s \in [t, T].$$

Then the Riccati equation (6.83) reads

$$\begin{cases} \dot{P}(s) + P(s)A(s) + A(s)^T P(s) + Q(s) = 0, & s \in [t, T], \\ P(T) = G, \end{cases}$$

which is linear. Such an equation is called a *Lyapunov equation*. It always admits a unique solution, given by

$$P(s) = \Phi(T, s)^T G \Phi(T, s) + \int_s^T \Phi(r, s)^T Q(r) \Phi(r, s) dr, \quad s \in [t, T],$$

where $\Phi(\cdot, \cdot)$ is the fundamental matrix of $A(\cdot)$.

Case 2. Let

$$Q(s) = 0, \qquad s \in [t, T].$$

Then the Riccati equation (6.83) becomes

$$\begin{cases} \dot{P}(s) + P(s)A(s) + A(s)^T P(s) - P(s)M(s)P(s) = 0, & s \in [t, T], \\ P(T) = G. \end{cases}$$

Then the above is equivalent to the following integral equation:

$$P(s) = \Phi(T, s)^T G \Phi(T, s) - \int_s^T \Phi(r, s)^T P(r) M(r) P(r) \Phi(r, s) dr, \quad s \in [t, T].$$

Consequently, noting $\Phi(t, s)^{-1} = \Phi(s, t)$,

$$\Phi(s, T)^T P(s) \Phi(s, T)$$

$$= G - \int_s^T \Phi(s, T)^T \Phi(r, s)^T P(r) M(r) P(r) \Phi(r, s) \Phi(s, T) dr$$

$$= G - \int_s^T \Phi(r, T)^T P(r) \Phi(r, T) \Phi(T, r) M(r) \Phi(T, r)^T \Phi(r, T)^T P(r) \Phi(r, T) dr.$$

If we denote

$$\widetilde{P}(s) = \Phi(s, T)^T P(s) \Phi(s, T), \quad \widetilde{M}(s) = \Phi(T, s) M(s) \Phi(T, s)^T, \quad s \in [t, T],$$

then the above becomes

$$\widetilde{P}(s) = G - \int_s^T \widetilde{P}(r)\widetilde{M}(r)\widetilde{P}(r)\,dr, \qquad s \in [t,T],$$

which is equivalent to the following:

$$\begin{cases} \dot{\widetilde{P}}(s) = \widetilde{P}(s)\widetilde{M}(s)\widetilde{P}(s), & s \in [t,T], \\ \widetilde{P}(T) = G. \end{cases} \qquad (6.85)$$

Since $G \in \mathbb{S}^n$, we may write

$$G = G_0^T \widetilde{G} G_0, \qquad \widetilde{G} \in \mathbb{S}^n, \qquad \det \widetilde{G} \neq 0.$$

In fact, we have

$$\begin{aligned} G &= \begin{pmatrix} \Gamma_1 & \Gamma_2 \\ \Gamma_3 & \Gamma_4 \end{pmatrix} \begin{pmatrix} \Gamma_0 & 0 \\ 0 & 0 \end{pmatrix} \begin{pmatrix} \Gamma_1^T & \Gamma_3^T \\ \Gamma_2^T & \Gamma_4^T \end{pmatrix} = \begin{pmatrix} \Gamma_1 \Gamma_0 \Gamma_1^T & \Gamma_1 \Gamma_0 \Gamma_3^T \\ \Gamma_3 \Gamma_0 \Gamma_1^T & \Gamma_3 \Gamma_0 \Gamma_3^T \end{pmatrix} \\ &= \begin{pmatrix} \Gamma_1 & \Gamma_2 \\ \Gamma_3 & 0 \end{pmatrix} \begin{pmatrix} \Gamma_0 & 0 \\ 0 & I \end{pmatrix} \begin{pmatrix} \Gamma_1^T & \Gamma_3^T \\ \Gamma_2^T & 0 \end{pmatrix} \equiv G_0^T \widetilde{G} G_0, \end{aligned}$$

with

$$G_0 = \begin{pmatrix} \Gamma_1^T & \Gamma_3^T \\ \Gamma_2^T & 0 \end{pmatrix}, \qquad \widetilde{G} = \begin{pmatrix} \Gamma_0 & 0 \\ 0 & I \end{pmatrix}, \qquad \det \widetilde{G} = \det \Gamma_0 \neq 0.$$

Note that $\widetilde{G}$ may be indefinite, and G_0 may be singular. If G itself is invertible, we may take $\widetilde{G} = G$ and $G_0 = I$. Now, the Riccati equation (6.85) becomes

$$\begin{cases} \dot{\widetilde{P}}(s) = \widetilde{P}(s)\widetilde{M}(s)\widetilde{P}(s), & s \in [t,T], \\ \widetilde{P}(T) = G_0^T \widetilde{G} G_0. \end{cases}$$

We claim that the solution is given by

$$\widetilde{P}(s) = G_0^T \left(\widetilde{G}^{-1} + \int_s^T G_0 \widetilde{M}(r) G_0^T\, dr \right)^{-1} G_0, \qquad s \in [t,T],$$

provided the right-hand side of the above is well-defined. In fact,

$$\begin{aligned} \dot{\widetilde{P}}(s) &= G_0^T \left(\widetilde{G}^{-1} + \int_s^T G_0 \widetilde{M}(r) G_0^T\, dr \right)^{-1} G_0 \widetilde{M}(s) G_0^T \\ &\quad \cdot \left(\widetilde{G}^{-1} + \int_s^T G_0 \widetilde{M}(r) G_0^T\, dr \right)^{-1} G_0 \\ &= \widetilde{P}(s)\widetilde{M}(s)\widetilde{P}(s). \end{aligned}$$

Consequently,

$$P(s)= \Phi(T,s)^T G_0^T \Big(\widetilde{G}^{-1} + \int_s^T G_0 \Phi(T,r)M(r)\Phi(T,r)^T G_0^T \, dr \Big)^{-1} G_0 \Phi(T,s),$$

$$s \in [t,T].$$

Clearly, the above $P(\cdot)$ is well-defined on $[t,T]$ if

$$\det \Big(\widetilde{G}^{-1} + \int_s^T G_0 \Phi(T,r)M(r)\Phi(T,r)^T G_0^T \, dr \Big) \neq 0, \qquad s \in [t,T].$$

Case 3. Let A, M, and Q be constant matrices, and the following algebraic Riccati equation admits a solution $P_0 \in \mathbb{S}^n$:

$$P_0 A + A^T P_0 - P_0 M P_0 + Q = 0.$$

Let

$$\bar{P} = P - P_0.$$

Then

$$\begin{aligned}
\dot{\bar{P}} = \dot{P} &= -PA - A^T P + PMP - Q \\
&= -(P - P_0)A - A^T(P - P_0) + PMP - P_0 M P_0 \\
&= -\bar{P}A - A^T \bar{P} + (\bar{P} + P_0)M(\bar{P} + P_0) - P_0 M P_0 \\
&= -\bar{P}A - A^T \bar{P} + \bar{P}M\bar{P} + P_0 M \bar{P} + \bar{P}M P_0 \\
&= -\bar{P}(A - MP_0) - (A - MP_0)^T \bar{P} + \bar{P}M\bar{P}.
\end{aligned}$$

Hence,

$$\begin{cases}
\dot{\bar{P}} + \bar{P}(A - MP_0) + (A - MP_0)^T \bar{P} - \bar{P}M\bar{P} = 0, \qquad s \in [t,T], \\
\bar{P}(T) = G - P_0.
\end{cases}$$

Then it is reduced to the case of $Q = 0$. Hence, a representation of $P(\cdot)$ can be obtained as the above Case 2.

6.6.2 *One-dimensional case*

In this subsection, we will carry out one-dimensional cases for which the results are much more complete. We consider the following one-dimensional controlled linear system:

$$\begin{cases}
\dot{X}(s) = AX(s) + B_1 u_1(s) + B_2 u_2(s), \qquad s \in [t,T], \\
X(t) = x,
\end{cases}$$

with the performance functional:

$$J(t, x; u_1(\cdot), u_2(\cdot))$$
$$= \int_t^T \left[QX(s)^2 + R_1 u_1(s)^2 + R_2 u_2(s)^2 \right] ds + GX(T)^2,$$

where $A, B_1, B_2, A, R_1, R_2, G \in \mathbb{R}$. We assume that

$$R_1 > 0, \quad R_2 < 0.$$

Note that in terms of Chapter 4, for the current case, $\mu = 2$. One has

$$H^{\pm}(t, x, p) = H(t, x, p) = \inf_{u_1} \sup_{u_2} \left[pf(t, x, u_1, u_2) + g(t, x, u_1, u_2) \right]$$

$$= Apx + Qx^2 + \inf_{u_1} \left[R_1 u_1^2 + pB_1 u_1 \right] + \sup_{u_2} \left[R_2 u_2^2 + pB_2 u_2 \right]$$

$$= Apx + Qx^2 + \left(\frac{B_2^2}{4R_2} + \frac{B_1^2}{4R_1} \right) p^2.$$

Consequently, the upper and lower HJI equations have the same form:

$$\begin{cases} V_t(t, x) + AxV_x(t, x) + Qx^2 + \left(\dfrac{B_2^2}{4R_2} + \dfrac{B_1^2}{4R_1} \right) V_x(t, x)^2 = 0, \\ \qquad\qquad\qquad\qquad\qquad\qquad (t, x) \in [0, T] \times \mathbb{R}, \\ V(T, x) = Gx^2, \qquad x \in \mathbb{R}. \end{cases} \qquad (6.86)$$

If the above HJI equation has a viscosity solution, by the uniqueness, the solution has to be of the following form:

$$V(t, x) = p(t)x^2, \qquad (t, x) \in [0, T] \times \mathbb{R},$$

where $p(\cdot)$ is the solution to the following Riccati equation:

$$\begin{cases} \dot{p}(t) + 2Ap(t) + Q + \left(\dfrac{B_2^2}{R_2} + \dfrac{B_1^2}{R_1} \right) p(t)^2 = 0, \qquad t \in [0, T], \\ p(T) = G. \end{cases} \qquad (6.87)$$

In other words, the solvability of (6.86), in the viscosity sense, is equivalent to that of (6.87).

Our claim is that Riccati equation (6.87) is not always solvable for any $T > 0$. To state our result in a relatively neat way, let us rewrite equation (6.87) as follows:

$$\begin{cases} \dot{p} + \alpha p + \beta p^2 + \gamma = 0, \\ p(T) = g, \end{cases} \qquad (6.88)$$

with

$$\alpha = 2A, \qquad \beta = \frac{B_2^2}{R_2^2} + \frac{B_1^2}{R_1^2}, \qquad \gamma = Q, \qquad g = G.$$

Note that β could be positive, negative, or zero. We have the following result.

Proposition 6.6.1. *Riccati equation (6.88) admits a solution on $[0, T]$ for any $T > 0$ if and only if one of the following holds:*

$$\alpha^2 - 4\beta\gamma \geqslant 0, \quad 2\beta g + \alpha - \sqrt{\alpha^2 - 4\beta\gamma} \leqslant 0. \qquad (6.89)$$

Proof. We split the proof in several cases.

Case 1. $\beta = 0$. The Riccati equation reads

$$\begin{cases} \dot{p} + \alpha p + \gamma = 0, \\ p(T) = g. \end{cases}$$

This is an initial value problem for a linear equation, which admits a unique global solution $p(\cdot)$ on $[0, T]$.

Case 2. $\beta \neq 0$. Then Riccati equation reads

$$\begin{cases} \dot{p} + \beta\left[\left(p + \dfrac{\alpha}{2\beta}\right)^2 + \dfrac{4\beta\gamma - \alpha^2}{4\beta^2}\right] = 0, \\ p(T) = g. \end{cases}$$

Let

$$\kappa = \frac{\sqrt{|\alpha^2 - 4\beta\gamma|}}{2|\beta|} \geqslant 0.$$

There are three subcases.

Subcase 1. $\alpha^2 - 4\beta\gamma = 0$. The Riccati equation becomes

$$\begin{cases} \dot{p} + \beta\left(p + \dfrac{\alpha}{2\beta}\right)^2 = 0, \\ p(T) = g. \end{cases}$$

Therefore, in the case

$$2\beta g + \alpha = 0,$$

we have that $p(t) \equiv -\frac{\alpha}{2\beta}$ is the (unique) global solution on $[0, T]$. Now, let

$$2\beta g + \alpha \neq 0.$$

Then we have

$$\frac{dp}{(p + \frac{\alpha}{2\beta})^2} = -\beta dt,$$

which leads to

$$\frac{1}{p(t) + \frac{\alpha}{2\beta}} = \frac{1}{g + \frac{\alpha}{2\beta}} - \beta(T - t) = \frac{2\beta - \beta(2\beta g + \alpha)(T - t)}{2\beta g + \alpha}.$$

Thus,

$$p(t) = -\frac{\alpha}{2\beta} + \frac{2\beta g + \alpha}{2\beta - \beta(2\beta g + \alpha)(T - t)},$$

which is well-defined on $[0, T]$ if and only if

$$2 - (2\beta g + \alpha)(T - t) \neq 0, \qquad t \in [0, T].$$

This is equivalent to the following:

$$(2\beta g + \alpha)T < 2.$$

The above is true for all $T > 0$ if and only if

$$2\beta g + \alpha \leqslant 0.$$

Subcase 2. $\alpha^2 - 4\beta\gamma < 0$. The Riccati equation is

$$\dot{p} + \beta\left[\left(p + \frac{\alpha}{2\beta}\right)^2 + \kappa^2\right] = 0.$$

Hence,

$$\frac{dp}{(p + \frac{\alpha}{2\beta})^2 + \kappa^2} = -\beta dt,$$

which results in

$$\frac{1}{\kappa}\tan^{-1}\left[\frac{1}{\kappa}\left(p(t) + \frac{\alpha}{2\beta}\right)\right] = -\beta t + C.$$

By the terminal condition,

$$C = \beta T + \frac{1}{\kappa}\tan^{-1}\left[\frac{1}{\kappa}\left(g + \frac{\alpha}{2\beta}\right)\right].$$

Consequently,

$$\tan^{-1}\left[\frac{1}{\kappa}\left(p(t) + \frac{\alpha}{2\beta}\right)\right] = \kappa\beta(T - t) + \tan^{-1}\left[\frac{1}{\kappa}\left(g + \frac{\alpha}{2\beta}\right)\right].$$

Then

$$p(t) = \frac{\alpha}{2\beta} + \kappa\tan\left\{\kappa\beta(T - t) + \tan^{-1}\left(\frac{2\beta g + \alpha}{2\kappa\beta}\right)\right\}.$$

The above is well-defined for $t \in [0, T]$ if and only if

$$-\frac{\pi}{2} < \tan^{-1}\frac{2\beta g + \alpha}{2\kappa\beta} + \kappa\beta T < \frac{\pi}{2},$$

which is true for all $T > 0$ if and only if $\beta = 0$.

Subcase 3. $\alpha^2 - 4\beta\gamma > 0$. The Riccati equation becomes

$$\dot{p} + \beta\left[\left(p + \frac{\alpha}{2\beta}\right)^2 - \kappa^2\right] = 0.$$

If

$$(2\beta g + \alpha - 2\kappa\beta)(2\beta g + \alpha + 2\kappa\beta) \equiv 4\beta^2\left(g + \frac{\alpha}{2\beta} - \kappa\right)\left(g + \frac{\alpha}{2\beta} + \kappa\right) = 0, \quad (6.90)$$

then one of the following

$$p(t) \equiv -\frac{\alpha}{2\beta} \pm \kappa, \qquad t \in [0, T],$$

is the unique global solution to the Riccati equation. We now let

$$(2\beta g + \alpha - 2\kappa\beta)(2\beta g + \alpha + 2\kappa\beta) \equiv 4\beta^2\left(g + \frac{\alpha}{2\beta} - \kappa\right)\left(g + \frac{\alpha}{2\beta} + \kappa\right) \neq 0.$$

Then

$$\frac{dp}{(p + \frac{\alpha}{2\beta})^2 - \kappa^2} = -\beta dt.$$

Hence,

$$\frac{1}{2\kappa} \ln\left|\frac{p(t) + \frac{\alpha}{2\beta} - \kappa}{p(t) + \frac{\alpha}{2\beta} + \kappa}\right| = -\beta t + \widetilde{C},$$

which implies

$$\frac{p(t) + \frac{\alpha}{2\beta} - \kappa}{p(t) + \frac{\alpha}{2\beta} + \kappa} = Ce^{-2\kappa\beta t},$$

with

$$C = e^{2\kappa\beta T}\frac{g + \frac{\alpha}{2\beta} - \kappa}{g + \frac{\alpha}{2\beta} + \kappa} = e^{2\kappa\beta T}\frac{2\beta g + \alpha - 2\kappa\beta}{2\beta g + \alpha + 2\kappa\beta}.$$

Then

$$\frac{p(t) + \frac{\alpha}{2\beta} - \kappa}{p(t) + \frac{\alpha}{2\beta} + \kappa} = e^{2\kappa\beta(T-t)}\frac{2\beta g + \alpha - 2\kappa\beta}{2\beta g + \alpha + 2\kappa\beta}.$$

Consequently,

$$p(t) + \frac{\alpha}{2\beta} - \kappa = e^{2\kappa\beta(T-t)}\frac{2\beta g + \alpha - 2\kappa\beta}{2\beta g + \alpha + 2\kappa\beta}\left[p(t) + \frac{\alpha}{2\beta} + \kappa\right].$$

Thus, $p(\cdot)$ globally exists on $[0, T]$ if and only if

$$e^{2\kappa\beta(T-t)}\frac{2\beta g + \alpha - 2\kappa\beta}{2\beta g + \alpha + 2\kappa\beta} - 1 \neq 0, \qquad \forall t \in [0, T],$$

which is equivalent to

$$\psi(t) \equiv e^{2\kappa\beta(T-t)}(2\beta g + \alpha - 2\kappa\beta) - (2\beta g + \alpha + 2\kappa\beta) \neq 0, \quad \forall t \in [0, T].$$

Since $\psi'(t)$ does not change sign on $[0, T]$, the above is equivalent to the following:

$$0 < \psi(0)\psi(T) = \left[e^{2\kappa\beta T}(2\beta g + \alpha - 2\kappa\beta) - (2\beta g + \alpha + 2\kappa\beta) \right](-4\kappa\beta),$$

which is equivalent to

$$\left[e^{2\kappa\beta T}(2\beta g + \alpha - 2\kappa\beta) - (2\beta g + \alpha + 2\kappa\beta) \right]\beta < 0.$$

Note when (6.90) holds, the above is true. In the case $\beta > 0$, the above reads

$$e^{2\kappa\beta T}(2\beta g + \alpha - 2\kappa\beta) < 2\beta g + \alpha + 2\kappa\beta,$$

which is true for all $T > 0$ if and only if

$$2\beta g + \alpha - 2\kappa\beta \leqslant 0. \tag{6.91}$$

Finally, if $\beta < 0$, then

$$\begin{aligned}
0 &< e^{2\kappa\beta T}(2\beta g + \alpha - 2\kappa\beta) - (2\beta g + \alpha + 2\kappa\beta) \\
&= e^{-2\kappa|\beta|T}(-2|\beta|g + \alpha + 2\kappa|\beta|) - (-2|\beta|g + \alpha - 2\kappa|\beta|) \\
&= e^{-2\kappa|\beta|T}\left[-\left(2|\beta|g - \alpha - 2\kappa|\beta|\right) + e^{2\kappa|\beta|T}\left(2|\beta|g - \alpha + 2\kappa|\beta|\right) \right],
\end{aligned}$$

which is true for all $T > 0$ if and only if

$$0 \leqslant 2|\beta|g - \alpha + 2\kappa|\beta| = -(2\beta g + \alpha - 2\kappa|\beta|).$$

Thus,

$$2\beta g + \alpha - 2\kappa|\beta| \leqslant 0,$$

which has the same form as (6.91). This completes the proof. $\qquad\square$

It is clear that there are a lot of cases for which the Riccati equation is not solvable. For example,

$$\alpha = \beta = \gamma = 1,$$

which violates (6.89). Also, the case

$$\alpha = 0, \quad \beta = -1, \quad \gamma = 1, \quad g = -2,$$

which also violates (6.89). For the above two cases, Riccati equation (6.88) does not have a global solution on $[0, T]$ for some $T > 0$. Correspondingly we have some two-person zero-sum differential game with unbounded controls for which (H2) introduced in Chapter 4 fails and the upper and lower value functions are not defined on the whole time interval $[0, T]$, or equivalently, the corresponding upper/lower HJI equations have no viscosity solutions on $[0, T]$.

6.6.3 *Open-loop values and saddle point*

Now, we consider Problem (LQG) under open-loop controls. First we present the following result concerning the finiteness of lower value function.

Theorem 6.6.2. *Let* (LQG1)–(LQG3) *hold. Suppose that for any initial pair* $(t, x) \in [0, T] \times \mathbb{R}^n$, *the lower open-loop value* $V^-(t, x)$ *is uniquely achievable. Then the following two Riccati equations admit unique solutions* $P_1(\cdot)$ *and* $P(\cdot)$ *on* $[0, T]$, *respectively:*

$$
\begin{cases}
\dot{P}_1(s) + P_1(s)A(s) + A(s)^T P_1(s) - P_1(s)M_1(s)P_1(s) + Q(s) = 0, \\
\qquad\qquad\qquad\qquad\qquad\qquad\qquad\qquad s \in [t, T], \qquad (6.92) \\
P_1(T) = G,
\end{cases}
$$

with $M_1(\cdot) = B_1(\cdot)R_1(\cdot)^{-1}B_1(\cdot)^T$, *and*

$$
\begin{cases}
\dot{P}(s) + P(s)A(s) + A(s)^T P(s) - P(s)M(s)P(s) + Q(s) = 0, \\
\qquad\qquad\qquad\qquad\qquad\qquad\qquad\qquad s \in [t, T], \qquad (6.93) \\
P(T) = G,
\end{cases}
$$

with $M(\cdot)$ *defined by* (6.84). *Consequently, Problem* (LQG) *admits a closed-loop saddle point. Moreover,* $V^-(t, x)$ *is achieved by* $(\bar{u}_1(\cdot), \bar{u}_2(\cdot))$ *with*

$$
\begin{cases}
\bar{u}_1(s) = -R_1(s)^{-1}B_1(s)P(s)\bar{X}(s), \\
\bar{u}_2(s) = R_2(s)^{-1}B_2(s)P(s)\bar{X}(s),
\end{cases}
\qquad s \in [t, T], \qquad (6.94)
$$

where $\bar{X}(\cdot)$ *is the solution to*

$$
\begin{cases}
\dot{\bar{X}}(s) = \big[A(s) - M(s)P(s)\big]\bar{X}(s), \qquad s \in [t, T], \\
\bar{X}(t) = x.
\end{cases}
\qquad (6.95)
$$

Finally,

$$
V^-(t, x) = \frac{1}{2}\langle P(t)x, x \rangle, \qquad \forall (t, x) \in [0, T] \times \mathbb{R}^n. \qquad (6.96)
$$

The interesting point of the above result is that the finiteness of the lower open-loop value $V^-(t, x)$ of the game for the initial pair $(t, x) \in [0, T) \times \mathbb{R}^n$ implies the existence of the closed-loop saddle point of the game. From this, we should expect that the existence of closed-loop saddle point should not even imply the existence of the open-loop value function. We will present a simple example about this shortly.

Proof. For any initial pair $(t, x) \in [0, T) \times \mathbb{R}^n$, let $\bar{\mu}_1 : \mathcal{U}_2[t, T] \to \mathcal{U}_1[t, T]$ and $\bar{u}_2(\cdot) \in \mathcal{U}_2[t, T]$ such that $(\bar{\mu}_1(\cdot), \bar{u}_2(\cdot))$ achieves $V^-(t, x)$ in the following sense:

$$J(t, x; \bar{\mu}_1[u_2(\cdot)], u_2(\cdot)) = \inf_{u_1(\cdot) \in \mathcal{U}_1[t, T]} J(t, x; u_1(\cdot), u_2(\cdot)), \qquad (6.97)$$

for any $u_2(\cdot) \in \mathcal{U}_2[t, T]$ such that the right-hand side of the above is finite, and

$$J(t, x; \bar{\mu}_1[\bar{u}_2(\cdot)], \bar{u}_2(\cdot)) = \sup_{u_2(\cdot) \in \mathcal{U}_2[t, T]} J(t, x; \bar{\mu}_1[u_2(\cdot)], u_2(\cdot))$$

$$= \sup_{u_2(\cdot) \in \mathcal{U}_2[t, T]} \inf_{u_1(\cdot) \in \mathcal{U}_1[t, T]} J(t, x; u_1(\cdot), u_2(\cdot)) = V^-(t, x).$$

Let us now find $\bar{\mu}_1(\cdot)$ and $\bar{u}_2(\cdot)$. For any $u_2(\cdot) \in \mathcal{U}_2[t, T]$ fixed such that (6.97) makes sense, consider state equation

$$\begin{cases} \dot{X}_1(s) = A(s)X_1(s) + B_1(s)u_1(s) + B_2(s)u_2(s), & s \in [t, T], \\ X_1(t) = x, \end{cases}$$

with cost functional

$$J_1(t, x; u_1(\cdot)) = \int_t^T \Big[\langle Q(s)X(s), X(s) \rangle + \langle R_1(s)u_1(s), u_1(s) \rangle \Big] ds$$

$$+ \langle GX(T), X(T) \rangle,$$

regarding $B_2(\cdot)u_2(\cdot)$ as a nonhomogeneous term. Then $\bar{u}_1(\cdot) \equiv \bar{\mu}_1[u_2(\cdot)] \in \mathcal{U}_1[t, T]$ is the unique optimal control of the corresponding LQ problem. By Theorem 6.1.13, we have that the optimal control $\bar{u}_1(\cdot) \equiv \bar{\mu}_1[u_2(\cdot)]$ admits the following representation:

$$\bar{u}_1(s) = -R_1(s)^{-1} \big[B_1(s)^T P_1(s) \bar{X}_1(s) + B_1(s)^T \eta_1(s) \big], \qquad s \in [t, T],$$

where $P_1(\cdot)$ solves (6.92) and $\eta_1(\cdot)$ solves

$$\begin{cases} \dot{\eta}_1(s) = -[A(s) - M_1(s)P_1(s)]^T \eta_1(s) - P_1(s)B_2(s)u_2(s), \\ \hspace{6cm} s \in [t, T], \qquad (6.98) \\ \eta_1(T) = 0. \end{cases}$$

The closed-loop system reads:

$$\begin{cases} \dot{\bar{X}}_1(s) = \big[A(s) - M_1(s)P_1(s) \big] \bar{X}_1(s) + B_2(s)u_2(s) - M_1(s)\eta_1(s), \\ \hspace{6cm} s \in [t, T], \\ \bar{X}_1(t) = x. \end{cases}$$

Further, the optimal value of the cost functional is given by

$$J_1(t, x; \bar{u}_1(\cdot)) = \inf_{u_1(\cdot) \in \mathcal{U}_1[t,T]} J_1(t, x; u(\cdot)) = \frac{1}{2} \langle P_1(t)x, x \rangle + \langle \eta_1(t), x \rangle$$

$$- \frac{1}{2} \int_t^T \left\{ \langle R_1^{-1}(s) B_1(s)^T \eta_1(s), B_1(s)^T \eta_1(s) \rangle - 2 \langle \eta_1(s), B_2(s) u_2(s) \rangle \right\} ds$$

$$= \frac{1}{2} \langle P_1(t)x, x \rangle + \langle \eta_1(t), x \rangle$$

$$- \frac{1}{2} \int_t^T \left\{ \langle M_1(s)\eta_1(s), \eta_1(s) \rangle - 2 \langle B_2(s)^T \eta_1(s), u_2(s) \rangle \right\} ds.$$

Hence,

$$\inf_{u_1(\cdot) \in \mathcal{U}_1[t,T]} J(t, x; u_1(\cdot), u_2(\cdot)) = \frac{1}{2} \langle P_1(t)x, x \rangle + \langle \eta_1(t), x \rangle$$

$$- \frac{1}{2} \int_t^T \left\{ \langle M_1(s)\eta_1(s), \eta_1(s) \rangle - 2 \langle B_2(s)^T \eta_1(s), u_2(s) \rangle \right.$$

$$\left. + \langle R_2(s)u_2(s), u_2(s) \rangle \right\} ds \stackrel{\Delta}{=} J_2(t, x; u_2(\cdot)).$$

Therefore, we end up with an LQ problem with state equation (6.98) and payoff functional $J_2(t, x; u_2(\cdot))$ (to be maximized). By our assumption, this LQ problem admits a unique optimal control $\bar{u}_2(\cdot)$, with the corresponding optimal state trajectory $\bar{\eta}_1(\cdot)$. Then

$$0 = \langle \eta_1(t), x \rangle - \int_t^T \left\{ \langle M_1(s)\bar{\eta}_1(s), \eta_1(s) \rangle - \langle B_2(s)^T \bar{\eta}_1(s), u_2(s) \rangle \right.$$

$$\left. - \langle B_2(s)^T \eta_1(s), \bar{u}_2(s) \rangle + \langle R_2(s)\bar{u}_2(s), u_2(s) \rangle \right\} ds$$

$$= \langle \eta_1(t), x \rangle - \int_t^T \left\{ \langle M_1(s)\bar{\eta}_1(s) - B_2(s)\bar{u}_2(s), \eta_1(s) \rangle \right.$$

$$\left. + \langle -B_2(s)^T \bar{\eta}_1(s) + R_2(s)\bar{u}_2(s), u_2(s) \rangle \right\} ds,$$

where $(\eta_1(\cdot), u_2(\cdot))$ is any state-control pair of (6.98). Now, let $\psi_1(\cdot)$ be undetermined with $\psi_1(t) = x$. Observe the following:

$$- \langle \eta_1(t), x \rangle = \int_t^T \left[\langle -[A(s) - M_1(s)P_1(s)]^T \eta_1(s) \right.$$

$$\left. - P_1(s) B_2(s) u_2(s), \psi_1(s) \rangle + \langle \eta_1(s), \dot{\psi}_1(s) \rangle \right] ds.$$

Thus,

$$
\begin{aligned}
0 &= \langle\, \eta_1(t), x \,\rangle - \int_t^T \Big\{ \langle\, M_1(s)\bar\eta_1(s) - B_2(s)\bar u_2(s), \eta_1(s) \,\rangle \\
&\qquad + \langle\, -B_2(s)^T \bar\eta_1(s) + R_2(s)\bar u_2(s), u_2(s) \,\rangle \,\Big\} ds \\
&= \int_t^T \Big[\langle\, ([A(s) - M_1(s)P_1(s)]^T \eta_1(s) + P_1(s)B_2(s)u_2(s), \psi_1(s) \,\rangle \\
&\qquad - \langle\, \eta_1(s), \dot\psi_1(s) \,\rangle - \Big\{ \langle\, M_1(s)\bar\eta_1(s) - B_2(s)\bar u_2(s), \eta_1(s) \,\rangle \\
&\qquad + \langle\, -B_2(s)^T \bar\eta_1(s) + R_2(s)\bar u_2(s), u_2(s) \,\rangle \,\Big\} \Big] ds \\
&= \int_t^T \Big[\langle\, \eta_1(s), -\dot\psi_1(s) + [A(s) - M_1(s)P_1(s)]\psi_1(s) - M_1(s)\bar\eta_1(s) + B_2(s)\bar u_2(s) \,\rangle \\
&\qquad + \langle\, u_2(s), B_2(s)^T P_1(s)\psi_1(s) + B_2(s)^T \bar\eta_1(s) - R_2(s)\bar u_2(s) \,\rangle \Big] ds.
\end{aligned}
$$

Hence, we let $\psi_1(\cdot)$ solve the following:

$$
\begin{cases}
\dot\psi_1(s) = [A(s) - M_1(s)P_1(s)]\psi_1(s) - M_1(s)\bar\eta_1(s) + B_2(s)\bar u_2(s), \\
\psi_1(t) = x.
\end{cases}
$$

Then we must have

$$
B_2(s)^T P_1(s)\psi_1(s) + B_2(s)^T \bar\eta_1(s) - R_2(s)\bar u_2(s) = 0.
$$

Thus,

$$
\bar u_2(s) = R_2(s)^{-1}\Big(B_2(s)^T P_1(s)\psi_1(s) + B_2(s)^T \bar\eta_1(s) \Big).
$$

Consequently,

$$
\begin{aligned}
\dot\psi_1(s) &= \Big[A(s) - M_1(s)P_1(s)\Big]\psi_1(s) - M_1(s)\bar\eta_1(s) \\
&\qquad + B_2(s)R_2(s)^{-1}\Big(B_2(s)^T P_1(s)\psi_1(s) + B_2(s)^T \bar\eta_1(s) \Big) \\
&= \Big[A(s) - M(s)P_1(s)\Big]\psi_1(s) - M(s)\bar\eta_1(s),
\end{aligned}
$$

and

$$
\begin{aligned}
\dot{\bar\eta}_1(s) &= -\Big[A(s) - M_1(s)P_1(s)\Big]^T \bar\eta_1(s) \\
&\qquad - P_1(s)B_2(s)R_2(s)^{-1}\Big(B_2(s)^T P_1(s)\psi_1(s) + B_2(s)^T \bar\eta_1(s) \Big) \\
&= -\Big[A(s) - M(s)P_1(s)\Big]^T \bar\eta_1(s) - P_1(s)M_2(s)P_1(s)\psi_1(s).
\end{aligned}
$$

Hence, we obtain

$$
\begin{cases}
\dot{\psi}_1(s) = \Big[A(s) - M(s)P_1(s)\Big]\psi_1(s) - M(s)\bar{\eta}_1(s), \\
\dot{\bar{\eta}}_1(s) = -P_1(s)M_2(s)P_1(s)\psi_1(s) - \Big[A(s) - M(s)P_1(s)\Big]^T \bar{\eta}_1(s), \\
\psi_1(t) = x, \quad \bar{\eta}_1(T) = 0.
\end{cases}
$$

The above admits a unique solution $(\psi_1(\cdot), \bar{\eta}_1(\cdot))$ for any $(t, x) \in [0, T) \times \mathbb{R}^n$. Therefore, similar to the proof of Theorem 6.1.13, the following Riccati equation admits a solution $\bar{P}_2(\cdot)$:

$$
\begin{cases}
\dot{\bar{P}}_2(s) + \bar{P}_2(s)\big[A(s) - M(s)P_1(s)\big] + \big[A(s) - M(s)P_1(s)\big]^T \bar{P}_2(s) \\
\qquad -\bar{P}_2(s)M(s)\bar{P}_2(s) + P_1(s)M_2(s)P_1(s) = 0, \qquad s \in [0, T], \\
\bar{P}_2(T) = 0.
\end{cases}
$$

Let $P(\cdot) = P_1(\cdot) + \bar{P}_2(\cdot)$. Then (suppressing s)

$$
\begin{aligned}
0 &= \dot{P} + PA + A^T P + Q - \bar{P}_2 M P_1 - P_1 M \bar{P}_2 - \bar{P}_2 M \bar{P}_2 + P_1 M_2 P_1 - P_1 M_1 P_1 \\
&= \dot{P} + PA + A^T P + Q - \bar{P}_2 M P_1 - P_1 M \bar{P}_2 - \bar{P}_2 M \bar{P}_2 - P_1 M P_1 \\
&= \dot{P} + PA + A^T P + Q - PMP.
\end{aligned}
$$

Hence, $P(\cdot)$ is the solution to (6.93).

Note that one can check directly,

$$
\bar{\eta}_1(s) = \bar{P}_2(s)\psi_1(s), \qquad s \in [t, T].
$$

Hence, the optimal control $\bar{u}_2(\cdot)$ can be written as

$$
\bar{u}_2(s) = R_2(s)^{-1} B_2(s)^T P(s)\psi_1(s), \qquad s \in [t, T].
$$

Then

$$
\dot{\psi}_1(s) = \big[A(s) - M(s)P(s)\big]\psi_1(s), \qquad s \in [t, T].
$$

Also,

$$
\begin{aligned}
\dot{\bar{X}}_1(s) = \big[A(s) - M_1(s)P_1(s)\big]\bar{X}_1(s) + M_2(s)P(s)\psi_1(s) \\
- M_1(s)\bar{P}_2(s)\psi_1(s).
\end{aligned}
$$

Consequently,

$$
\dot{\bar{X}}(s) - \dot{\psi}_1(s) = \big[A(s) - M_1(s)P_1(s)\big]\big[\bar{X}_1(s) - \psi_1(s)\big].
$$

Together with $\bar{X}_1(t) - \psi_1(t) = 0$, we get

$$
\bar{X}_1(s) = \psi_1(s), \qquad s \in [t, T].
$$

Hence, we obtain the representation of $(\bar{u}_1(\cdot), \bar{u}_2(\cdot))$. Also, under such a pair, we have a representation (6.96) for $V^-(t, x)$.   □

Symmetrically, we can prove a similar result concerning $V^+(\cdot, \cdot)$, by considering $-J(t, x; u_1(\cdot), u_2(\cdot))$, which is stated here.

Theorem 6.6.3. *Let (LQG1)–(LQG3) hold. Suppose for any $(t, x) \in [0, T) \times \mathbb{R}^n$, the upper open-loop value $V^+(t, x)$ is uniquely achievable. Then the following Riccati equation admits a unique solution $P_2(\cdot)$*

$$
\begin{cases}
\dot{P}_2(s) + P_2(s)A(s) + A(s)^T P_2(s) - P_2(s)M_2(s)P_2(s) - Q(s) = 0, \\
\qquad\qquad\qquad\qquad\qquad\qquad\qquad\qquad\qquad s \in [t, T], \\
P_2(T) = -G,
\end{cases}
$$

with $M_2(\cdot) = B_2(\cdot)R_2(\cdot)^{-1}B_2(\cdot)^T$, and Riccati equation (6.93) also admits a unique solution $P(\cdot)$. Consequently, Problem (LQG) admits a closed-loop saddle point. Moreover, $V^+(t, x)$ is achieved by $(\bar{u}_1(\cdot), \bar{u}_2(\cdot))$ defined in (6.94) with $\bar{X}(\cdot)$ being the solution to (6.95). Finally,

$$
V^+(t, x) = \frac{1}{2}\langle P(t)x, x \rangle, \qquad \forall (t, x) \in [0, T] \times \mathbb{R}^n.
$$

The following corollary is interesting.

Corollary 6.6.4. *Let (LQG1)–(LQG3) hold. Suppose for any $(t, x) \in [0, T) \times \mathbb{R}^n$, the upper and lower open-loop values are uniquely achievable. Then it is necessary that*

$$
V^+(t, x) = V^-(t, x), \qquad \forall (t, x) \in [0, T] \times \mathbb{R}^n.
$$

Further, the controls $(\bar{u}_1(\cdot), \bar{u}_2(\cdot))$ defined by (6.94) is an open-loop saddle point of Problem (LQG).

Proof. According to Theorems 6.6.1 and 6.6.2, we have

$$
V^+(t, x) = \frac{1}{2}\langle P(t), x, x \rangle = V^-(t, x), \qquad (t, x) \in [0, T] \times \mathbb{R}^n.
$$

By Proposition 1.2.13, we see that $(\bar{u}_1(\cdot), \bar{u}_2(\cdot))$ is an open-loop saddle point of Problem (LQG).   □

We now present an example which shows that a closed-loop saddle point exists, whereas the open-loop value function does not exist.

Example 6.6.5. Consider the following state equation:

$$
\begin{cases}
\dot{X}(s) = u_1(s) + u_2(s), \qquad s \in [t, T], \\
X(t) = x,
\end{cases}
$$

and performance functional

$$J(t, x; u_1(\cdot), u_2(\cdot)) = \int_t^T \left(|u_1(s)|^2 - |u_2(s)|^2 \right) ds + |X(T)|^2.$$

Note that in the current case,

$$M(s) = 0, \qquad s \in [0, T].$$

Thus, the Riccati equation for closed-loop problem reads

$$\begin{cases} \dot{P}(s) = 0, & s \in [0, T], \\ P(T) = 1. \end{cases}$$

Hence, $P(s) \equiv 1$, consequently, a closed-loop saddle point exists, by Theorem 6.1.11. On the other hand, for any given $u_2(\cdot) \in \mathcal{U}_2[t, T]$, the Riccati equation for Player 1 reads

$$\begin{cases} \dot{P}_1(s) - P_1(s)^2 = 0, & s \in [0, T], \\ P_1(T) = 1, \end{cases}$$

whose solution is given by

$$P_1(s) = \frac{1}{1 + (T - s)}, \qquad s \in [0, T].$$

Hence,

$$V^-(t, x) = x^2, \qquad (t, x) \in [0, T] \times \mathbb{R}.$$

Now, the Riccati equation for Player 2 reads

$$\begin{cases} \dot{P}_2(s) - P_2(s)^2 = 0, \\ P_2(T) = -1. \end{cases}$$

Then the solution is given by

$$P_2(s) = \frac{-1}{1 - (T - s)}.$$

Hence, if $T > 1$, then $P_2(\cdot)$ cannot exist on $[t, T]$, with $0 \le t \le T - 1$. In fact, one can check directly that

$$V^+(t, x) = \infty, \qquad t \in [0, T - 1], \ x \in \mathbb{R} \setminus \{0\}.$$

This example shows that the closed-loop saddle point exists. But the open-loop value does not exist.

6.7   Fredholm Integral Equation

In this subsection, we will look at an equivalent form of the Riccati equation. Sometimes, such an alternative might be useful. Consider the following Riccati equation:

$$
\begin{cases}
\dot{P}(s) + A_1(s)^T P(s) + P(s)A_2(s) - P(s)M(s)P(s) + Q(s) = 0, \\
\qquad\qquad\qquad\qquad\qquad\qquad\qquad s \in [0,T], \qquad (6.99) \\
P(T) = G.
\end{cases}
$$

Here, we allow $A_1(\cdot)$ and $A_2(\cdot)$ to have different orders, and $M(\cdot)$ and $Q(\cdot)$ are not necessarily square matrix valued. Hence, $P(\cdot)$ is allowed to be non-square! (So no symmetry is assumed, of course.) Let $\Phi_1(t,s)$ and $\Phi_2(t,s)$ be the evolution operator generated by $A_1(\cdot)$ and $A_2(\cdot)$, respectively, i.e., $(i = 1,2)$

$$
\begin{cases}
\dfrac{\partial}{\partial s}\Phi_i(s,t) = A_i(s)\Phi_i(s,t), \qquad s \in [t,T], \\
\Phi_i(t,t) = I.
\end{cases}
$$

We define

$$
\Gamma(s,t) = \Phi_1(T,s)^T G \Phi_2(T,t) + \int_{s\vee t}^{T} \Phi_1(r,s)^T Q(r)\Phi_2(r,t)dr, \qquad (6.100)
$$
$$
s,t \in [0,T].
$$

Then we introduce the following family of *Fredholm integral equations* of the second kind, parameterized by $t \in [0,T]$:

$$
H(s,t) = \Gamma(s,t) - \int_{t}^{T} \Gamma(s,\tau)M(\tau)H(\tau,t)d\tau, \qquad s \in [t,T]. \qquad (6.101)
$$

We have the following equivalence result for (6.99) and (6.101):

Theorem 6.7.1. (i) *If Riccati equation (6.99) admits a solution $P(\cdot)$, then for any $t \in [0,T]$, Fredholm integral equation (6.101) has a solution given by*

$$
H(s,t) = P(s)\Psi(s,t), \qquad 0 \le t \le s \le T, \qquad (6.102)
$$

where $\Psi(\cdot,\cdot)$ is the evolution operator generated by $A_2(\cdot) - M(\cdot)P(\cdot)$.

(ii) *If for any $t \in [0,T]$, Fredholm integral equation (6.101) admits a unique solution $H(\cdot,t)$, then Riccati equation (6.99) has a solution. Moreover, the solution is given by*

$$
P(t) = H(t,t), \qquad t \in [0,T]. \qquad (6.103)
$$

Proof. (i) Let $P(\cdot)$ be a solution of Riccati equation (6.99). We let $\Psi(\cdot,\cdot)$ be the evolution operator generated by $A_2(\cdot) - M(\cdot)P(\cdot)$ and define $H(\cdot,\cdot)$ by (6.102). We calculate the following:

$$\frac{\partial}{\partial s}H(s,t) = \dot{P}(s)\Psi(s,t) + P(s)\Big[A_2(s) - M(s)P(s)\Big]\Psi(s,t)$$

$$= \Big[-A_1(s)^T P(s) - Q(s)\Big]\Psi(s,t) = -A_1(s)^T H(s,t) - Q(s)\Psi(s,t),$$

$$0 \le s \le t \le T.$$

Thus,

$$H(s,t) = \Phi_1(T,s)^T G\Psi(T,t) + \int_s^T \Phi_1(\tau,s)^T Q(\tau)\Psi(\tau,t)d\tau, \qquad (6.104)$$

$$0 \le t \le s \le T.$$

On the other hand, from

$$\begin{cases} \dfrac{\partial}{\partial s}\Psi(s,t) = A_2(s)\Psi(s,t) - M(s)P(s)\Psi(s,t) \\ \qquad\quad = A_2(s)\Psi(s,t) - M(s)H(s,t), \\ \Psi(t,t) = I, \end{cases}$$

we have

$$\Psi(s,t) = \Phi_2(s,t) - \int_t^s \Phi_2(s,\tau)M(\tau)H(\tau,t)d\tau, \qquad 0 \le t \le s \le T. \quad (6.105)$$

Thus, substituting (6.105) into (6.104), one obtains

$$H(s,t) = \Phi_1(T,s)^T G\Big[\Phi_2(T,t) - \int_t^T \Phi_2(T,\tau)M(\tau)H(\tau,t)d\tau\Big]$$

$$+ \int_s^T \Phi_1(\tau,s)^T Q(\tau)\Big[\Phi_2(\tau,t) - \int_t^\tau \Phi_2(\tau,r)M(r)H(r,t)dr\Big]d\tau$$

$$= \Gamma(s,t) - \int_t^T \Gamma(s,\tau)M(\tau)H(\tau,t)d\tau.$$

This means that $H(s,t)$ defined by (6.102) is a solution of (6.101).

(ii) Let $H(\cdot,t)$ be the unique solution of (6.101) (for given $t \in [0,T]$). We first claim that

$$\frac{\partial}{\partial t}H(s,t) = H(s,t)\Big[M(t)H(t,t) - A_2(t)\Big], \qquad 0 \le t \le s \le T. \quad (6.106)$$

In fact, for $0 \le t \le s \le T$,

$$\Gamma(s,t) = \Phi_1(T,s)^T G\Phi_2(T,t) + \int_s^T \Phi_1(r,s)^T Q(r)\Phi_2(r,t)dr.$$

Thus, we have

$$\frac{\partial}{\partial t}H(s,t) = \frac{\partial}{\partial t}\Gamma(s,t) + \Gamma(s,t)M(t)H(t,t) - \int_t^T \Gamma(s,\tau)M(\tau)\frac{\partial}{\partial t}H(\tau,t)d\tau$$

$$= -\Gamma(s,t)A_2(t) + \Gamma(s,t)M(t)H(t,t) - \int_t^T \Gamma(s,\tau)M(\tau)\frac{\partial}{\partial t}H(\tau,t)d\tau$$

$$= -\Gamma(s,t)\Big[A_2(t) - M(t)H(t,t)\Big] - \int_t^T \Gamma(s,\tau)M(\tau)\frac{\partial}{\partial t}H(\tau,t)d\tau.$$

Consequently,

$$\frac{\partial}{\partial t}H(s,t) + H(s,t)\Big[A_2(t) - M(t)H(t,t)\Big]$$

$$= \Big[H(s,t) - \Gamma(s,t)\Big]\Big[A_2(t) - M(t)H(t,t)\Big] - \int_t^T \Gamma(s,\tau)M(\tau)\frac{\partial}{\partial t}H(\tau,t)d\tau$$

$$= -\int_t^T \Gamma(t,\tau)M(\tau)\Big\{\frac{\partial}{\partial t}H(\tau,t) + H(\tau,t)\Big[A_2(t) - M(t)H(t,t)\Big]\Big\}d\tau.$$

Then, by the uniqueness of the solutions to (6.101) (for any given $t \in [0,T]$), we obtain (6.106). Next, we let

$$P(t) = H(t,t) = \Gamma(t,t) - \int_t^T \Gamma(t,\tau)M(\tau)H(\tau,t)d\tau$$

$$= \Phi_1(T,t)^T G\Phi_2(T,t) + \int_t^T \Phi_1(r,t)^T Q(r)\Phi_2(r,t)dr$$

$$- \int_t^T \Big[\Phi_1(T,t)^T G\Phi_2(T,\tau) + \int_\tau^T \Phi_1(r,t)^T Q(r)\Phi_2(r,\tau)dr\Big]M(\tau)H(\tau,t)d\tau,$$

$$t \in [0,T].$$

Then

$$\dot{P}(t) = -A_1(t)^T \Phi_1(T,t)^T G\Phi_2(T,t) - \Phi_1(T,t)^T G\Phi_2(T,t)A_2(s) - Q(t)$$

$$- \int_t^T \Big[A_1(t)^T \Phi_1(r,t)^T Q(r)\Phi_2(r,t) + \Phi_1(r,t)^T Q(r)\Phi_2(r,t)A_2(t)\Big]dr$$

$$+ \Big[\Phi_1(T,t)^T G\Phi_2(T,t) + \int_t^T \Phi_1(r,t)^T Q(r)\Phi_2(r,t)dr\Big]M(t)H(t,t)$$

$$- \int_t^T \Big\{ -A_1(t)^T \Big[\Phi_1(T,t)^T G\Phi_2(T,\tau)$$

$$+ \int_\tau^T \Phi_1(r,t)^T Q(r)\Phi_2(r,\tau)dr\Big]M(\tau)H(\tau,t)$$

$$+ \Big[\Phi_1(T,t)^T G\Phi_2(T,\tau) + \int_\tau^T \Phi_1(r,t)^T Q(r)\Phi_2(r,\tau)dr\Big]$$

$$\cdot M(\tau)H(\tau,t)\Big[M(t)H(t,t) - A_2(t)\Big]\Big\}d\tau$$

$$= -A_1(t)^T \Big\{ \Phi_1(T,t)^T G \Phi_2(T,t) + \int_t^T \Phi_1(r,t)^T Q(r) \Phi_2(r,t) dr$$

$$- \int_t^T \Big[\Phi_1(T,t)^T G \Phi_2(T,\tau) + \int_\tau^T \Phi_1(r,t)^T Q(r) \Phi_2(r,\tau) dr \Big] M(\tau) H(\tau,t) d\tau \Big\}$$

$$- \Big\{ \Phi_1(T,t)^T G \Phi_2(T,t) + \int_t^T \Phi_1(r,t)^T Q(r) \Phi_2(r,t) dr$$

$$- \int_t^T \Big[\Phi_1(T,t)^T G \Phi_2(T,\tau) + \int_\tau^T \Phi_1(r,t)^T Q(r) \Phi_2(r,\tau) dr \Big] M(\tau) H(\tau,t) d\tau \Big\} A_2(t)$$

$$- Q(t) + \Big[\Phi_1(T,t)^T G \Phi_2(T,t) + \int_t^T \Phi_1(r,t)^T Q(r) \Phi_2(r,t) dr \Big] M(t) H(t,t)$$

$$- \Big(\int_t^T \Big[\Phi_1(T,t)^T G \Phi_2(T,\tau) + \int_\tau^T \Phi_1(r,t)^T Q(r) \Phi_2(r,\tau) dr \Big]$$

$$\cdot M(\tau) H(\tau,t) d\tau \Big) M(t) H(t,t)$$

$$= -A_1(t)^T P(t) - P(t) A_2(t) - Q(t) + P(t) M(t) P(t).$$

Thus, $P(\cdot)$ defined by (6.103) is a solution of Riccati equation (6.99). ☐

Note that Riccati equation (6.99) is nonlinear, whereas Fredholm integral equation (6.101) is linear. Therefore, the above result gives us an equivalence between a nonlinear equation and a linear equation. Let us look at the corresponding Fredholm integral equation for the problem in Example 6.1.15.

In the current case, the following hold:
$$A_1(\cdot) = A_2(\cdot) = Q(\cdot) = 0, \quad M(\cdot) = 1, \quad G = g.$$
Thus, (6.100) leads to
$$\Gamma(s,t) = g, \quad s,t \in [0,T].$$
Hence, the Fredholm integral equation becomes
$$H(s,t) = g - \int_t^T g H(\tau,t) d\tau, \quad s \in [t,T].$$
Clearly, $H(s,t) \equiv H(t)$ is independent of s. Hence, the above is equivalent to
$$H(t) = g - g H(t)(T - t).$$
Consequently,
$$H(t) = \frac{g}{1 + g(T - t)}.$$
Then
$$P(t) = H(t,t) = \frac{g}{1 + g(T - t)},$$
which coincides with the result presented in Example 6.1.15.

6.8   Brief Historic Remarks

The history of deterministic linear-quadratic optimal control problems can be traced back to the works of Bellman–Glicksberg–Gross ([7]) in 1958, Kalman ([62]), and Letov ([67]) in 1960.

The study of linear-quadratic two-person zero sum differential games can be traced back to the work of Ho–Bryson–Baron [56] in 1965. In 1970, Schmitendorf studied both open-loop and closed-loop strategies for Problem (LQG) ([105], [106], see also [107]); among other things, it was shown that the existence of a closed-loop saddle point may not imply that of an open-loop saddle point. In 1979, Bernhard carefully investigated Problem (LQG) from a closed-loop point of view ([15]); see also the book by Basar and Bernhard [3] in this aspect. In 2005, Zhang [131] proved that for a special Problem (LQG), the existence of the open-loop value is equivalent to the finiteness of the corresponding open-loop lower and upper values, which is also equivalent to the existence of an open-loop saddle point. Along this line, there were follow-up extensions by Delfour ([35]) and Delfour–Sbarbar ([36]).

Most of the material presented in this chapter is based on the work of Sun–Yong ([112]) and Mou–Yong [78] (see also [128] and [129] for some relevant results). More precisely, Sections 6.1.2, 6.3, and 6.6.3 are based on some results from [78]. Corollaries 6.1.6 and 6.3.2 seem to be new. By using a property of Fredholm operator, the equivalence between the existence of an open-loop saddle point and the finiteness of the upper and lower values (respectively, the equivalence between the existence of an open-loop optimal control and the finiteness of the value function, for LQ optimal control problem) becomes very simple and transparent. The original proof of this fact presented by Zhang ([131]) was very technical (see also the extension by Deflour [35] and Delfour-Sbarbar [36]). Sections 6.1.3–6.1.4, 6.2, 6.4 and 6.5 are a modification of some relevant material from [112], where the notion of closed-loop strategy presented here is adopted. Section 6.6.1 is based on [105], [106], with proper modification and extension. Section 6.6.2 is taken from Qiu–Yong ([102]). Section 6.7 is mainly based on a work by Chen ([27]).

Chapter 7

Differential Games with Switching Strategies

7.1 Optimal Switching Control Problems

In this section, we will introduce a different type of controls called switching controls, and related optimal control problems.

7.1.1 *Switching controls*

Let $\mathbb{M} = \{1, 2, \cdots, m\}$. Consider m controlled systems:

$$\dot{X}_a(s) = f(s, X_a(s), a), \qquad s \in [t, T], \quad a \in \mathbb{M},$$

where $f : [0, T] \times \mathbb{R}^n \times \mathbb{M} \to \mathbb{R}^n$. For an initial pair $(t, x) \in [0, T) \times \mathbb{R}^n$, let $\{(\theta_i, a_i)\}_{i=0}^k \subseteq [t, T] \times \mathbb{M}$ be a finite sequence with the following properties:

$$\begin{cases} a_{i+1} \neq a_i, & 0 \leqslant i \leqslant k - 1, \\ t = \theta_0 \leqslant \theta_1 \leqslant \theta_2 \leqslant \cdots \leqslant \theta_k = T. \end{cases} \tag{7.1}$$

Here, $k \geqslant 0$ is a nonnegative integer. We now describe the following *switching process* driven by $\{(\theta_i, a_i)\}_{i=0}^k$. On $[t, \theta_1)$, the system a_0 is running, i.e., we have

$$\begin{cases} \dot{X}(s) = f(s, X(s), a_0), & s \in [\theta_0, \theta_1), \\ X(\theta_0) = x. \end{cases}$$

At time θ_1, the system is switched from a_0 to a_1, and system a_1 will be running on $[\theta_1, \theta_2)$, i.e.,

$$\begin{cases} \dot{X}(s) = f(s, X(s), a_1), & s \in [\theta_1, \theta_2), \\ X(\theta_1 + 0) = X(\theta_1 - 0). \end{cases}$$

This procedure can be continued. We call $\{(\theta_i, a_i)\}_{i=0}^k$ a *switching control*. We now identify $\{(\theta_i, a_i)\}_{i=0}^k$ with the map $a(\cdot)$ defined by the following:

$$a(s) = \sum_{i=0}^{k-1} a_i I_{[\theta_i,\theta_{i+1})}(s) + a_k I_{[\theta_{k-1},\theta_k]}(s), \qquad s \in [t,T]. \qquad (7.2)$$

When $k = 0$, the corresponding switching control has no switching. For this case, $\{(\theta_0, a_0)\}$ is a singleton, and we identify such a (trivial) switching control with a_0, i.e.,

$$a(s) = a_0, \qquad s \in [0,T].$$

For any $a_0 \in \mathbb{M}$, let

$$\mathcal{S}^{a_0}[t,T] = \Big\{ a(\cdot) \equiv \{(\theta_i, a_i)\}_{i=0}^k \mid \theta_0 = t, \ (7.1) \text{ holds} \Big\}.$$

Now for any $(t, x, a_0) \in [0,T) \times \mathbb{R}^n \times \mathbb{M}$ and $a(\cdot) \in \mathcal{S}^{a_0}[t,T]$, the system under the above described switching can be written as follows:

$$\begin{cases} \dot{X}(s) = f(s, X(s), a(s)), \qquad s \in [t,T], \\ X(t) = x, \end{cases}$$

which is a standard form of control system, with a piecewise constant control $a(\cdot)$.

Next, we introduce the following cost functional:

$$J^{a_0}(t,x;a(\cdot)) = \int_t^T g(s, X(s), a(s))ds + h(X(T)) \\ + \sum_{i=0}^{k-1} \kappa(\theta_{i+1}, a_i, a_{i+1}), \qquad (7.3)$$

with the convention that $\sum_{i=0}^{-1}\{\cdots\} = 0$. On the right-hand side of the above, the first term is called a *running cost* and the second term is called a *switching cost*. The term $\kappa(\theta_{i+1}, a_i, a_{i+1})$ represents the cost of switching (which is positive) from system a_i to system a_{i+1} at θ_{i+1}. When $k = 0$, the switching cost term is absent.

Note that in the definition of switching control $a(\cdot) = \{(\theta_i, a_i)\}_{i=0}^k$, we allow $\theta_{i+1} = \theta_i$. When this happens, it means that at θ_i, two switchings are made (from a_{i-1} to a_i, then from a_i to a_{i+1}). As far as the state trajectory $X(\cdot)$ is concerned, this is the same as the switching control that is switched from a_{i-1} directly to a_{i+1}, at time θ_i. However, these two controls are different when the switching cost is concerned. Therefore, although for simplicity, we make the identification of $a(\cdot)$ defined by (7.2) with $\{(\theta_i, a_i)\}_{i=0}^k$, we should keep in mind that $\theta_i = \theta_{i+1}$ could happen.

Next, for any $0 \leqslant t < \tau \leqslant T$ and $a(\cdot) \equiv \{(\theta_i, a_i)\}_{i=0}^k \in \mathcal{S}^{a_0}[\tau, T]$, we define its *extension* on $[t, T]$ as follows:

$$\begin{cases} a_0 I_{[t,\tau)}(\cdot) \oplus a(\cdot) = \{(\bar\theta_i, a_i)\}_{i=0}^k \in \mathcal{S}^{a_0}[t, T], \\ \bar\theta_0 = t, \qquad \bar\theta_i = \theta_i, \quad 1 \leqslant i \leqslant k. \end{cases}$$

Therefore, under $a_0 I_{[t,\tau)}(\cdot) \oplus a(\cdot)$, there is no switching on $[t, \tau)$. On the other hand, for any $a(\cdot) \equiv \{(\theta_i, a_i)\}_{i=0}^k \in \mathcal{S}^{a_0}[t, T]$, we define its *compression* on $[\tau, T]$ as follows:

$$\begin{cases} a(\cdot)\big|_{[\tau,T]} \equiv \{(\bar\theta_i, a_i)\}_{i=0}^k \in \mathcal{S}^{a_0}[\tau, T], \\ \bar\theta_i = \theta_i \vee \tau, \qquad 0 \leqslant i \leqslant k. \end{cases} \tag{7.4}$$

This amounts to moving all the switchings made before τ in $a(\cdot) \in \mathcal{S}^{a_0}[t, T]$ to the moment τ. Since the switching cost is positive, the cost functional depends not only on the initial pair (t, x), but also on the initial value of the switching control $a(\cdot)$. Such a dependence of the cost functional on $a_0 = a(t)$ is indicated on the left-hand side of (7.3). We now pose the following problem.

Problem (S). For any given $(t, x, a) \in [0, T) \times \mathbb{R}^n \times \mathbb{M}$, find an $a^*(\cdot) \in \mathcal{S}^a[t, T]$ such that

$$J^a(t, x; a^*(\cdot)) = \inf_{a(\cdot) \in \mathcal{S}^a[t,T]} J(t, x; a(\cdot)) \equiv V^a(t, x).$$

We denote

$$V(t, x) = (V^1(t, x), V^2(t, x), \cdots, V^m(t, x)), \quad (t, x) \in [0, T] \times \mathbb{R}^n,$$

and call it the *value function* of Problem (S).

Let us now introduce the following assumptions.

(S1) The map $f : [0, T] \times \mathbb{R}^n \times \mathbb{M} \to \mathbb{R}^n$ is continuous and there exists a constant $L > 0$ such that

$$\begin{cases} |f(s, x_1, a) - f(s, x_2, a)| \leqslant L|x_1 - x_2|, \\ \qquad\qquad \forall (s, a) \in [0, T] \times \mathbb{M}, \ x_1, x_2 \in \mathbb{R}^n, \\ |f(s, 0, a)| \leqslant L, \qquad \forall (s, a) \in [0, T] \times \mathbb{M}. \end{cases}$$

(S2) The maps $g : [0, T] \times \mathbb{R}^n \times \mathbb{M} \to \mathbb{R}$ and $h : \mathbb{R}^n \to \mathbb{R}$ are continuous and there exists a continuous increasing function $\theta : \mathbb{R}_+ \to \mathbb{R}_+$ such that

$$\begin{cases} |g(t, x_1, a) - g(t, x_2, a)| + |h(x_1) - h(x_2)| \leqslant \theta(|x_1| \vee |x_2|)|x_1 - x_2|, \\ \qquad\qquad (t, a) \in [0, T] \times \mathbb{M}, \ x_1, x_2 \in \mathbb{R}^n, \\ |g(t, 0, a)| + |h(0)| \leqslant \theta(0), \qquad (t, a) \in [0, T] \times \mathbb{M}. \end{cases}$$

(S2)′ The maps $g : [0, T] \times \mathbb{R}^n \times \mathbb{M} \to \mathbb{R}$ and $h : \mathbb{R}^n \to \mathbb{R}$ are continuous and there exists a local modulus of continuity $\omega(\cdot, \cdot)$ such that

$$
\begin{cases}
|g(t, x_1, a) - g(t, x_2, a)| + |h(x_1) - h(x_2)| \leqslant \omega(|x_1| \vee |x_2|, |x_1 - x_2|), \\
\qquad\qquad\qquad (t, a) \in [0, T] \times \mathbb{M}, \ x_1, x_2 \in \mathbb{R}^n, \\
|g(t, 0, a)| + |h(0)| \leqslant L, \qquad (t, a) \in [0, T] \times \mathbb{M}.
\end{cases}
$$

(S3) The map $\kappa : [0, T] \times \mathbb{M} \times \mathbb{M} \to (0, \infty)$ is continuous and

$$
\begin{cases}
\kappa(t, a, \widetilde{a}) < \kappa(t, a, \bar{a}) + \kappa(t, \bar{a}, \widetilde{a}), \\
\qquad \forall t \in [0, T], \ a, \widetilde{a}, \bar{a} \in \mathbb{M}, \ a \neq \bar{a} \neq \widetilde{a}, \\
\kappa(t, a, a) = 0, \qquad \forall (t, a) \in [0, T] \times \mathbb{M}, \\
\kappa(t_2, a, \bar{a}) \leqslant \kappa(t_1, a, \bar{a}), \qquad 0 \leqslant t_1 \leqslant t_2 \leqslant T, \ a, \bar{a} \in \mathbb{M}.
\end{cases}
$$

(S3)′ The map $\kappa : [0, T] \times \mathbb{M} \times \mathbb{M} \to (0, \infty)$ satisfies (S3) and is independent of t.

Before going further, let us make some remarks. It is easy to see that under (S2), $x \mapsto (g(t, x, a), h(x))$ is locally Lipschitz continuous uniformly in t, which is stronger than (S2)′. Fortunately, in the case, say, $x \mapsto (g(t, x, a), h(x))$ is differentiable with bounded gradient, (S2) holds. Hence, (S2) is still very general. On the other hand, (S3) is much more restrictive than (S3). Recall that in Sections 2.4–2.5, we are able to discuss the corresponding optimal control problem, including the uniqueness of viscosity solution to the HJB equation, under conditions similar to (S1) and (S2)′. However, for the current Problem (S), due to the appearance of the switching cost, the situation becomes very subtle. We will see that the uniqueness of viscosity solution for the corresponding HJB equation for Problem (S) will be guaranteed under conditions (S1), (S2)′ and (S3)′, or under (S1)–(S3). In other words, either g and h are general with κ independent of t, or g and h are less general allowing κ to depend on t. Finally, we note that under (S3) of (S3)′, there exists a $\kappa_0 > 0$ such that

$$
\kappa(t, a, \widetilde{a}) \geqslant \kappa_0, \qquad \forall t \in [0, T], \ a, \widetilde{a} \in \mathbb{M}, \ a \neq \widetilde{a}.
$$

We now present the continuity of the value function.

Theorem 7.1.1. *Let (S1)–(S3) hold. Then the value function* $V(\cdot, \cdot)$ *is continuous. In addition, if (S2)′ holds, then there exists a continuous increasing function* $\widetilde{\theta} : \mathbb{R}_+ \to \mathbb{R}_+$ *such that*

$$
|V(t_1, x_1) - V(t_2, x_2)| \leqslant \widetilde{\theta}(|x_1| \vee |x_2|)\Big(|t_1 - t_2| + |x_1 - x_2|\Big), \tag{7.5}
$$
$$
\forall (t_1, x_1), (t_2, x_2) \in [0, T] \times \mathbb{R}^n.
$$

Proof. First of all, under (S1), for any $(t, x) \in [0, T) \times \mathbb{R}^n$ and $a(\cdot) \in \mathcal{S}^a[t, T]$, the state equation admits a unique trajectory denoted by $X(\cdot; t, x, a(\cdot))$. Similar to Proposition 2.1.1, we have

$$
\begin{cases}
|X(s; t, x, a(\cdot))| \leqslant e^{L(s-t)}(1 + |x|) - 1, \\
|X(s; t, x, a(\cdot)) - x| \leqslant \left[e^{L(s-t)} - 1 \right](1 + |x|), \\
\qquad\qquad\qquad\qquad s \in [t, T], \ a(\cdot) \in \mathcal{S}^a[t, T].
\end{cases}
$$

Further, for any $t \in [0, T)$, $x_1, x_2 \in \mathbb{R}^n$, and $a(\cdot) \in \mathcal{S}^a[t, T]$,

$$
|X(s; t, x_1, a(\cdot)) - X(s; t, x_2, a(\cdot))| \leqslant e^{L(s-t)}|x_1 - x_2|, \quad \forall s \in [t, T].
$$

Now, for any $t \in [0, T)$, $x_1, x_2 \in \mathbb{R}^n$ and $a(\cdot) \in \mathcal{S}^a[t, T]$, let us denote $X_i(\cdot) = X(\cdot; t, x_i, a(\cdot))$, $i = 1, 2$. Then

$$
\begin{aligned}
&|J^a(t, x_1; a(\cdot)) - J^a(t, x_2; a(\cdot))| \\
&\leqslant \int_t^T \omega\big(|X_1(s)| \vee |X_2(s)|, |X_1(s) - X_2(s)|\big) ds \\
&\quad + \omega\big(|X_1(T)| \vee |X_2(T)|, |X_1(T) - X_2(T)|\big) \\
&\leqslant \omega\big(e^{LT}(1 + |x_1| \vee |x_2|), e^{LT}|x_1 - x_2|\big)(T + 1) \equiv \bar{\omega}\big(|x_1| \vee |x_2|, |x_1 - x_2|\big).
\end{aligned}
$$

This implies

$$
|V^a(t, x_1) - V^a(t, x_2)| \leqslant \bar{\omega}\big(|x_1| \vee |x_2|, |x_1 - x_2|\big).
$$

Next, we let $0 \leqslant t_1 < t_2 \leqslant T$ and $x \in \mathbb{R}^n$. For any $a(\cdot) \equiv \{\theta_i, a_i\}_{i \geqslant 0} \in \mathcal{S}^a[t_1, T]$, let $a_c(\cdot) \equiv a(\cdot)|_{[t_2, T]} \in \mathcal{S}^a[t_2, T]$ be the compression of $a(\cdot)$ on $[t_2, T]$, $X_c(\cdot) = X(\cdot; t_2, x, a_c(\cdot))$, and $X(\cdot) = X(\cdot; t_1, x, a(\cdot))$. Then

$$
\begin{aligned}
V^a(t_2, x) &\leqslant J^a\big(t_2, x; a_c(\cdot)\big) = \int_{t_2}^T g(s, X_c(s), a_c(s)) ds + h(X_c(T)) \\
&\quad + \sum_{\theta_{i+1} < t_2} \kappa(t_2, a_i, a_{i+1}) + \sum_{\theta_{i+1} \geqslant t_2} \kappa(\theta_{i+1}, a_i, a_{i+1}) \\
&\leqslant \int_{t_2}^T g(s, X_c(s), a_c(s)) ds + h(X_c(T)) \\
&\quad + \sum_{\theta_{i+1} < t_2} \kappa(\theta_{i+1}, a_i, a_{i+1}) + \sum_{\theta_{i+1} \geqslant t_2} \kappa(\theta_{i+1}, a_i, a_{i+1})
\end{aligned}
$$

$$= J^a(t_1, x; a(\cdot)) + \int_{t_2}^{T} \Big[g(s, X_c(s), a_c(s)) - g(s, X(s), a(s)) \Big] ds$$

$$- \int_{t_1}^{t_2} g(s, X(s), a(s)) ds + h(X_c(T)) - h(X(T))$$

$$\leqslant J^a(t_1, x; a(\cdot)) + \int_{t_2}^{T} \omega\big(|X_c(s)| \vee |X(s)|, |X_c(s) - X(s)| \big) ds$$

$$+ \int_{t_1}^{t_2} \Big(L + \omega\big(|X(s)|, X(s)| \big) \Big) ds + \omega\big(|X(T)| \vee |X_c(T)|, |X(T) - X_c(T)| \big)$$

$$\leqslant J^a(t_1, x; a(\cdot)) + (T - t_2 + 1)\omega\big(e^{LT}(1 + |x|), e^{LT}|x - X(t_2)| \big)$$

$$+ (t_2 - t_1)\Big(L + \omega\big(e^{LT}(1 + |x|), e^{LT}(1 + |x|) \big) \Big)$$

$$\leqslant J^a(t_1, x; a(\cdot)) + (T + 1)\omega\big(e^{LT}(1 + |x|), e^{LT}(e^{L(t_2 - t_1)} - 1)(1 + |x|) \big)$$

$$+ (t_2 - t_1)\Big(L + \omega\big(e^{LT}(1 + |x|), e^{LT}(1 + |x|) \big) \Big)$$

$$\equiv J^a(t_1, x; a(\cdot)) + \widetilde{\omega}(|x|, t_2 - t_1).$$

Then, one has

$$V^a(t_2, x) - V^a(t_1, x) \leqslant \widetilde{\omega}(|x|, t_2 - t_1).$$

On the other hand, for any $a(\cdot) \in \mathcal{S}^a[t_2, T]$, let $a_e(\cdot) = aI_{[t_1, t_2]}(\cdot) \oplus a(\cdot)$ $\in \mathcal{S}^a[t_1, T]$ be the natural extension of $a(\cdot)$ over $[t_1, T]$, and

$$X_e(\cdot) = X(\cdot; t_1, x; a_e(\cdot)), \qquad X(\cdot) = X(\cdot; t_2, x, a(\cdot)).$$

Then

$$V^a(t_1, x) \leqslant J^a(t_1, x; a_e(\cdot))$$

$$= \int_{t_1}^{T} g(s, X_e(s), a_e(s)) ds + h(X_e(T)) + \sum_{i \geqslant 0} \kappa(\theta_{i+1}, a_i, a_{i+1})$$

$$= J^a(t_2, x; a(\cdot)) + \int_{t_2}^{T} \Big(g(s, X_e(s), a_e(s)) - g(s, X(s), a(s)) \Big) ds$$

$$+ \int_{t_1}^{t_2} g(s, X_e(s), a) ds + h(X_e(T)) - h(X(T))$$

$$\leqslant J^a(t_2, x; a(\cdot)) + \int_{t_2}^{T} \omega\big(|X(s)| \vee |X_e(s)|, |X(s) - X_e(s)| \big) ds$$

$$+ \int_{t_1}^{t_2} \Big(L + \omega\big(|X_e(s)|, |X_e(s)| \big) \Big) ds$$

$$+ \omega\big(|X(T)| \vee |X_e(T)|, |X(T) - X_e(T)| \big)$$

$$\leqslant J^a\big(t_2, x; a(\cdot)\big) + (T+1)\omega\big(e^{LT}(1+|x|), e^{LT}(e^{L(t_2-t_1)}-1)(1+|x|)\big)$$
$$+(t_2 - t_1)\Big(L + \omega\big(e^{LT}(1+|x|), e^{LT}(1+|x|)\big)\Big)\Big)$$
$$\equiv J^a\big(t_2, x; a(\cdot)\big) + \widetilde{\omega}\big(|x|, t_2 - t_1\big).$$

Hence,

$$V^a(t_1, x) - V^a(t_2, x) \leqslant \widetilde{\omega}\big(|x|, t_2 - t_1\big).$$

Then we obtain the continuity of $V(\cdot, \cdot)$.

The proof of (7.5) under conditions (S1), (S2)′ and (S3) is very similar. We leave the details to the readers. □

7.1.2 Dynamic programming and quasi-variational inequality

We now present a principle of optimality.

Theorem 7.1.2. *Let* (S1)–(S3) *hold. Then the following hold.*

$$V^a(t, x) \leqslant \int_t^\tau g(s, X(s), a)ds + V^a(\tau, X(\tau)), \qquad \tau \in (t, T], \qquad (7.6)$$

$$V^a(t, x) \leqslant M^a[V](t, x) \equiv \min_{\bar{a} \neq a}\Big[V^a(t, x) + \kappa(t, a, \bar{a})\Big] \qquad (7.7)$$

and if at some $(t, x) \in [0, T) \times \mathbb{R}^n$, *the inequality in* (7.7) *is strict, then there exists a* $\bar{\tau} \in (t, T]$ *such that*

$$V^a(t, x) = \int_t^\tau g(s, X(s), a)ds + V^a(\tau, X(\tau)), \qquad \tau \in (t, \bar{\tau}). \qquad (7.8)$$

Proof. For any $\bar{a}_0 \in \mathbb{M}$, $\bar{a}_0 \neq a_0$, and $\bar{a}(\cdot) \equiv \{(\bar{\theta}_i, \bar{a}_i)\}_{i=0}^k \in \mathcal{S}^{\bar{a}_0}[t, T]$, let

$$\begin{cases} \widetilde{a}(\cdot) = \{(\widetilde{\theta}_i, \widetilde{a}_i)\}_{i \geqslant 0}, \\ \widetilde{\theta}_0 = \bar{\theta}_0 = t, \qquad \widetilde{\theta}_i = \bar{\theta}_{i-1}, \quad i \geqslant 1, \\ \widetilde{a}_0 = a_0, \qquad \widetilde{a}_i = \bar{a}_{i-1}, \quad i \geqslant 1. \end{cases}$$

Then

$$V^{a_0}(t, x) \leqslant J^{a_0}(t, x; \widetilde{a}(\cdot)) = J^{\bar{a}_0}(t, x; \bar{a}(\cdot)) + \kappa(t, a_0, \bar{a}_0).$$

Hence,

$$V^{a_0}(t, x) \leqslant \min_{\bar{a}_0 \neq a_0}\Big(V^{\bar{a}_0}(t, x) + \kappa(t, a_0, \bar{a}_0)\Big) \equiv M^{a_0}[V](t, x).$$

This proves (7.6). Next, fix any $(t,x,a) \in [0,T) \times \mathbb{R}^n \times \mathbb{M}$ and any $\tau \in (t,T]$, take any $a(\cdot) \equiv \{(\theta_i,a_i)\}_{i \geqslant 0} \in \mathcal{S}^a[\tau,T]$, denote $a_e(\cdot) = a I_{[t,\tau)}(\cdot) \oplus a(\cdot) \in \mathcal{S}^a[t,T]$, the extension of $a(\cdot)$ over $[t,T]$. Then

$$V^a(t,x) \leqslant J^a(t,x;a_e(\cdot)) = \int_t^\tau g(s,X(s),a)ds + J^a(\tau,X(\tau);a(\cdot)).$$

Hence,

$$V^a(t,x) \leqslant \int_t^\tau g(s,X(s),a)ds + V^a(\tau,X(\tau)),$$

proving (7.7). Now, let

$$V^a(t,x) < M^a[V](t,x) \equiv \min_{\bar{a} \neq a}\left(V^{\bar{a}}(t,x) + \kappa(t,a,\bar{a})\right). \tag{7.9}$$

For any $\varepsilon > 0$, let $a^\varepsilon(\cdot) \equiv \{(\theta_i^\varepsilon,a_i^\varepsilon)\}_{i \geqslant 0} \in \mathcal{S}^a[t,T]$ such that

$$V^a(t,x) + \varepsilon > J^a(t,x;a^\varepsilon(\cdot)).$$

We first claim that

$$\theta_1^\varepsilon > t, \qquad \forall \varepsilon > 0. \tag{7.10}$$

In fact, if for some $\varepsilon > 0$,

$$\theta_1^\varepsilon = \theta_0^\varepsilon = t,$$

then letting

$$\begin{cases} \widetilde{a}^\varepsilon(\cdot) = \{\widetilde{\theta}_i^\varepsilon,\widetilde{a}_i^\varepsilon\}_{i \geqslant 0} \in \mathcal{S}^{a_1^\varepsilon}[t,T], \\ \widetilde{\theta}_i^\varepsilon = \theta_{i+1}^\varepsilon, \quad \widetilde{a}_i^\varepsilon = a_{i+1}^\varepsilon, \quad i \geqslant 0, \end{cases}$$

we have

$$V^a(t,x) + \varepsilon > J^a(t,x;a^\varepsilon(\cdot)) = J^{a_1^\varepsilon}(t,x;\widetilde{a}^\varepsilon(\cdot)) + \kappa(t,a,\widetilde{a}_1^\varepsilon)$$
$$\geqslant V^{a_1^\varepsilon}(t,x) + \kappa(t,a,\widetilde{a}_1^\varepsilon) \geqslant M^a[V](t,x),$$

which contradicts (7.9). Thus, (7.10) holds. Next, we claim that there exists a $\bar{\tau} \in (t,T)$ such that

$$\theta_1^\varepsilon > \bar{\tau}, \qquad \forall \varepsilon > 0. \tag{7.11}$$

If not, then along a sequence $\varepsilon\downarrow 0$, one has $\theta_1^\varepsilon \to t$. We let $X^\varepsilon(\cdot) = X(\cdot\,;t,x,a^\varepsilon(\cdot))$. Then

$$V^a(t,x) + \varepsilon > J^a(t,x;a^\varepsilon(\cdot))$$

$$= \int_t^T g(s,X^\varepsilon(s),a^\varepsilon(s))ds + h(X^\varepsilon(T)) + \sum_{i\geqslant 0}\kappa(\theta_{i+1}^\varepsilon,a_i^\varepsilon,a_{i+1}^\varepsilon)$$

$$= \int_t^{\theta_1^\varepsilon} g(s,X^\varepsilon(s),a)ds + \int_{\theta_1^\varepsilon}^T g(s,X^\varepsilon(s),a^\varepsilon(s))ds + h(X^\varepsilon(T))$$

$$+ \sum_{i\geqslant 1}\kappa(\theta_{i+1}^\varepsilon,a_i^\varepsilon,a_{i+1}^\varepsilon) + \kappa(\theta_1^\varepsilon,a,a_1^\varepsilon)$$

$$= J^{a_1^\varepsilon}(\theta_1^\varepsilon,X^\varepsilon(\theta_1^\varepsilon);\tilde{a}^\varepsilon(\cdot)) + \int_t^{\theta_1^\varepsilon} g(s,X^\varepsilon(s)a)ds + \kappa(\theta_1^\varepsilon,a,a_1^\varepsilon)$$

$$\geqslant V^{a_1^\varepsilon}(\theta_1^\varepsilon,X^\varepsilon(\theta_1^\varepsilon)) - \int_t^{\theta_1^\varepsilon}\Big(L + \omega\big(|X^\varepsilon(s)|,|X^\varepsilon(s)|\big)\Big)ds + \kappa(\theta_1^\varepsilon,a,a_1^\varepsilon).$$

Note that for fixed $x \in \mathbb{R}^n$,

$$\lim_{\varepsilon\to 0}\sup_{s\in[t,\theta_1^\varepsilon]}|X^\varepsilon(s;t,x,a(\cdot)) - x| \leqslant \lim_{\varepsilon\to 0}\sup_{s\in[t,\theta_1^\varepsilon]}\big[e^{L(s-t)} - 1\big](1 + |x|) = 0,$$

uniformly in $a(\cdot) \in \mathcal{S}^a[t,T]$. Also, $\mathbb{M}$ is a finite set, we may assume that along a sequence, $a_1^\varepsilon = a_1 \in \mathbb{M}$ is independent of ε (with $a_1 \neq a$). Then passing to the limit in the above, we obtain

$$V^a(t,x) \geqslant V^{a_1}(t,x) + \kappa(t,a,a_1) \geqslant M^a[V](t,x),$$

which is a contradiction again. Hence, (7.11) holds. Consequently, for any $\tau \in (t,\bar{\tau})$, one has

$$V^a(t,x) + \varepsilon > J^a(t,x;a^\varepsilon(\cdot))$$

$$= \int_t^\tau g(s,X(s),a)ds + J^a(\tau,X^\varepsilon(\tau);a_c^\varepsilon(\cdot))$$

$$\geqslant \int_t^\tau g(s,X(s),a)ds + V^a(\tau,X(\tau)).$$

Sending $\varepsilon \to 0$, we obtain

$$V^a(t,x) \geqslant \int_t^\tau g(s,X(s),a)ds + V^a(\tau,X(\tau)).$$

Then combining (7.6), we get (7.8). $\qquad\qquad\square$

The following is a verification theorem.

Theorem 7.1.3. *Let (S1)–(S3) hold. Suppose $V(\cdot,\cdot) \equiv (V^1(\cdot,\cdot),\cdots,V^m(\cdot,\cdot))$ is a continuous function such that the conclusions of Theorem*

7.1.2 hold. Then, for any $(t, x, a) \in [0, T) \times \mathbb{R}^n \times \mathbb{M}$, an $a^*(\cdot) \in \mathcal{S}^a[t, T]$ can be constructed such that

$$V^a(t, x) = J^a(t, x; a^*(\cdot)). \tag{7.12}$$

Consequently, if the conclusions of Theorem 7.1.2 characterizes the value function of Problem (S), then the constructed $a^*(\cdot)$ is an optimal switching control of Problem (S).

Proof. Let $(t, x, a) \in [0, T) \times \mathbb{R}^n \times \mathbb{M}$ be given. Suppose

$$V^a(t, x) < M^a[V](t, x).$$

The other case can be treated similarly (see below). In this case, we solve the a-th state equation

$$\begin{cases} \dot{X}(s) = f(s, X(s), a), & s \geqslant t, \\ X(t) = x. \end{cases}$$

Let

$$\theta_1^* = \inf \big\{ s \in (t, T] \mid V^a(s, X(s)) = M^a[V](s, X(s)) \big\},$$

with the convention that $\inf \phi = T$. If $\theta_1^* = T$, then let $a^*(\cdot) \equiv a$, and we have

$$V^a(s, X(s)) < M^a[V](s, X(s)), \qquad s \in [t, T).$$

Then by (7.8), we have

$$V^a(t, x) = \int_t^\tau g(s, X(s), a)ds + V^a(\tau, X(\tau)), \qquad \forall \tau \in [t, T).$$

Sending $\tau \to T$, we obtain

$$V^a(t, x) = \int_t^T g(s, X(s), a)ds + h(X(T)) = J^a(t, x; a^*(\cdot)).$$

If $\theta_1^* < T$, then we have

$$V^a(\theta_1^*, X(\theta_1^*)) = M^a[V](\theta_1^*, X(\theta_1^*)) = V^{a_1^*}(\theta_1^*, X(\theta_1^*)) + \kappa(\theta_1^*, a, a_1^*),$$

for some $a_1^* \in \mathbb{M} \setminus \{a\}$. In general, such an a_1^* might not be unique. For definiteness, we take the smallest $a_1^* \in \mathbb{M}$ such that the above holds. Next, we claim that

$$V^{a_1^*}(\theta_1^*, X(\theta_1^*)) < M^{a_1^*}[V](\theta_1^*, X(\theta_1^*)). \tag{7.13}$$

In fact, if not, then for some $a_2 \neq a_1^*$,

$$V^{a_1^*}(\theta_1^*, X^*(\theta_1^*)) = M^{a_1^*}[V](\theta_1^*, X^*(\theta_1^*)) = V^{a_2}(\theta_1^*, X^*(\theta_1^*)) + \kappa(\theta_1^*, a_1^*, a_2),$$

which leads to

$$V^a\left(\theta_1^*, X^*(\theta_1^*)\right) = V^{a_2}\left(\theta_1^*, X^*(\theta_1^*)\right) + \kappa(\theta_1^*, a_1^*, a_2) + \kappa(\theta_1^*, a, a_1^*)$$
$$> V^{a_2}\left(\theta_1^*, X^*(\theta_1^*)\right) + \kappa(\theta_1^*, a, a_2) \geqslant M^a[V]\left(\theta_1^*, X^*(\theta_1^*)\right).$$

This is a contradiction. Hence, (7.13) holds and we arrive at the same case as at the beginning. By induction, we can construct a switching control $a^*(\cdot) = \{(\theta_i^*, a_i^*)\}_{i \geqslant 0} \in \mathcal{S}^a[t, T]$ such that

$$\begin{cases} V^{a_i^*}\left(\theta_i^*, X^*(\theta_i^*)\right) = \displaystyle\int_{\theta_i^*}^{\theta_{i+1}^*} g(s, X^*(s), a_i^*) ds + V^{a_i^*}\left(\theta_{i+1}^*, X^*(\theta_{i+1}^*)\right), \\ V^{a_i^*}\left(\theta_{i+1}^*, X^*(\theta_{i+1}^*)\right) = V^{a_{i+1}^*}\left(\theta_{i+1}^*, X^*(\theta_{i+1}^*)\right) + \kappa(\theta_{i+1}^*, a_i^*, a_{i+1}^*), \end{cases} \quad i \geqslant 0.$$

Since

$$V^a(T, x) = h(x) < M^a[h](x) = M^a[V](T, x), \qquad x \in \mathbb{R}^n.$$

By the continuity of the value function, we see that there exists a $k \geqslant 0$ such that

$$V^{a_k^*}(\theta_k^*, X^*(\theta_k^*)) = \int_{\theta_k^*}^T g(s, X^*(s), a_k^*) ds + h(X^*(T)).$$

That is k is the total number of switchings in $a^*(\cdot)$. Then we have

$$V^a(t, x) = \int_t^T g(s, X^*(s), a^*(s)) ds + h(X^*(T))$$
$$+ \sum_{i \geqslant 0} \kappa(\theta_{i+1}^*, a_i^*, a_{i+1}^*) = J^a(t, x; a^*(\cdot)).$$

This proves (7.12). Now, in the case that the function $V(\cdot, \cdot)$ characterized by Theorem 7.1.2 has to be the value function of Problem (S), then (7.12) implies that the constructed switching control $a^*(\cdot) \in \mathcal{S}^a[t, T]$ is optimal for Problem (S). $\qquad\square$

Our next goal is to investigate if the conclusions of Theorem 7.1.2 uniquely characterizes the value function of Problem (S). The following result introduces the corresponding Hamilton-Jacobi-Bellman equation for Problem (S).

Proposition 7.1.4. *Let* (S1)–(S3) *hold. Suppose the value function* $V(\cdot, \cdot)$ *of Problem* (S) *is continuously differentiable. Then the following equation is satisfied:*

$$\begin{cases} \min\left\{V_t^a(t, x) + H^a(t, x, V_x^a(t, x)), M^a[V](t, x) - V^a(t, x)\right\} = 0, \\ \qquad\qquad\qquad\qquad (t, x, a) \in [0, T] \times \mathbb{R}^n \times \mathbb{M}, \qquad (7.14) \\ V^a(T, x) = h(x), \qquad (x, a) \in \mathbb{R}^n \times \mathbb{M}, \end{cases}$$

where

$$H^a(t, x, p) = \langle p, f(t, x, a) \rangle + g(t, x, a).$$

Proof. From (7.6), we have

$$
0 \leqslant \frac{1}{\tau - t}\left[V^a(\tau, X(\tau)) - V^a(t, x) + \int_t^\tau g(s, X(s), a)ds\right]
$$

$$
= \frac{1}{\tau - t}\int_t^\tau \left[V_t^a(s, X(s)) + \langle V_x^a(s, X(s)), f(s, X(s), a) \rangle + g(s, X(s), a)\right]ds
$$

$$
= \frac{1}{\tau - t}\int_t^\tau \left[V_t(s, X(s)) + H^a(s, X(s), V_x^a(s, X(s)))\right]ds.
$$

Hence, we obtain

$$
\min\left\{V_t^a(t, x) + H^a(t, x, V_x^a(t, x)), M^a[V](t, x) - V^a(t, x)\right\} \geqslant 0.
$$

Now, on the set $V^a(t, x) < M^a[V](t, x)$, we have some $\bar{\tau} > t$ such that (7.8) holds, which leads to

$$
V_t^a(t, x) + H^a(t, x, V_x^a(t, x)) = 0.
$$

Hence, (7.14) holds. □

Note that when $V(\cdot, \cdot)$ is C^1, (7.14) is equivalent to the following:

$$
\begin{cases}
V_t^a(t, x) + H^a(t, x, V_x^a(t, x)) \geqslant 0, \quad M^a[V](t, x) - V^a(t, x) \geqslant 0, \\
\left(V_t^a(t, x) + H^a(t, x, V_x^a(t, x))\right)\left(M^a[V](t, x) - V^a(t, x)\right) = 0, \\
\qquad\qquad (t, x, a) \in [0, T] \times \mathbb{R}^n \times \mathbb{M}, \\
V^a(T, x) = h(x), \qquad (x, a) \in \mathbb{R}^n \times \mathbb{M}.
\end{cases}
$$

This is called a *quasi-variational inequality*, and $M^a[V]$ is called a *switching obstacle*.

7.1.3 *Viscosity solutions of quasi-variational inequalities*

Proposition 7.1.4 shows that when the value function $V(\cdot, \cdot)$ is continuously differentiable, it is a classical solution to (7.14). Now, if we can show that (7.14) admits at most one solution and the value function $V(\cdot, \cdot)$ is differentiable, then the solution has to be the value function. Hence, (7.14) characterizes the value function $V(\cdot, \cdot)$ of Problem (S), and via the value function, one can construct an optimal switching control for Problem (S). However, similar to the classical optimal control problem, the value function $V(\cdot, \cdot)$ of Problem (S) is not necessarily differentiable. Hence, we need to realize the above idea of solving Problem (S) in the framework of viscosity solutions. Note that since $V(\cdot,\cdot)$ is a vector-valued function, the corresponding definition is slightly different from that for classical optimal control problems.

Definition 7.1.5. (i) A continuous function $V(\cdot,\cdot) \equiv (V^1(\cdot,\cdot), \cdots, V^m(\cdot,\cdot))$ is called a viscosity sub-solution of (7.14) if

$$V^a(T,x) \leqslant h(x), \qquad \forall (x,a) \in \mathbb{R}^n \times \mathbb{M}, \qquad (7.15)$$

and for any continuous differentiable function $\varphi(\cdot,\cdot)$, as long as $V^a(\cdot,\cdot) - \varphi(\cdot,\cdot)$ attains a local maximum at $(t_0,x_0) \in [0,T) \times \mathbb{R}^n$, the following holds:

$$\min \Big\{ \varphi_t(t_0,x_0) + H^a(t_0,x_0,\varphi_x(t_0,x_0)), \\ M^a[V](t_0,x_0) - V^a(t_0,x_0) \Big\} \geqslant 0. \qquad (7.16)$$

(ii) A continuous function $V(\cdot,\cdot) \equiv (V^1(\cdot,\cdot), \cdots, V^m(\cdot,\cdot))$ is called a viscosity super-solution of (7.14) if

$$V^a(T,x) \geqslant h(x), \qquad \forall (x,a) \in \mathbb{R}^n \times \mathbb{M}, \qquad (7.17)$$

and for any continuous differentiable function $\varphi(\cdot,\cdot)$, as long as $V^a(\cdot,\cdot) - \varphi(\cdot,\cdot)$ attains a local minimum at $(t_0,x_0) \in [0,T) \times \mathbb{R}^n$, the following holds:

$$\min \Big\{ \varphi_t(t_0,x_0) + H^a(t_0,x_0,\varphi_x(t_0,x_0)), \\ M^a[V](t_0,x_0) - V^a(t_0,x_0) \Big\} \leqslant 0. \qquad (7.18)$$

(iii) If $V(\cdot,\cdot)$ is both viscosity sub-solution and viscosity super-solution to (7.14), it is called a viscosity solution to (7.14).

The following result is a rigorous version of Proposition 7.1.4.

Theorem 7.1.6. *Let* (S1), (S2)' *and* (S3) *hold. Then the value function* $V(\cdot,\cdot)$ *of Problem* (S) *is a viscosity solution of* (7.14).

Proof. Fix an $a \in \mathbb{M}$, let $V^a(\cdot,\cdot) - \varphi(\cdot,\cdot)$ attain a local maximum at $(t_0,x_0) \in [0,T) \times \mathbb{R}^n$. Let $X(\cdot) = X(\cdot\,;t_0,x_0,a)$. By Theorem 7.1.2, we have

$$0 \leqslant \int_{t_0}^{\tau} g(s,X(s),a)ds + V^a(\tau,X(\tau)) - V^a(t_0,x_0)$$

$$\leqslant \int_{t_0}^{\tau} g(s,X(s),a)ds + \varphi(\tau,X(\tau)) - \varphi(t_0,x_0).$$

Then dividing by $\tau - t_0$ and sending $\tau \downarrow t_0$, we obtain

$$0 \leqslant \varphi_t(t_0,x_0) + H^a(t_0,x_0,\varphi_x(t_0,x_0)).$$

On the other hand, Theorem 7.1.2 tells us that one always has

$$M^a[V](t_0, x_0) \geqslant V^a(t_0, x_0).$$

Thus, (7.16) holds, and by definition, $V(\cdot, \cdot)$ is a viscosity sub-solution of (7.14).

Next, let $V^a(\cdot, \cdot) - \varphi(\cdot, \cdot)$ attain a local minimum at $(t_0, x_0) \in [0, T) \times \mathbb{R}^n$. If

$$M^a[V](t_0, x_0) - V^a(t_0, x_0) = 0,$$

we have (7.18). If the following holds:

$$M^a[V](t_0, x_0) > V^a(t_0, x_0).$$

Then by Theorem 7.1.2, there exists a $\bar{\tau} \in (t_0, T]$ such that

$$
\begin{aligned}
0 &= \int_{t_0}^{\tau} g(s, X(s), a)ds + V^a(\tau, X(\tau)) - V^a(t_0, x_0) \\
&\geqslant \int_{t_0}^{\tau} g(s, X(s), a)ds + \varphi(\tau, X(\tau)) - \varphi(t_0, x_0), \quad \forall \tau \in (t_0, \bar{\tau}).
\end{aligned}
$$

Hence, dividing by $\tau - t_0$ and sending $\tau \downarrow t_0$, we obtain

$$0 \geqslant \varphi_t(t_0, x_0) + H^a(t_0, x_0, \varphi_x(t_0, x_0)).$$

Consequently, (7.18) holds and by definition, $V(\cdot, \cdot)$ is a viscosity super-solution of (7.14). Therefore, $V(\cdot, \cdot)$ is a viscosity solution of (7.14). □

We now consider the uniqueness of viscosity solutions to (7.14). Let $\mathcal{V}$ be the set of all continuous functions $v : [0, T] \times \mathbb{R}^n \to \mathbb{R}^m$ such that for some increasing continuous function $\theta : \mathbb{R}_+ \to \mathbb{R}_+$,

$$|v(t_1, x) - v(t_2, x)| \leqslant \theta(|x|)|t_1 - t_2|, \quad \forall t_1, t_2 \in [0, T], x \in \mathbb{R}^n.$$

We have the following comparison theorem which will lead to the uniqueness of viscosity solutions.

Theorem 7.1.7. (i) *Let (S1), (S2) and (S3)' hold. Let $V(\cdot, \cdot), \widehat{V}(\cdot, \cdot) \in C([0, T] \times \mathbb{R}^n; \mathbb{R}^m)$ be a viscosity sub-solution and a viscosity super-solution of variational inequality (7.14), respectively. Then*

$$V^a(t, x) \leqslant \widehat{V}^a(t, x), \qquad \forall (t, x, a) \in [0, T[\times \mathbb{R}^n \times \mathrm{M}. \tag{7.19}$$

(ii) *Let (S1)–(S3) hold. Let $V(\cdot, \cdot), \widehat{V}(\cdot, \cdot) \in \mathcal{V}$ be a viscosity sub-solution and a viscosity super-solution of the variational inequality (7.14), respectively. Then (7.19) holds.*

For (i), since the switching cost $\kappa(\cdot)$ is independent of t, the proof is almost the same as that of Theorem 2.5.3, with a little modification which one can see in the proof of part (ii) below. For (ii), due to the dependence of $\kappa(\cdot)$ on the time variable t, the method used in the proof of Theorem 2.5.3 does not work. Hence, we need to introduce some new method. But for that new method to work, we strengthen (S2)' to (S2) so that $t \mapsto V(t, x)$ is Lipschitz. To provide a proof to part (ii) of the above theorem, we need to make some preparations.

For function $v : [0, T] \times \mathbb{R}^n \to [-\infty, +\infty]$, $(s, z) \in [0, T) \times \mathbb{R}^n$, let

$$D_{t,x}^{1,+} v(t, x) = \Big\{ (q, p) \in \mathbb{R} \times \mathbb{R}^n \mid v(s, y) \leqslant v(t, x) + q(s - t) + \langle p, y - x \rangle$$
$$+ o\big(|t - s| + |y - x| \big) \Big\},$$

$$D_{t,x}^{1,-} v(t, x) = \Big\{ (q, p) \in \mathbb{R} \times \mathbb{R}^n \mid v(s, y) \geqslant v(t, x) + q(s - t) + \langle p, y - x \rangle$$
$$+ o\big(|t - s| + |y - x| \big) \Big\}.$$

We call $D_{t,x}^{1,+} v(t, x)$ and $D_{t,x}^{1,-} v(t, x)$ the *super-* and *sub-gradient* of $v(\cdot, \cdot)$ at (t, x), respectively. We further define

$$\overline{D}_{t,x}^{1,+} v_1(t, x) = \Big\{ (q, p) \in \mathbb{R} \times \mathbb{R}^n \mid \exists (t_k, x_k, q_k, p_k) \to (t, x, b, p),$$
$$(q_k, p_k) \in D_{t,x}^{1,+} v(t_k, x_k) \Big\},$$

$$\overline{D}_{t,x}^{1,-} v_1(t, x) = \Big\{ (q, p) \in \mathbb{R} \times \mathbb{R}^n \mid \exists (t_k, x_k, q_k, p_k) \to (t, x, q, p),$$
$$(q_k, p_k) \in D_{t,x}^{1,-} v(t_k, x_k) \Big\}.$$

It is clear that for any $v(\cdot, \cdot) \in C([0, T] \times \mathbb{R}^n)$,

$$\begin{cases} D_{t,x}^{1,+}(-v)(t, x) = -D_{t,x}^{1,-} v(t, x), \\ \overline{D}_{t,x}^{1,+}(-v)(t, x) = -\overline{D}_{t,x}^{1,-} v(t, x), \\ D_{t,x}^{1,+} v(t, x) \subseteq \overline{D}_{t,x}^{1,+} v(t, x), \\ D_{t,x}^{1,-} v(t, x) \subseteq \overline{D}_{t,x}^{1,-} v(t, x), \end{cases} \quad (t, x) \in [0, T] \times \mathbb{R}^n. \quad (7.20)$$

The following result gives a representation of elements in $D_{t,x}^{1,\pm} v(t, x)$.

Lemma 7.1.8. *Let* $v(\cdot, \cdot) \in C([0, T] \times \mathbb{R}^n)$ *and* $(t_0, x_0) \in [0, T) \times \mathbb{R}^n$ *be given. Then*

(i) $(q, p) \in D_{t,x}^{1,+} v(t_0, x_0)$ *if and only if there exists a* $\varphi \in C_0^1(\mathbb{R} \times \mathbb{R}^n)$, *such that* $v - \varphi$ *attains a strict maximum at* (t_0, x_0) *and*

$$\big(\varphi(t_0, x_0), \varphi_t(t_0, x_0), \varphi_x(t_0, x_0) \big) = (v(t_0, x_0), q, p). \quad (7.21)$$

(ii) $(q,p) \in D_{t,x}^{1,-} v(t_0, x_0)$ if and only if there exists a $\varphi \in C_0^1(\mathbb{R} \times \mathbb{R}^n)$, such that $v - \varphi$ attains a strict minimum at (t_0, x_0) and (7.21) holds.

Proof. We prove (i). By taking into account (7.20), one can prove (ii). Suppose $(q,p) \in D_{t,x}^{1,+} v(t_0, x_0)$. Define

$$\Phi(t,x) = \begin{cases} \dfrac{\big(v(t,x) - v(t_0,x_0) - q(t-t_0) - \langle p, x - x_0 \rangle\big) \vee 0}{|t - t_0| + |x - x_0|}, \\ \qquad \text{if } (t_0,x_0) \neq (t,x) \in [0,T] \times \mathbb{R}^n, \\ 0, \qquad \text{otherwise,} \end{cases}$$

and

$$\varepsilon(r) = \sup\{\Phi(t,x) : (t,x) \in [0,T] \times \mathbb{R}^n, |s - t| + |y - x| \leqslant r\}.$$

Then it is seen that $\varepsilon : \mathbb{R} \to [0, \infty)$ is a continuous nondecreasing function with $\varepsilon(0) = 0$. Further,

$$v(t,x) - v(t_0,x_0) - q(t-t_0) - \langle p, x - x_0 \rangle$$
$$\leqslant (|t - t_0| + |x - x_0|)\varepsilon(|t - t_0| + |x - x_0|), \quad \forall (t,x) \in [0,T] \times \mathbb{R}^n.$$

Set

$$\psi(t,x) = \int_0^{2(|t-t_0|+|x-x_0|)} \varepsilon(\rho)d\rho + \big(|t-t_0|^2 + |x-x_0|^2\big)^2, \quad (t,x) \in [0,T] \times \mathbb{R}^n.$$

Clearly, $\psi \in C^1(\mathbb{R} \times \mathbb{R}^n)$ with

$$\psi(t_0,x_0) = 0, \quad \psi_t(t_0,x_0) = 0, \quad \psi_x(t_0,x_0) = 0,$$

and

$$\psi(t,x) \geqslant \int_{|t-t_0|+|x-x_0|}^{2(|t-t_0|+|x-x_0|)} \varepsilon(\rho)d\rho + \big(|t-t_0|^2 + |x-x_0|^2\big)^2$$
$$> (|t-t_0| + |x-x_0|)\varepsilon(|t-t_0| + |x-x_0|)$$
$$\geqslant v(t,x) - v(t_0,x_0) - q(t-t_0) - \langle p, x - x_0 \rangle, \quad \forall (t,x) \in [0,T] \times \mathbb{R}^n.$$

By defining

$$\varphi(t,x) = v(t_0,x_0) + q(t-t_0) + \langle p, x - x_0 \rangle + \psi(t,x), \quad (t,x) \in \mathbb{R} \times \mathbb{R}^n,$$

we have

$$v(t,x) - \varphi(t,x) \leqslant v(t_0,x_0) - \varphi(t_0,x_0) = 0, \qquad \forall (t,x) \in [0,T] \times \mathbb{R}^n,$$

and (7.21) holds. This proves the "only if" part.

Conversely, if there exists a $\varphi(\cdot,\cdot) \in C_0^1(\mathbb{R} \times \mathbb{R}^n)$ such that $v - \varphi$ attains a strict maximum at (t_0, x_0), and (7.21) holds. Then

$$
\begin{aligned}
v(s,y) &\leqslant v(t_0, x_0) + \varphi(s,y) - \varphi(t_0, x_0) \\
&\leqslant v(t_0, x_0) + \varphi_t(t_0, x_0)(s - t_0) + \langle \varphi_x(t_0, x_0), y - x_0 \rangle \\
&\quad + o(|s - t_0| + |y - x_0|) \\
&\leqslant v(t_0, x_0) + q(s - t_0) + \langle p, y - x_0 \rangle + o(|s - t_0| + |y - x_0|).
\end{aligned}
$$

This implies that

$$
(q, p) \in D_{t,x}^{1,+} v(t_0, x_0),
$$

proving the "if" part. □

The following gives an equivalent definition of viscosity solutions to the quasi-variational inequality (7.14), whose proof is obvious by applying Lemma 7.1.8.

Proposition 7.1.9. *Let (S1)–(S3) hold. Then function* $V(\cdot,\cdot) \in C([0,T] \times \mathbb{R}^n; \mathbb{R}^m)$ *is a viscosity sub-solution (resp. super-solution) of (7.14) if and only if (7.15) (resp. (7.17)) is satisfied and the following holds:* $\forall (t,x) \in [0,T) \times \mathbb{R}^n$,

$$
\min\{q + H^a(t,x,p), M^a[V](t,x) - V^a(t,x)\} \geqslant 0 \ (resp. \leqslant 0),
$$

$$
\forall (q,p) \in \overline{D}_{t,x}^{1,+} V^a(t,x) \quad (resp. \ \overline{D}_{t,x}^{1,-} V^a(t,x)), \ a \in \mathbb{M}.
$$

Proof. We prove the viscosity sub-solution case. The viscosity super-solution case can be proved similarly. Let $V(\cdot,\cdot)$ be a viscosity sub-solution of (7.14), then (7.15) holds, and for any $(t_0, x_0, a) \in [0,T) \times \mathbb{R}^n \times \mathbb{M}$, any $(q_0, p_0) \in \overline{D}_{t,x}^{1,+} V^a(t_0, x_0)$, we have some sequence $(t_k, x_k, q_k, p_k) \to (t_0, x_0, q_0, p_0)$ such that

$$
(q_k, p_k) \in D_{t,x}^{1,+} V^a(t_k, x_k), \qquad k \geqslant 1.
$$

Next, for each $k \geqslant 1$, by Lemma 7.1.8, we can find a $\varphi(\cdot,\cdot) \in C_0^1(\mathbb{R} \times \mathbb{R}^n)$ such that $V^a(\cdot,\cdot) - \varphi(\cdot,\cdot)$ attains a strict maximum at (t_k, x_k) and

$$
\big(\varphi(t_k, x_k), \varphi_t(t_k, x_k), \varphi_x(t_k, x_k)\big) = \big(V^a(t_k, x_k), q_k, p_k\big).
$$

Then

$$
\min\{q_k + H^a(t_k, x_k, p_k), M^a[V](t_k, x_k) - V^a(t_k, x_k)\} \geqslant 0.
$$

Letting $k \to \infty$, we obtain

$$
\min\{q_0 + H^a(t_0, x_0, p_0), M^a[V](t_0, x_0) - V^a(t_0, x_0)\} \geqslant 0.
$$

The converse is clear. ∎

The following result will play an essential role below.

Lemma 7.1.10. *Suppose* $v_1(\cdot,\cdot), v_2(\cdot,\cdot) \in \mathcal{V}$, *and* $\varphi : [0,T] \times \mathbb{R}^n \times \mathbb{R}$ *is continuously differentiable such that* $(t,x,y) \mapsto v_1(t,x) + v_2(t,y) - \varphi(t,x,y)$ *attains a local maximum at* $(\bar{t}, \bar{x}, \bar{y})$. *Then there exist* $q_1, q_2 \in \mathbb{R}$ *such that*

$$\begin{cases} \left(q_1, \varphi_x(\bar{t}, \bar{x}, \bar{y})\right) \in \overline{D}_{t,x}^{1,+} v_1(\bar{t}, \bar{x}), \\ \left(q_2, \varphi_y(\bar{t}, \bar{x}, \bar{y})\right) \in \overline{D}_{t,x}^{1,+} v_2(\bar{t}, \bar{x}), \\ q_1 + q_2 = \varphi_t(\bar{t}, \bar{x}, \bar{y}). \end{cases}$$

Proof. By assumption, we let $\varepsilon > 0$ such that

$$v_1(\bar{t}, \bar{x}) + v_2(\bar{t}, \bar{y}) - \varphi(\bar{t}, \bar{x}, \bar{y}) \geqslant v_1(t, x) + v_2(t, y) - \varphi(t, x, y),$$
$$(t, x, y) \in \bar{B}_\varepsilon(\bar{t}, \bar{x}, \bar{y}).$$

Now, for any $\delta \in (0, \varepsilon)$, let

$$\Phi_\delta(t, x, s, y) = v_1(t, x) + v_2(s, y) - \varphi(t, x, y)$$
$$-\frac{1}{\delta}|t - s|^2 - |t - \bar{t}|^2 - |x - \bar{x}|^2 - |y - \bar{y}|^2,$$
$$\forall (t, x, s, y) \in \Gamma_\varepsilon,$$

with

$$\Gamma_\varepsilon \equiv \left\{(t, x, s, y) \mid (t, x, y), (s, x, y) \in B_\varepsilon(\bar{t}, \bar{x}, \bar{y})\right\}.$$

Then for $\delta > 0$ small enough, $\Phi_\delta(t, x, s, y)$ attains its maximum over $\bar{\Gamma}_\varepsilon$ at some $(t_\delta, x_\delta, s_\delta, y_\delta) \in \bar{\Gamma}_\varepsilon$. Hence,

$$v_1(t_\delta, x_\delta) + v_2(s_\delta, y_\delta) - \varphi(t_\delta, x_\delta, y_\delta)$$
$$-\frac{1}{\delta}|t_\delta - s_\delta|^2 - |t_\delta - \bar{t}|^2 - |x_\delta - \bar{x}|^2 - |y_\delta - \bar{y}|^2 \qquad (7.22)$$
$$= \Phi_\delta(t_\delta, x_\delta, s_\delta, y_\delta) \geqslant \Phi_\delta(\bar{t}, \bar{x}, \bar{t}, \bar{y})$$
$$= v_1(\bar{t}, \bar{x}) + v_2(\bar{t}, \bar{y}) - \varphi(\bar{t}, \bar{x}, \bar{y}).$$

Therefore,

$$\frac{1}{\delta}|t_\delta - s_\delta|^2 + |t_\delta - \bar{t}|^2 + |x_\delta - \bar{x}|^2 + |y_\delta - \bar{y}|^2$$
$$\leqslant v_1(t_\delta, x_\delta) + v_2(s_\delta, y_\delta) - \varphi(t_\delta, x_\delta, y_\delta)$$
$$-v_1(\bar{t}, \bar{x}) - v_2(\bar{t}, \bar{y}) + \varphi(\bar{t}, \bar{x}, \bar{y}) \leqslant K.$$

This leads to

$$|t_\delta - s_\delta|^2 \leqslant K\delta \to 0, \qquad \delta \to 0.$$

Since $(t_\delta, x_\delta, y_\delta) \in \bar{B}_\varepsilon(\bar{t}, \bar{x}, \bar{y})$, we may let

$$(t_\delta, x_\delta, y_\delta) \to (\tilde{t}, \tilde{x}, \tilde{y}), \qquad \delta \to 0.$$

Then from (7.22), one also has

$$v_1(t_\delta, x_\delta) - v_2(s_\delta, y_\delta) - \varphi(t_\delta, x_\delta, y_\delta) \geqslant \Phi_\delta(t_\delta, x_\delta, s_\delta, y_\delta)$$
$$\geqslant v_1(\bar{t}, \bar{x}) - v_2(\bar{t}, \bar{y}) - \varphi(\bar{t}, \bar{x}, \bar{y}) - |t_\delta - \bar{t}|^2 - |x_\delta - \bar{x}|^2 - |y_\delta - \bar{y}|^2.$$

Hence, letting $\delta \to 0$, we get

$$v_1(\tilde{t}, \tilde{x}) + v_2(\tilde{t}, \tilde{y}) - \varphi(\tilde{t}, \tilde{x}, \tilde{y}) - |\tilde{t} - \bar{t}|^2 - |\tilde{x} - \bar{x}|^2 - |\tilde{y} - \bar{y}|^2$$
$$\geqslant v_1(\bar{t}, \bar{x}) + v_2(\bar{t}, \bar{y}) - \varphi(\bar{t}, \bar{x}, \bar{y}) \geqslant v_1(\tilde{t}, \tilde{x}) + v_2(\tilde{t}, \tilde{y}) - \varphi(\tilde{t}, \tilde{x}, \tilde{y}).$$

Thus, it is necessary that

$$(\tilde{t}, \tilde{x}, \tilde{y}) = (\bar{t}, \bar{x}, \bar{y}).$$

On the other hand, for any $(t, x) \in \bar{\Gamma}_\varepsilon$,

$$v_1(t_\delta, x_\delta) + v_2(s_\delta, y_\delta) - \varphi(t_\delta, x_\delta, y_\delta)$$
$$-\frac{1}{\delta}|t_\delta - s_\delta|^2 - |t_\delta - \bar{t}|^2 - |x_\delta - \bar{x}|^2 - |y_\delta - \bar{y}|^2$$
$$= \Phi_\delta(t_\delta, x_\delta, s_\delta, y_\delta) \geqslant \Phi_\delta(t, x, s_\delta, y_\delta)$$
$$= v_1(t, x) + v_2(s_\delta, y_\delta) - \varphi(t, x, y_\delta)$$
$$-\frac{1}{\delta}|t - s_\delta|^2 - |t - \bar{t}|^2 - |x - \bar{x}|^2 - |y_\delta - \bar{y}|^2,$$

which leads to

$$0 \geqslant v_1(t, x) - v_1(t_\delta, x_\delta) - \left[\varphi(t, x, y_\delta) - \varphi(t_\delta, x_\delta, y_\delta)\right]$$
$$-\frac{1}{\delta}\left(|t - s_\delta|^2 - |t_\delta - s_\delta|^2\right) - \left(|t - \bar{t}|^2 - |t_\delta - \bar{t}|^2\right)$$
$$-\left(|x - \bar{x}|^2 - |x_\delta - \bar{x}|^2\right)$$
$$= v_1(t, x) - v_1(t_\delta, x_\delta) - \left[\varphi_t(t_\delta, x_\delta, y_\delta) + \frac{2}{\delta}(t_\delta - s_\delta) + 2(t_\delta - \bar{t})\right](t - t_\delta)$$
$$- \langle \varphi_x(t_\delta, x_\delta, y_\delta) + 2(x_\delta - \bar{x}), x - x_\delta \rangle + o(|t - t_\delta| + |x - x_\delta|).$$

Consequently,

$$(q_1^\delta, \varphi_x(t_\delta, x_\delta, y_\delta) + 2(x_\delta - \bar{x})) \in D_{t,x}^{1,+} v_1(t_\delta, x_\delta),$$

with

$$q_1^\delta = \varphi_t(t_\delta, x_\delta, y_\delta) + \frac{2}{\delta}(t_\delta - s_\delta) + 2(t_\delta - \bar{t}).$$

Also,

$$v_1(t_\delta, x_\delta) + v_2(s_\delta, y_\delta) - \varphi(t_\delta, x_\delta, y_\delta) - \frac{1}{\delta}|t_\delta - s_\delta|^2 - |y_\delta - \bar{y}|^2$$

$$= \Phi_\delta(t_\delta, x_\delta, s_\delta, y_\delta) \geqslant \Phi_\delta(t_\delta, x_\delta, s, y)$$

$$= v_1(t_\delta, x_\delta) + v_2(s, y) - \varphi(t_\delta, x_\delta, y) - \frac{1}{\delta}|t_\delta - s|^2 - |y - \bar{y}|^2,$$

which leads to

$$0 \geqslant v_2(s, y) - v_2(s_\delta, y_\delta) - \left[\varphi(t_\delta, x_\delta, y) - \varphi(t_\delta, x_\delta, y_\delta)\right]$$

$$- \frac{1}{\delta}\left(|t_\delta - s|^2 - |t_\delta - s_\delta|^2\right) - \left(|y_\delta - \bar{y}|^2 - |y - \bar{y}|^2\right)$$

$$= v_2(s, y) - v_2(s_\delta, y_\delta) - \frac{2}{\delta}(s_\delta - t_\delta)(s - s_\delta)$$

$$- \langle \varphi_y(t_\delta, x_\delta, y_\delta) + 2(y_\delta - \bar{y}), y - y_\delta \rangle + o(|s - s_\delta| + |y - y_\delta|).$$

Therefore,

$$\left(q_2^\delta, \varphi_y(t_\delta, x_\delta, y_\delta) + 2(y_\delta - \bar{y})\right) \in D_{t,x}^{1,+} v_2(s_\delta, y_\delta),$$

with

$$q_2^\delta = \frac{2}{\delta}(s_\delta - t_\delta).$$

We see that

$$q_1^\delta + q_2^\delta = \varphi_t(t_\delta, x_\delta, y_\delta) + 2(t_\delta - \bar{t}).$$

Note that

$$v_1(t, x_\delta) \leqslant v_1(t_\delta, x_\delta) + q_1^\delta(t - t_\delta) + o(|t - t_\delta|),$$

which leads to (since $v_1(\cdot, \cdot) \in \mathcal{V}$)

$$|q_1^\delta| \leqslant \frac{|v_1(s, x_\delta) - v_1(t_\delta, x_\delta)|}{|s - t_\delta|} + o(1) \leqslant \theta(|x_\delta|) + 1.$$

Hence, $\{q_1^\delta\}_{\delta \in (0,\varepsilon)}$ is bounded. Similarly, $\{q_2^\delta\}_{\delta \in (0,\varepsilon)}$ is also bounded. Therefore, we may choose a sequence $\delta \downarrow 0$ such that

$$(q_1^\delta, q_2^\delta) \to (q_1, q_2),$$

with

$$q_1 + q_2 = \varphi_t(\bar{t}, \bar{x}, \bar{y}),$$

and

$$\varphi_x(t_\delta, x_\delta, y_\delta) + 2(x_\delta - \bar{x}) \to \varphi_x(\bar{t}, \bar{x}, \bar{y}),$$
$$\varphi_y(t_\delta, x_\delta, y_\delta) + 2(y_\delta - \bar{y}) \to \varphi_y(\bar{t}, \bar{x}, \bar{y}), \qquad \delta \to 0.$$

By definition, we have

$$(q_1, \varphi_x(\bar{t}, \bar{x}, \bar{y})) \in \overline{D}_{t,x}^{1,+} v_1(\bar{t}, \bar{x}),$$
$$(q_2, \varphi_y(\bar{t}, \bar{x}, \bar{y})) \in \overline{D}_{t,x}^{1,-} v_2(\bar{t}, \bar{y}).$$

This completes the proof. □

Lemma 7.1.11. *Let* (S1), (S2)′ *and* (S3) *hold. Suppose that* $V(\cdot, \cdot)$ *is a viscosity solution to* (7.14). *Then*

$$V^a(t, x) \leqslant M^a[V](t, x), \qquad \forall (t, x) \in [0, T] \times \mathbb{R}^n. \tag{7.23}$$

Proof. Suppose at some point $(t_0, x_0, a) \in [0, T) \times \mathbb{R}^n \times \mathbb{M}$, it holds

$$V^a(t_0, x_0) > M^a[V](t_0, x_0).$$

By continuity, we can find a $\delta > 0$ such that

$$V^a(t, x) > M^a[V](t, x) + \delta,$$
$$(t, x) \in [0, T) \times \mathbb{R}^n, \ |t - t_0| + |x - x_0| < \delta. \tag{7.24}$$

Let $\zeta(\cdot, \cdot)$ be smooth satisfying

$$\begin{cases} \operatorname{supp} \zeta \subseteq \{(t, x) \in [0, T] \times \mathbb{R}^n \mid |t - t_0| + |x - x_0| \leqslant \delta\}, \\ 0 \leqslant \zeta(t, x) \leqslant 1, \qquad \forall (t, x) \in [0, T] \times \mathbb{R}^n, \\ \zeta(t_0, x_0) = 1, \qquad 0 \leqslant \zeta(t, x) < 1, \quad \forall (t, x) \neq (t_0, x_0). \end{cases}$$

Let

$$\Phi^a(t, x) = V^a(t, x) + 2R\zeta(t, x), \qquad (t, x) \in [0, T] \times \mathbb{R}^n,$$

with

$$R > \max_{|t - t_0| + |x - x_0| \leqslant \delta} |V^a(t, x)|.$$

Then for any (t, x) with $|t - t_0| + |x - x_0| = \delta$,

$$\Phi^a(t, x) = V^a(t, x) < R \leqslant \Phi^a(t_0, x_0).$$

Hence, there exists a point (t_1, x_1) with $|t_1 - t_0| + |x_1 - x_0| < \delta$ at which $\Phi^a(\cdot, \cdot)$ attains its local maximum. Then by the definition of viscosity solution, one has

$$\min \Big\{ - 2R\zeta_t(t_1, x_1) + H^a(t_1, x_1, -2R\zeta_x(t_1, x_1)),$$
$$M^a[V](t_1, x_1) - V^a(t_1, x_1) \Big\} \geqslant 0.$$

This implies

$$M^a[V](t_1, x_1) \geqslant V^a(t_1, x_1),$$

which contradicts (7.24). Hence, (7.23) holds.   □

Now, we are ready to present a proof of Theorem 7.1.7, part (ii), which is a careful modification of that for Theorem 2.5.3, with the aid of the above preparation.

Proof of Theorem 7.1.7, Part (ii). Let $V(\cdot,\cdot)$ be a viscosity sub-solution of (7.14) and $\widehat{V}(\cdot,\cdot)$ be a viscosity super-solution of (7.14). We are going to prove (7.19), or equivalently,

$$\max_{a\in M}\left[V^a(t,x)-\widehat{V}^a(t,x)\right]\leqslant 0,\qquad \forall(t,x)\in[0,T]\times\mathbb{R}^n. \tag{7.25}$$

We split the proof into several steps.

Step 1. A reduction.

Let $T_0=(T-\frac{1}{2L})^+$. Then

$$0<T-T_0=T-(T-\frac{1}{2L})^+=T\wedge\frac{1}{2L}.$$

For any $x_0\in\mathbb{R}^n$, let

$$L_0=2L(1+|x_0|),$$

and define

$$\Delta(x_0)=\Big\{(t,x)\in[T_0,T]\times\mathbb{R}^n \ \big| \ |x-x_0|<L_0(t-T_0)\Big\}.$$

We are going to show that for any $x_0\in\mathbb{R}^n$,

$$\sup_{(t,x)\in\Delta(x_0)}\max_{a\in M}\left[V^a(t,x)-\widehat{V}^a(t,x)\right]\leqslant 0. \tag{7.26}$$

Since

$$[T_0,T]\times\mathbb{R}^n=\bigcup_{x_0\in\mathbb{R}^n}\overline{\Delta(x_0)},$$

from (7.26), we get

$$\max_{a\in M}\left[V^a(t,x)-\widehat{V}^a(t,x)\right]\leqslant 0,\qquad \forall(t,x)\in[T_0,T]\times\mathbb{R}^n.$$

Then, one may replace T by T_0 and continue the procedure. Repeating the procedure at most $[2LT]+1$ times, (7.25) will be proved.

Step 2. Construction of an auxiliary function.

To prove (7.25) by contradiction, we suppose

$$\sup_{(t,x)\in\Delta(x_0)}\max_{a\in M}\left[V^a(t,x)-\widehat{V}^a(t,x)\right]=\bar{\sigma}>0.$$

Note that under (S1)–(S2), for any $(t, x_1), (t, x_2) \in \Delta(x_0)$, $a \in \mathbb{M}$, and $p_1, p_2 \in \mathbb{R}^n$,

$$|H^a(t, x_1, p_1) - H^a(t, x_2, p_2)|$$

$$\leqslant \Big(L|p_1| + \theta(|x_1| \vee |x_2|)\Big)|x_1 - x_2| + L(1 + |x_2|)|p_1 - p_2|$$

$$\leqslant \Big(L|p_1| + \theta(|x_1| \vee |x_2|)\Big)|x_1 - x_2| + L\Big(1 + |x_0| + L_0(T - T_0)\Big)|p_1 - p_2|$$

$$\leqslant \Big(L|p_1| + \theta(|x_1| \vee |x_2|)\Big)|x_1 - x_2| + L\Big(1 + |x_0| + \frac{L_0}{2L}\Big)|p_1 - p_2|$$

$$= \Big(L|p_1| + \theta(|x_2| \vee |x_2|)\Big)|x_1 - x_2| + L_0|p_1 - p_2|.$$

Take small $\varepsilon, \delta > 0$ satisfying

$$\varepsilon + 2\delta < L_0(T - T_0),$$

and define

$$\Delta_{\varepsilon, \delta}(x_0) = \big\{(t, x) \in \Delta(x_0) \mid \langle x \rangle_\varepsilon < L_0(t - T_0) - \delta\big\},$$

with $\langle x \rangle_\varepsilon = \sqrt{|x - x_0|^2 + \varepsilon^2}$. We may assume that

$$\sup_{(t,x) \in \Delta_{\varepsilon, 2\delta}(x_0)} \max_{a \in \mathbb{M}} \big[V^a(t, x) - \widehat{V}^a(t, x)\big] \geqslant \frac{\bar{\sigma}}{2} > 0.$$

Let $K > 0$ be sufficiently large so that

$$K > \sup_{(t,x,y) \in \Gamma(x_0)} \max_{a \in \mathbb{M}} \big[V^a(t, x) - \widehat{V}^a(t, y)\big],$$

where

$$\Gamma(x_0) = \big\{(t, x, y) \in [0, T] \times \mathbb{R}^{2n} \mid (t, x), (t, y) \in \Delta(x_0)\big\}.$$

Introduce $\zeta_\delta(\cdot) \in C^\infty(\mathbb{R})$ satisfying

$$\zeta_\delta(r) = \begin{cases} 0, & r \leqslant -2\delta, \\ -K, & r \geqslant -\delta, \end{cases} \qquad \zeta_\delta'(r) \leqslant 0, \quad \forall r \in \mathbb{R}.$$

Define

$$\Phi^a(t, x, y) = V^a(t, x) - \widehat{V}^a(t, y) - \frac{1}{\beta}|x - y|^2 + \zeta_\delta\big(\langle x \rangle_\varepsilon - L_0(t - T_0)\big)$$

$$+ \zeta_\delta\big(\langle y \rangle_\varepsilon - L_0(t - T_0)\big) + \sigma(t - T), \quad (t, x, y) \in \overline{\Gamma(x_0)}.$$

Let

$$\Gamma_{\varepsilon, \delta}(x_0) = \big\{(t, x, y) \in [0, T] \times \mathbb{R}^{2n} \mid (t, x), (t, y) \in \Delta_{\varepsilon, \delta}(x_0)\big\},$$

and let $(\bar{t}, \bar{x}, \bar{y}) \in \overline{\Gamma_{\varepsilon,\delta}(x_0)}$ satisfy the following:

$$\Phi^{\bar{a}}(\bar{t}, \bar{x}, \bar{y}) = \max_{(t,x,y) \in \overline{\Gamma_{\varepsilon,\delta}(x_0)}} \max_{a \in \mathbb{M}} \Phi^a(t, x, y).$$

Keep in mind that $(\bar{t}, \bar{x}, \bar{y}, \bar{a})$ depends on β, as well as other parameters $\varepsilon, \delta, \sigma$.

Step 3. We may assume that $\bar{a}$ is independent of the parameters $\beta, \varepsilon, \delta$, etc. and

$$\widehat{V}^{\bar{a}}(\bar{t}, \bar{y}) < M^{\bar{a}}[\widehat{V}](\bar{t}, \bar{y}). \tag{7.27}$$

In fact, if

$$\widehat{V}^{\bar{a}}(\bar{t}, \bar{y}) = M^{\bar{a}}[\widehat{V}](\bar{t}, \bar{y}) = \widehat{V}^{\hat{a}}(\bar{t}, \bar{y}) + \kappa(\bar{t}, \bar{a}, \hat{a}),$$

for some $\hat{a} \in \mathbb{M} \setminus \{\bar{a}\}$, then

$$V^{\bar{a}}(\bar{t}, \bar{x}) - \widehat{V}^{\bar{a}}(\bar{t}, \bar{y}) \leqslant M^{\bar{a}}[V](\bar{t}, \bar{y}) - \widehat{V}^{\hat{a}}(\bar{t}, \bar{y}) - \kappa(\bar{t}, \bar{a}, \hat{a})$$
$$\leqslant V^{\hat{a}}(\bar{t}, \bar{x}) - \widehat{V}^{\hat{a}}(\bar{t}, \bar{y}),$$

which implies

$$\Phi^{\bar{a}}(\bar{t}, \bar{x}, \bar{y}) \leqslant \Phi^{\hat{a}}(\bar{t}, \bar{x}, \bar{y}).$$

By the definition of $(\bar{t}, \bar{x}, \bar{y})$ and $\bar{a}$, it is necessary that

$$\Phi^{\hat{a}}(\bar{t}, \bar{x}, \bar{y}) = \Phi^{\bar{a}}(\bar{t}, \bar{x}, \bar{y}) = \max_{(t,x,y) \in \overline{\Gamma_{\varepsilon,\delta}(x_0)}} \max_{a \in \mathbb{M}} \Phi^a(t, x, y).$$

On the other hand, we must have

$$\widehat{V}^{\hat{a}}(\bar{t}, \bar{y}) < M^{\hat{a}}[\widehat{V}](\bar{t}, \bar{y}).$$

In fact, if for some $\tilde{a} \in \mathbb{M} \setminus \{\hat{a}\}$,

$$\widehat{V}^{\hat{a}}(\bar{t}, \bar{y}) = M^{\hat{a}}[\widehat{V}](\bar{t}, \bar{y}) = \widehat{V}^{\tilde{a}}(\bar{t}, \bar{y}) + \kappa(\bar{t}, \hat{a}, \tilde{a}).$$

Then

$$\widehat{V}^{\bar{a}}(\bar{t}, \bar{y}) = \widehat{V}^{\tilde{a}}(\bar{t}, \bar{y}) + \kappa(\bar{t}, \hat{a}, \tilde{a}) + \kappa(\bar{t}, \bar{a}, \hat{a}) > \begin{cases} \widehat{V}^{\bar{a}}(\bar{t}, \bar{y}), & \tilde{a} = \bar{a}, \\ M^{\bar{a}}[\widehat{V}](\bar{t}, \bar{y}), & \tilde{a} \neq \bar{a}. \end{cases}$$

This is a contradiction. Hence, we may assume that (7.27) holds. On the other hand, since $\mathbb{M}$ is a finite set, there must be one $\bar{a}$ appearing infinitely many times that (7.27) holds. By choosing such an $\bar{a}$ (corresponding to a sequence $\beta \downarrow 0$), we have the independence of $\bar{a}$ on the parameters.

Step 4. It holds

$$\frac{1}{\beta} |\bar{x} - \bar{y}|^2 \leqslant \omega_0\big(\sqrt{\beta \bar{\omega}}\big) \to 0, \qquad \beta \to 0, \tag{7.28}$$

where

$$\omega_0(r) = \frac{1}{2} \sup_{\substack{|x-y| \leqslant r \\ (t,x,y) \in \Gamma(x_0)}} \left(|V^{\bar{a}}(t,x) - V^{\bar{a}}(t,y)| + |\widehat{V}^{\bar{a}}(t,x) - \widehat{V}^{\bar{a}}(t,y)| \right).$$

In fact, from

$$\Phi^{\bar{a}}(\bar{t},\bar{x},\bar{x}) + \Phi^{\bar{a}}(\bar{t},\bar{y},\bar{y}) \leqslant 2\Phi^{\bar{a}}(\bar{t},\bar{x},\bar{y}),$$

we have

$$V^{\bar{a}}(\bar{t},\bar{x}) - \widehat{V}^{\bar{a}}(\bar{t},\bar{x}) + 2\zeta_\delta \big(\langle \bar{x} \rangle_\varepsilon - L_0(\bar{t} - T_0) \big) + \sigma(\bar{t} - T)$$

$$+ V^{\bar{a}}(\bar{t},\bar{y}) - \widehat{V}^{\bar{a}}(\bar{t},\bar{y}) + 2\zeta_\delta \big(\langle \bar{y} \rangle_\varepsilon - L_0(\bar{t} - T_0) \big) + \sigma(\bar{t} - T)$$

$$\leqslant 2V^{\bar{a}}(\bar{t},\bar{x}) - 2\widehat{V}^{\bar{a}}(\bar{t},\bar{y}) - \frac{2}{\beta}|\bar{x} - \bar{y}|^2$$

$$+ 2\zeta_\delta \big(\langle \bar{x} \rangle_\varepsilon - L_0(\bar{t} - T_0) \big) + 2\zeta_\delta \big(\langle \bar{y} \rangle_\varepsilon - L_0(\bar{t} - T_0) \big) + 2\sigma(\bar{t} - T),$$

which results in

$$\frac{2}{\beta}|\bar{x} - \bar{y}|^2 \leqslant V^{\bar{a}}(\bar{t},\bar{x}) - V^{\bar{a}}(\bar{t},\bar{y}) + \widehat{V}^{\bar{a}}(\bar{t},\bar{x}) - \widehat{V}^{\bar{a}}(\bar{t},\bar{y})$$

$$\leqslant 2\omega_0 \big(|\bar{x} - \bar{y}| \big), \tag{7.29}$$

where $\omega_0(\cdot)$ is defined in the above. Clearly,

$$\lim_{r \to 0} \omega_0(r) = 0, \qquad \bar{\omega}_0 \equiv \sup_{r \geq 0} \omega_0(r) < \infty.$$

Hence, (7.29) implies

$$|\bar{x} - \bar{y}| \leqslant \sqrt{\beta \bar{\omega}_0},$$

and thus, (7.28) holds.

Step 5. It holds that

$$\langle \bar{x} \rangle_\varepsilon < L_0(\bar{t} - T_0) - \delta, \quad \langle \bar{y} \rangle_\varepsilon < L_0(\bar{t} - T_0) - \delta, \tag{7.30}$$

and when $\beta, \sigma > 0$ are small,

$$\bar{t} < T. \tag{7.31}$$

In fact, if (7.30) fails, then

$$\zeta_\delta \big(\langle \bar{x} \rangle_\varepsilon - L_0(\bar{t} - T_0) \big) + \zeta_\delta \big(\langle \bar{y} \rangle_\varepsilon - L_0(\bar{t} - T_0) \big) \leqslant -K.$$

Consequently,

$$0 = V^{\bar{a}}(T,x_0) - \widehat{V}^{\bar{a}}(T,x_0) + 2\zeta_\delta(\varepsilon - L_0(T - T_0))$$

$$= \Phi^{\bar{a}}(T,x_0,x_0) \leqslant \Phi^{\bar{a}}(\bar{t},\bar{x},\bar{y})$$

$$= V^{\bar{a}}(\bar{t},\bar{x}) - \widehat{V}^{\bar{a}}(\bar{t},\bar{y}) - \frac{1}{\beta}|\bar{x} - \bar{y}|^2 + \zeta_\delta \big(\langle \bar{x} \rangle_\varepsilon - L_0(\bar{t} - T_0) \big)$$

$$+ \zeta_\delta(\langle \bar{y} \rangle_\varepsilon - L_0(\bar{t} - T_0)) + \sigma(\bar{t} - T)$$

$$< K - K + \sigma\bar{t} - \sigma T \leqslant 0,$$

a contradiction. Thus, (7.30) holds. Next, if instead of (7.31), one has $\bar{t} = T$, then with $\sigma > 0$ small enough,

$$\theta(|\bar{x}| \vee |\bar{y}|)|\bar{x} - \bar{y}| \geqslant h(\bar{x}) - h(\bar{y}) = V^{\bar{a}}(T, \bar{x}) - \widehat{V}^{\bar{a}}(T, \bar{y})$$

$$= \Phi^{\bar{a}}(T, \bar{x}, \bar{y}) + \frac{1}{\beta}|\bar{x} - \bar{y}|^2 - \zeta_\delta(\langle\, \bar{x}\, \rangle_\varepsilon - L_0(T - T_0))$$

$$-\zeta_\delta(\langle\, \bar{y}\, \rangle_\varepsilon - L_0(T - T_0))$$

$$\geqslant \sup_{(t,x)\in\Delta_{\varepsilon,2\delta}(x_0)} \Phi^{\bar{a}}(t, x, x)$$

$$\geqslant \sup_{(t,x)\in\Delta_{\varepsilon,2\delta}(x_0)} \left[V^{\bar{a}}(t, x) - \widehat{V}^{\bar{a}}(t, x) + \sigma(t - T)\right] \geqslant \frac{\bar{\sigma}}{4} > 0,$$

which will lead to a contradiction when $\beta > 0$ is small.

Step 6. Completion of the proof.

Now, let us denote

$$\varphi(t, x, y) = \frac{1}{\beta}|x - y|^2 - \zeta_\delta(\langle\, x\, \rangle_\varepsilon - L_0(t - T_0))$$

$$-\zeta_\delta(\langle\, y\, \rangle_\varepsilon - L_0(t - T_0)) + \sigma(T - t).$$

Then

$$\begin{cases} \varphi_t(t, x, y) = -\sigma + L_0\left[\zeta'_\delta(X_\varepsilon) + \zeta'_\delta(Y_\varepsilon)\right], \\ \varphi_x(t, x, y) = \frac{2}{\beta}(x - y) - \zeta'_\delta(X_\varepsilon)\dfrac{x - x_0}{\langle\, x\, \rangle_\varepsilon}, \\ \varphi_y(t, x, y) = \frac{2}{\beta}(y - x) - \zeta'_\delta(Y_\varepsilon)\dfrac{y - x_0}{\langle\, y\, \rangle_\varepsilon}, \end{cases}$$

where

$$X_\varepsilon = \langle\, x\, \rangle_\varepsilon - L_0(t - T_0), \quad Y_\varepsilon = \langle\, \bar{y}\, \rangle_\varepsilon - L_0(\bar{t} - T_0).$$

Applying Lemma 7.1.10 to the function

$$V^{\bar{a}}(t, x) + (-\widehat{V}^{\bar{a}})(t, y) - \varphi(t, x, y)$$

at point $(\bar{t}, \bar{x}, \bar{y})$, we can find $q_1, q_2 \in \mathbb{R}$ such that

$$\begin{cases} (q_1, \varphi_x(\bar{t}, \bar{x}, \bar{y})) \in \overline{D}_{t,x}^{1,+} V^{\bar{a}}(\bar{t}, \bar{x}), \\ (q_2, \varphi_y(\bar{t}, \bar{x}, \bar{y})) \in \overline{D}_{t,x}^{1,+} (-\widehat{V}^{\bar{a}})(\bar{t}, \bar{y}), \\ q_1 + q_2 = \varphi_t(\bar{t}, \bar{x}, \bar{y}). \end{cases}$$

By Proposition 7.1.9, we have

$$\min\Big\{q_1 + H^{\bar{a}}\Big(\bar{t}, \bar{x}, \frac{2}{\beta}(\bar{x} - \bar{y}) - \zeta'_\delta(X_\varepsilon)\frac{\bar{x} - x_0}{\langle\, \bar{x}\, \rangle_\varepsilon}\Big),$$

$$M^{\bar{a}}[V](\bar{t}, \bar{x}) - V^{\bar{a}}(\bar{t}, \bar{x})\Big\} \geqslant 0,$$

and

$$\min \left\{ -q_2 + H^{\bar{a}}\left(\bar{t}, \bar{x}, \frac{2}{\beta}(\bar{x} - \bar{y}) + \zeta'_\delta(Y_\varepsilon)\frac{\bar{y} - x_0}{\langle \bar{y} \rangle_\varepsilon}\right), \right.$$
$$\left. M^{\bar{a}}[\widehat{V}](\bar{t}, \bar{y}) - \widehat{V}^{\bar{a}}(\bar{t}, \bar{y}) \right\} \leqslant 0.$$

Thus, noting $M^{\bar{a}}[\widehat{V}](\bar{t}, \bar{y}) > \widehat{V}^{\bar{a}}(\bar{t}, \bar{y})$, we obtain

$$\begin{cases} q_1 + H^{\bar{a}}\left(\bar{t}, \bar{x}, \frac{2}{\beta}(\bar{x} - \bar{y}) - \zeta'_\delta(X_\varepsilon)\frac{\bar{x} - x_0}{\langle \bar{x} \rangle_\varepsilon}\right) \geqslant 0, \\ -q_2 + H^{\bar{a}}\left(\bar{t}, \bar{y}, \frac{2}{\beta}(\bar{x} - \bar{y}) + \zeta'_\delta(Y_\varepsilon)\frac{\bar{y} - x_0}{\langle \bar{y} \rangle_\varepsilon}\right) \leqslant 0. \end{cases}$$

Consequently,

$$\sigma = -\varphi_t(\bar{t}, \bar{x}, \bar{y}) + L_0\left[\zeta'_\delta(X_\varepsilon) + \zeta'_\delta(Y_\varepsilon)\right]$$
$$= -q_1 - q_2 + L_0\left[\zeta'_\delta(X_\varepsilon) + \zeta'_\delta(Y_\varepsilon)\right]$$
$$\leqslant L_0\left[\zeta'_\delta(X_\varepsilon) + \zeta'_\delta(Y_\varepsilon)\right] + H^{\bar{a}}\left(\bar{t}, \bar{x}, \frac{2}{\beta}(\bar{x} - \bar{y}) - \zeta'_\delta(X_\varepsilon)\frac{\bar{x} - x_0}{\langle \bar{x} \rangle_\varepsilon}\right)$$
$$\qquad - H^{\bar{a}}\left(\bar{t}, \bar{y}, \frac{2}{\beta}(\bar{x} - \bar{y}) + \zeta'_\delta(Y_\varepsilon)\frac{\bar{y} - x_0}{\langle \bar{y} \rangle_\varepsilon}\right)$$
$$\leqslant L_0\left[\zeta'_\delta(X_\varepsilon) + \zeta'_\delta(Y_\varepsilon)\right] + \left[L\left(\frac{2}{\beta}|\bar{x} - \bar{y}| + |\zeta'_\delta(X_\varepsilon)|\right) + \omega\big(|\bar{x}| \vee |\bar{y}|\big)\right]|\bar{x} - \bar{y}|$$
$$\qquad + L_0\Big(|\zeta'_\delta(X_\varepsilon)| + |\zeta'_\delta(Y_\varepsilon)|\Big)$$
$$\leqslant 2L\frac{|\bar{x} - \bar{y}|^2}{\beta} + \Big(L|\zeta'_\delta(X_\varepsilon)| + \omega\big(|\bar{x}| \vee |\bar{y}|\big)\Big)|\bar{x} - \bar{y}|.$$

In the above, we have used the fact that

$$\zeta'_\delta(r) \leqslant 0, \qquad \forall r \in \mathbb{R}.$$

Now, let $\beta \to 0$, by (7.28), we obtain

$$0 < \sigma \leqslant 0,$$

a contradiction. This completes the proof. $\qquad\square$

The above comparison theorem leads to the uniqueness of viscosity solution of the quasi-variational inequality (7.14), under two sets of conditions: either (S1), (S2)', (S3)', or (S1), (S2), (S3). Therefore, at least in principle, Problem (S) can be solved under these two sets of conditions.

7.2 Differential Games with Switching Controls

In this section, we are going to look at a two-person zero-sum differential game in which both players are using switching controls. We denote such a problem by Problem (SG). Let us make it precise now.

Let $M_i = \{1, 2, \cdots, m_i\}$, $i = 1, 2$. Similar to the previous section, we define $\mathcal{S}_1^a[t, T]$ and $\mathcal{S}_2^b[t, T]$ to be the sets of switching controls for Player 1 and Player 2, respectively, with $a \in M_1$ and $b \in M_2$. For any initial pair $(t, x) \in [0, T) \times \mathbb{R}^n$ and $a(\cdot) \equiv \{(\theta_i, a_i)\}_{i \geqslant 0} \in \mathcal{S}_1^{a_0}[t, T]$, $b(\cdot) \equiv \{(\tau_j, b_j)\}_{j \geqslant 0} \in \mathcal{S}_2^{b_0}[t, T]$, we consider the following controlled system:

$$\begin{cases} \dot{X}(s) = f(s, X(s), a(s), b(s)), & s \in [t, T], \\ X(t) = x. \end{cases} \tag{7.32}$$

Under certain conditions, the above state equation admits a unique solution $X(\cdot) = X(\cdot; t, x, a(\cdot), b(\cdot))$. In the game, the first player uses control $a(\cdot)$ from $\mathcal{S}^{a_0}[t, T]$ to minimize the payoff functional

$$J^{a_0, b_0}(t, x; a(\cdot), b(\cdot)) = \int_t^T g(s, X(s), a(s), b(s))ds + h(X(T))$$
$$+ \sum_{i \geqslant 1} \kappa_1(\theta_i, a_{i-1}, a_i) - \sum_{j \geqslant 1} \kappa_2(\tau_j, b_{j-1}, b_j),$$

and the second player uses control $b(\cdot)$ to maximize the above payoff. In the above, $\kappa_1(\cdot, \cdot, \cdot)$ and $\kappa_2(\cdot, \cdot, \cdot)$ are switching costs for Players 1 and 2, respectively. We now introduce the following hypotheses.

(SG1) The map $f : [0, T] \times \mathbb{R}^n \times M_1 \times M_2 \to \mathbb{R}^n$ is continuous and there exists a constant $L > 0$ such that

$$|f(t, x_1, a, b) - f(t, x_2, a, b)| \leqslant L|x_1 - x_2|,$$
$$(t, a, b) \in [0, T] \times M_1 \times M_2, \ x_1, x_2 \in \mathbb{R}^n,$$

and

$$|f(t, 0, a, b)| \leqslant L, \qquad (t, a, b) \in [0, T] \times M_1 \times M_2.$$

(SG2) The maps $g : [0, T] \times \mathbb{R}^n \times M_1 \times M_2 \to \mathbb{R}$ and $h : \mathbb{R}^n \to \mathbb{R}$ are continuous and there exists a constant $L > 0$ such that

$$|g(t, x_1, a, b) - g(t, x_2, a, b)| + |h(x_1) - h(x_2)| \leqslant L|x_1 - x_2|,$$
$$(t, a, b) \in [0, T] \times M_1 \times M_2, \ x_1, x_2 \in \mathbb{R}^n,$$

and

$$|g(t, 0, a, b)| + |h(0)| \leqslant L, \qquad (t, a, b) \in [0, T] \times M_1 \times M_2.$$

(SG3) The maps $\kappa_i : [0, T] \times \mathbb{M}_i \times \mathbb{M}_i \to \mathbb{R}$, $i = 1, 2$, are continuous and for all $a, \widehat{a}, \widetilde{a} \in \mathbb{M}_1$, $a \neq \widehat{a} \neq \widetilde{a}$, $b, \widehat{b}, \widetilde{b} \in \mathbb{M}_2$, $b \neq \widehat{b} \neq \widetilde{b}$ and $0 \leqslant t \leqslant s \leqslant T$,

$$
\begin{cases}
\kappa_1(t, a, \widetilde{a}) < \kappa_1(t, a, \widehat{a}) + \kappa_1(t, \widehat{a}, \widetilde{a}), \\
\kappa_1(t, a, \widehat{a}) > 0, \quad \kappa_1(t, a, a) = 0, \\
\kappa_1(s, a, \widetilde{a}) \leqslant \kappa_1(t, a, \widetilde{a}).
\end{cases}
$$

$$
\begin{cases}
\kappa_2(t, b, \widetilde{b}) < \kappa_2(t, b, \widehat{b}) + \kappa_2(t, \widehat{b}, \widetilde{b}), \\
\kappa_2(t, b, \widehat{b}) > 0, \quad \kappa_2(t, b, b) = 0, \\
\kappa_2(s, b, \widetilde{b}) \leqslant \kappa_2(t, b, \widetilde{b}).
\end{cases}
$$

Note that (SG2) is comparable with (S2). We now introduce the following definition.

Definition 7.2.1. For given $t \in [0, T)$ and $a \in \mathbb{M}_1$ (resp. $b \in \mathbb{M}_2$), an Elliott–Kalton strategy α_1 (resp. α_2) for player I (resp. II) on $[t, T]$ is a map $\alpha_1 : \bigcup_{b \in \mathbb{M}_2} \mathcal{S}_2^b[t, T] \to \mathcal{S}_1^a[t, T]$ (resp. $\alpha_2 : \bigcup_{a \in \mathbb{M}_1} \mathcal{S}_1^a[t, T] \to \mathcal{S}_2^b[t, T]$) such that

$$
b(s) = \widehat{b}(s) \quad (\text{resp. } a(t) = \widehat{a}(t)), \quad \forall s \in [t, \widehat{t}\,],
$$

implies

$$
\alpha_1[b(\cdot)](s) = \alpha_1[\widehat{b}(\cdot)](s) \quad (\text{resp. } \alpha_2[a(\cdot)](s) = \alpha_2[\widehat{a}(\cdot)](s)), \quad \forall s \in [t, \widehat{t}\,].
$$

We denote all Elliott–Kalton strategies for player I (resp. II) on $[t, T]$ by $\Gamma_1^a[t, T]$ (resp. $\Gamma_2^b[t, T]$). We make the convention that

$$
\begin{aligned}
\mathcal{S}_1^a[T, T] &= \{a\}, \quad \Gamma_1^a[T, T] = \{a\}, \\
\mathcal{S}_2^b[T, T] &= \{b\}, \quad \Gamma_2^b[T, T] = \{b\}.
\end{aligned}
$$

It is clear that for any $b(\cdot) \in \mathcal{S}_2^b[t, T]$ (resp. $a(\cdot) \in \mathcal{S}_1^a[t, T]$) and $\alpha_1 \in \Gamma_1^a[t, T]$ (resp. $\alpha_2 \in \Gamma_2^b[t, T]$), one has

$$
\alpha_1[b(\cdot)] \in \mathcal{S}_1^a[t, T] \quad (\text{resp. } \alpha_2[a(\cdot)] \in \mathcal{S}_2^b[t, T]).
$$

On the other hand, for any $(a, b) \in \mathbb{M}_1 \times \mathbb{M}_2$, $(t, x) \in [0, T) \times \mathbb{R}^n$, and $(a(\cdot), b(\cdot)) \in \mathcal{S}_1^a[t, T] \times \mathcal{S}_2^b[t, T]$, by (SG1), there exists a unique solution to (7.32). Then, we consider the following performance functional:

$$
\begin{aligned}
J^{a,b}(t, x; a(\cdot), b(\cdot)) &= \int_t^T g(s, X(s), a(s), b(s)) ds + h(X(T)) \\
&\quad + \sum_{i \geqslant 1} \kappa_1(\theta_i, a_{i-1}, a_i) - \sum_{j \geqslant 1} \kappa_2(\tau_j, b_{j-1}, b_j).
\end{aligned} \tag{7.33}
$$

In the above and sequel, whenever terms like the right-hand side of (7.33) appear together, we always understand that

$$a(\cdot) = \sum_{i \geqslant 1} a_{i-1} I_{[\theta_{i-1}, \theta_i)}(\cdot), \quad a_0 = a,$$

$$b(\cdot) = \sum_{j \geqslant 1} b_{j-1} I_{[\tau_{j-1}, \tau_j)}(\cdot), \quad b_0 = b,$$

i.e., $\{(\theta_i, a_i)\}_{i \geqslant 1}$ and $\{(\tau_j, b_j)\}_{j \geqslant 1}$ are associated with $a(\cdot)$ and $b(\cdot)$, respectively. Also, by our convention,

$$J^{a,b}(T, x; a(\cdot), b(\cdot)) = h(x), \quad \forall (x, a, b) \in \mathbb{R}^n \times \mathbb{M}_1 \times \mathbb{M}_2.$$

From the above analysis, we see that for any $(t, x, a, b) \in [0, T] \times \mathbb{R}^n \times \mathbb{M}_2$, $b(\cdot) \in \mathcal{S}_2^b[t, T]$ and $\alpha_1 \in \Gamma_1^a[t, T]$, the state equation admits a unique solution $X(\cdot) \equiv X(\cdot\,; t, x, \alpha_1[b(\cdot)], b(\cdot))$. Thus, the performance functional $J^{a,b}(t, x; \alpha_1[b(\cdot)], b(\cdot))$ is well-defined. Consequently, we can define

$$\begin{cases} V^{a,b}(t, x) = \displaystyle\inf_{\alpha_1 \in \Gamma_1^a[t,T]} \sup_{b(\cdot) \in \mathcal{S}_2^b[t,T]} J^{a,b}(t, x; \alpha_1[b(\cdot)], b(\cdot)), \\ V^{a,b}(T, x) = h(x). \end{cases}$$

Similarly, we define

$$\begin{cases} W^{a,b}(t, x) = \displaystyle\sup_{\alpha_2 \in \Gamma_2^b[t,T]} \inf_{a(\cdot) \in \mathcal{S}_1^a[t,T]} J^{a,b}(t, x; a(\cdot), \alpha_2[a(\cdot)]), \\ W^{a,b}(T, x) = h(x). \end{cases}$$

Let

$$V(\cdot, \cdot) = \begin{pmatrix} V^{1,1}(\cdot, \cdot) & \cdots & V^{1,m_2}(\cdot, \cdot) \\ \vdots & \ddots & \vdots \\ V^{m_1,1}(\cdot, \cdot) & \cdots & V^{m_1,m_2}(\cdot, \cdot) \end{pmatrix},$$

$$W(\cdot, \cdot) = \begin{pmatrix} W^{1,1}(\cdot, \cdot) & \cdots & W^{1,m_2}(\cdot, \cdot) \\ \vdots & \ddots & \vdots \\ W^{m_1,1}(\cdot, \cdot) & \cdots & W^{m_1,m_2}(\cdot, \cdot) \end{pmatrix}.$$

We call the $(m_1 \times m_2)$-matrix valued functions $V(\cdot, \cdot)$ and $W(\cdot, \cdot)$ Elliott–Kalton lower and upper value functions of our differential game, respectively. Now, let us present some basic properties of the lower and upper value functions.

Proposition 7.2.2. *Let* (SG1)–(SG3) *hold. Then* $V(\cdot,\cdot)$ *and* $W(\cdot,\cdot)$ *are continuous on* $[0,T] \times \mathbb{R}^n$. *Moreover, for some constant* $K > 0$,

$$|V(t,x_1) - V(t,x_2)| + |W(t,x_1) - W(t,x_2)| \leqslant K|x_1 - x_2|, \tag{7.34}$$
$$\forall t \in [0,T], \ x_1, x_2 \in \mathbb{R}^n,$$

$$|V(t_1,x) - V(t_2,x)| + |W(t_1,x) - W(t_2,x)| \leqslant K(1+|x|)|t_1 - t_2|, \tag{7.35}$$
$$\forall t_1, t_2 \in [0,T], \ x \in \mathbb{R}^n.$$

Proof. For any $t \in [0,T)$, $x, x_1, x_2 \in \mathbb{R}^n$, and $(a(\cdot), b(\cdot)) \in \mathcal{S}^{a_0}[t,T] \times \mathcal{S}^{b_0}[t,T]$, similar to Proposition 2.1.1, for any $s \in [t,T]$, we have

$$\begin{cases} |X(s;t,x,a(\cdot),b(\cdot))| \leqslant e^{L(s-t)}(1+|x|) - 1, \\ |X(s;t,x,a(\cdot),b(\cdot)) - x| \leqslant \left[e^{L(s-t)} - 1\right](1+|x|), \\ |X(s;t,x_1,a(\cdot),b(\cdot)) - X(s;t,x_2,a(\cdot),b(\cdot))| \leqslant e^{L(s-t)}|x_1 - x_2|. \end{cases}$$

Thus, by denoting $X_i(\cdot) = X(\cdot;t,x_i,a(\cdot),b(\cdot))$, $i = 1,2$, one has

$$|J^{a_0,b_0}(t,x_1;a(\cdot),b(\cdot)) - J^{a_0,b_0}(t,x_2;a(\cdot),b(\cdot))|$$
$$\leqslant \int_t^T |g(s,X_1(s),a(s),b(s)) - g(s,X_2(s),a(s),b(s))|ds$$
$$+ |h(X_1(T)) - h(X_2(T))|$$
$$\leqslant \int_t^T L|X_1(s) - X_2(s)|ds + L|X_1(T) - X_2(T)|$$
$$\leqslant (1+T)e^{LT}|x_1 - x_2|.$$

Consequently,

$$\begin{cases} |V^{a_0,b_0}(t,x_1) - V^{a_0,b_0}(t,x_2)| \leqslant (1+T)e^{LT}|x_1 - x_2|, \\ |W^{a_0,b_0}(t,x_1) - W^{a_0,b_0}(t,x_2)| \leqslant (1+T)e^{LT}|x_1 - x_2|. \end{cases}$$

This proves (7.34). Next, let $0 \leqslant t_1 < t_2 \leqslant T$. For any

$$a(\cdot) \in \mathcal{S}_1^{a_0}[t_1,T], \quad b(\cdot) \in \mathcal{S}_2^{b_0}[t_1,T],$$

let

$$a_c(\cdot) = a(\cdot)\big|_{[t_2,T]} \in \mathcal{S}_1^{a_0}[t_2,T], \quad b_c(\cdot) = b(\cdot)\big|_{[t_2,T]} \in \mathcal{S}_2^{b_0}[t_2,T],$$

be the compression of $a(\cdot)$ and $b(\cdot)$ on $[t_2,T]$ (see (7.4)), respectively. Let

$$X(\cdot) = X(\cdot;t,x,a(\cdot),b(\cdot)), \quad \bar{X}(\cdot) = X(\cdot;\bar{t},x,a_c(\cdot),b_c(\cdot)).$$

Then for any $s \in [t_2, T]$,

$$|X(s) - \bar{X}(s)| \leqslant \int_{t_1}^{t_2} f(r, X(r), a(r), b(r))|dr$$

$$+ \int_{t_2}^{s} |f(r, X(r), a(r), b(r)) - f(r, \bar{X}(r), a_c(r), b_c(r))|dr$$

$$\leqslant \int_{t_1}^{t_2} L(1 + |X(r)|)dr + \int_{t_2}^{s} L|X(r) - \bar{X}(r)|dr$$

$$\leqslant Le^{LT}(1 + |x|)(t_2 - t_1) + \int_{t_2}^{s} L|X(r) - \bar{X}(r)|dr.$$

Hence, by the Gronwall's inequality, we have

$$|X(s) - \bar{X}(s)| \leq K(1 + |x|)(t_2 - t_1), \qquad \forall s \in [t_2, T],$$

for some constant $K > 0$. Now, for any $b(\cdot) \in \mathcal{S}_2^{b_0}[t_1, T]$ and $\widehat{\alpha}_1 \in \Gamma_1^{a_0}[t_2, T]$, we define $b_c(\cdot) = b(\cdot)|_{[t_2, T]} \in \mathcal{S}_2^{b_0}[t_2, T]$ as the compression of $b(\cdot)$ on $[t_2, T]$, and

$$\alpha_1[b(\cdot)](s) = \begin{cases} a_0, & s \in [t_1, t_2), \\ \widehat{\alpha}_1[b_c(\cdot)](s), & s \in [t_2, T]. \end{cases}$$

Thus, α_1 is an extension of $\widehat{\alpha}_1$ on $[t_1, T]$. Then, we have

$$J^{a_0, b_0}(t_1, x; \alpha_1[b(\cdot)], b(\cdot)) = \int_{t_1}^{t_2} g(r, X(r), a_0, b(r))dr$$

$$+ J^{a_0, b_0}(t_2, X(t_2); \widehat{\alpha}_1[b_c(\cdot)], b_c(\cdot))$$

$$- \sum_{\tau_j < t_2} \left[\kappa_2(\tau_j, b_{j-1}, b_j) - \kappa_2(t_2, b_{j-1}, b_j) \right]$$

$$\leqslant J^{a_0, b_0}(t_2, x; \widehat{\alpha}_1[b_c(\cdot)], b_c(\cdot)) + \int_{t_1}^{t_2} L(1 + |X(r)|)dr$$

$$+ |J^{a_0, b_0}(t_2, X(t_2); \widehat{\alpha}_1[b_c(\cdot)], b_c(\cdot)) - J^{a_0, b_0}(t_2, x; \widehat{\alpha}_1[b_c(\cdot)], b_c(\cdot))|$$

$$\leqslant J^{a_0, b_0}(t_2, x; \widehat{\alpha}_1[b_c(\cdot)], b_c(\cdot)) + Le^{LT}(1 + |x|)(t_2 - t_1)$$

$$+ (1 + T)e^{LT}|X(t_2) - x|$$

$$\leqslant J^{a_0, b_0}(t_2, x; \widehat{\alpha}_1[b_c(\cdot)], b_c(\cdot)) + Le^{LT}(1 + |x|)(t_2 - t_1)$$

$$+ (1 + T)e^{LT}\left[e^{L(t_2 - t_1)} - 1\right](1 + |x|)$$

$$\leqslant J^{a_0, b_0}(t_2, x; \widehat{\alpha}_1[b_c(\cdot)], b_c(\cdot)) + K(1 + |x|)(t_2 - t_1).$$

Hence,

$$\sup_{b(\cdot)\in\mathcal{S}_2^{b_0}[t_1,T]} J^{a_0,b_0}(t_1,x;\alpha_1[b(\cdot)],b(\cdot))$$

$$\leqslant \sup_{b(\cdot)\in\mathcal{S}_2^{b_0}[t_1,T]} J^{a_0,b_0}(t_2,x;\widehat{\alpha}_1[b_c(\cdot)],b_c(\cdot)) + K(1+|x|)(t_2-t_1),$$

which implies

$$V^{a_0,b_0}(t_1,x) \leqslant V^{a_0,b_0}(t_2,x) + K(1+|x|)(t_2-t_1). \tag{7.36}$$

Conversely, for any $b(\cdot) \in \mathcal{S}_2^{b_0}[t_2,T]$ and $\alpha_1[\,\cdot\,] \in \Gamma_1^{a_0}[t_1,T]$, we let $b_e(\cdot) = b_0 I_{[t_1,t_2)}(\cdot) \oplus b(\cdot) \in \mathcal{S}_2^{b_0}[t_1,T]$ be the extension of $b(\cdot)$ on $[t_1,T]$, and define the compression $\widehat{\alpha}_1 \in \Gamma_1^{a_0}[t_2,T]$ of $\alpha_1[\,\cdot\,]$ on $[t_2,T]$ by the following:

$$\widehat{\alpha}_1[b(\cdot)](s) = \alpha_1[b_e(\cdot)](s), \quad s \in [t_2,T], \quad \forall b(\cdot) \in \mathcal{S}_2^{b_0}[t_2,T].$$

Let $X(\cdot) = X(\cdot\,;t_1,x,\alpha_1[b_e(\cdot)],b_e(\cdot))$. Then

$$J^{a_0,b_0}(t_1,x;\alpha_1[b_e(\cdot)],b_e(\cdot))$$

$$\geqslant \int_{t_1}^{t_2} g(r,X(r),\alpha_1[b_e(\cdot)](r),b_0)dr + J^{a_0,b_0}(t_2,x;\widehat{\alpha}_1[b(\cdot)],b(\cdot))$$

$$-|J^{a_0,b_0}(t_2,X(t_2);\widehat{\alpha}_1[b(\cdot)],b(\cdot)) - J^{a_0,b_0}(t_2,x;\widehat{\alpha}_1[b(\cdot)],b(\cdot))|$$

$$+ \sum_{\theta_i<t_2} \kappa_1(\theta_i,a_{i-1},a_i)$$

$$\geqslant J^{a_0,b_0}(t_2,x;\widehat{\alpha}_1[b(\cdot)],b(\cdot)) - \int_{t_1}^{t_2} L(1+|X(r)|)dr - (1+T)e^{LT}|X(t_2)-x|$$

$$\geqslant J^{a_0,b_0}(t_2,x;\widehat{\alpha}_1[b(\cdot)],b(\cdot)) - Le^{LT}(1+|x|)(t_2-t_1)$$

$$-(1+T)e^{LT}\big[e^{L(t_2-t_1)}-1\big](1+|x|)$$

$$\geqslant J^{a_0,b_0}(t_2,x;\widehat{\alpha}_1[b(\cdot)],b(\cdot)) - K(1+|x|)(t_2-t_1).$$

Then we see that

$$\sup_{\widehat{b}(\cdot)\in\mathcal{S}_2^{b_0}[t,T]} J^{a_0,b_0}(t_1,x;\alpha_1[\widehat{b}(\cdot)],\widehat{b}(\cdot))$$

$$\geqslant \sup_{b(\cdot)\in\mathcal{S}_2^{b_0}[t_2,T]} J^{a_0,b_0}(t_1,x;\alpha_1[b_e(\cdot)],b_e(\cdot))$$

$$\geqslant \sup_{b(\cdot)\in\mathcal{S}^{b_0}[t_2,T]} J^{a_0,b_0}(t_2,x;\widehat{\alpha}_1[b(\cdot)],b(\cdot)) - K(1+|x|)(t_2-t_1)$$

$$\geqslant V^{a_0,b_0}(t_2,x) - K(1+|x|)(t_2-t_1).$$

Therefore,

$$V^{a_0,b_0}(t_1,x) \geqslant V^{a_0,b_0}(t_2,x) - K(1+|x|)(t_2-t_1). \tag{7.37}$$

Combining (7.36) and (7.37) we obtain

$$|V^{a_0,b_0}(t_1,x) - V^{a_0,b_0}(t_2,x)| \leqslant K(1+|x|)|t_1-t_2|.$$

Using the similar argument, we may obtain the same result for $W^{a_0,b_0}(\cdot,\cdot)$. Then (7.35) follows. $\qquad\square$

7.2.1   Bilateral obstacle quasi-variational inequality

We define the following mappings: For any $(m_1 \times m_2)$ matrix valued function $V(\cdot, \cdot) = (V^{a,b}(\cdot, \cdot))$ defined on $[0, T] \times \mathbb{R}^n$,

$$M_1^{a,b}[V](t, x) = \min_{\bar{a} \neq a}\{V^{\bar{a},b}(t, x) + \kappa_1(t, a, \bar{a})\},$$

$$M_2^{a,b}[V](t, x) = \max_{\bar{b} \neq b}\{V^{a,\bar{b}}(t, x) - \kappa_2(t, b, \bar{b})\}.$$

These two mappings are called *upper* and *lower switching obstacle operators*, respectively. Similar to Theorem 7.1.2, we have the following principle of optimality.

Theorem 7.2.3. *Let* (SG1)–(SG3) *hold. Then the lower value function* $V(\cdot)$ *satisfies the following:*

(i) *For any* $(t, x, a, b) \in [0, T] \times \mathbb{R}^n \times \mathbb{M}_1 \times \mathbb{M}_2$,

$$M_2^{a,b}[V](t, x) \leqslant V^{a,b}(t, x) \leqslant M_1^{a,b}[V](t, x). \qquad (7.38)$$

(ii) *Suppose at* $(t, x, a, b) \in [0, T] \times \mathbb{R}^n \times \mathbb{M}_1 \times \mathbb{M}_2$,

$$V^{a,b}(t, x) < M_1^{a,b}[V](t, x).$$

Then, there exists a $\bar{\tau} \in (t, T)$, *such that for any* $\tau \in (t, \bar{\tau})$,

$$V^{a,b}(t, x) \geqslant \int_t^\tau g(r, X(r), a, b)dr + V^{a,b}(\tau, X(\tau)).$$

(iii) *Suppose at* $(t, x, a, b) \in [0, T] \times \mathbb{R}^n \times \mathbb{M}_1 \times \mathbb{M}_2$,

$$V^{a,b}(t, x) > M_2^{a,b}[V](t, x).$$

Then there exists a $\bar{\tau} \in (t, T)$, *such that for any* $\tau \in (t, \bar{\tau})$,

$$V^{a,b}(t, x) \leqslant \int_t^\tau g(r, X(r), a, b)ds + V^{a,b}(\tau, X(\tau)).$$

Proof. (i) For any $\bar{a} \in \mathbb{M}_1 \setminus \{a_0\}$, and any $\alpha_1 \in \Gamma_1^{a_0}[t, T]$, define $\bar{\alpha}_1 \in \Gamma_1^{\bar{a}}[t, T]$ by the following: For any $b(\cdot) \in \mathcal{S}_2^{b_0}[t, T]$,

$$\begin{cases} \bar{\alpha}_1[b(\cdot)] = \{(\bar{\theta}_i, \bar{a}_i)\}_{i \geqslant 0}, \\ (\bar{\theta}_0, \bar{a}_0) = (t, \bar{a}), \quad (\bar{\theta}_i, \bar{a}_i) = (\theta_{i-1}, a_{i-1}), \quad \forall i \geqslant 1, \end{cases}$$

where $\alpha_1[b(\cdot)] = \{(\theta_i, a_i)\}_{i \geqslant 0}$. Then for any $b(\cdot) \in \mathcal{S}_2^{b_0}[t, T]$,

$$J^{a_0, b_0}(t, x; \alpha_1[b(\cdot)], b(\cdot)) = J^{\bar{a}, b_0}(t, x; \bar{\alpha}_1[b(\cdot)], b(\cdot)) + \kappa_1(t, a_0, \bar{a}).$$

Hence,

$$V^{a_0, b_0}(t, x) \leqslant V^{\bar{a}, b_0}(t, x) + \kappa_1(t, a_0, \bar{a}), \qquad \forall \bar{a} \in \mathbb{M}_1 \setminus \{a_0\}.$$

This leads to

$$V^{a_0,b_0}(t,x) \leqslant M_1^{a_0,b_0}[V](t,x).$$

The other half of (7.38) can be proved similarly.

(ii) For any $\tau > t$, $\varepsilon > 0$ and $b(\cdot) \equiv b_0 \in \mathbb{M}_2$, there exists an $\alpha_1^{\tau,\varepsilon} \in \Gamma_1^{a_0}[t,T]$ such that

$$\begin{aligned}
V^{a_0,b_0}(t,x) + \varepsilon &\geqslant \sup_{b(\cdot) \in \mathcal{S}_2^{b_0}[t,T]} J^{a_0,b_0}(t,x;\alpha_1^{\tau,\varepsilon}[b(\cdot)], b(\cdot)) \\
&\geqslant \int_t^\tau g(r, X(r), \alpha_1^{\tau,\varepsilon}[b_0](r), b_0) dr \qquad (7.39) \\
&\quad + \sum_{\theta_i^\varepsilon < \tau} \kappa_1(\theta_i^\varepsilon, a_{i-1}^\varepsilon, a_i^\varepsilon) + V^{\alpha_1^{\tau,\varepsilon}[b_0](\tau), b_0}(\tau, X(\tau)),
\end{aligned}$$

where

$$\alpha_1^{\tau,\varepsilon}[b_0] = \{(\theta_i^\varepsilon, a_i^\varepsilon)\}_{i \geqslant 0} \in \mathcal{S}_1^{a_0}[t,T].$$

Then, similar to the proof of Theorem 7.1.2, one can show that for all small $\varepsilon > 0$ and $\tau > t$ with $\tau - t$ sufficiently small,

$$\theta_1^\varepsilon \equiv \theta_1^{\tau,\varepsilon} > \tau.$$

Hence, (7.39) becomes

$$V^{a_0,b_0}(t,x) + \varepsilon \geqslant \int_t^\tau g(r, X(r), a_0, b_0) dr + V^{a_0,b_0}(\tau, X(\tau)).$$

Now, fix a $\bar{t} > t$ with $\bar{t} - t$ small and let $\varepsilon \to 0$, we obtain (ii).

(iii) It is similar to (ii). $\qquad \square$

Now, we introduce the following Hamiltonian

$$H^{a,b}(t,x,p) = \langle p, f(t,x,a,b) \rangle + g(t,x,a,b). \qquad (7.40)$$

The following is a consequence of Theorem 7.2.3.

Theorem 7.2.4. *Suppose the lower value function $V(\cdot,\cdot)$ is C^1. Then, for any $(a,b) \in \mathbb{M}_1 \times \mathbb{M}_2$,*

$$\begin{cases}
M_2^{a,b}[V](t,x) \leqslant V^{a,b}(t,x) \leqslant M_1^{a,b}[V](t,x), \\
V_t^{a,b}(t,x) + H^{a,b}(t,x,V_x^{a,b}(t,x)) \leqslant 0, \\
\qquad \text{if } M_1^{a,b}[V](t,x) > V^{a,b}(t,x), \\
V_t^{a,b}(t,x) + H^{a,b}(t,x,V_x^{a,b}(s,x)) \geqslant 0, \\
\qquad \text{if } M_2^{a,b}[V](t,x) < V^{a,b}(t,x), \\
V^{a,b}(T,x) = h(x).
\end{cases} \qquad (7.41)$$

The above (7.41) is referred to as a *bilateral obstacle evolutionary quasi-variational inequality* system. This is the corresponding HJI equation for the lower value function $V(\cdot,\cdot)$. The following gives an equivalent form of (7.41).

Theorem 7.2.5. *Suppose $V(\cdot,\cdot)$ is C^1. Then it satisfies (7.41) if and only if it satisfies the following system:*

$$
\begin{cases}
\max\Big\{ \min\big\{ V_t^{a,b}(t,x) + H^{a,b}(t,x,V_x^{a,b}(t,x)), \\
\qquad\qquad M_1^{a,b}[V](t,x) - V^{a,b}(t,x)\big\}, \\
\qquad\qquad M_2^{a,b}[V](t,x) - V^{a,b}(t,x)\Big\} = 0, \\
\qquad (t,x,a,b) \in [0,T) \times \mathbb{R}^n \times \mathbb{M}_1 \times \mathbb{M}_2, \\
\min\Big\{ \max\big\{ V_t^{a,b}(t,x) + H^{a,b}(t,x,V_x^{a,b}(t,x)), \\
\qquad\qquad M_2^{a,b}[V](t,x) - V^{a,b}(t,x)\big\}, \\
\qquad\qquad M_1^{a,b}[V](t,x) - V^{a,b}(t,x)\Big\} = 0, \\
\qquad (t,x,a,b) \in [0,T) \times \mathbb{R}^n \times \mathbb{M}_1 \times \mathbb{M}_2, \\
V^{a,b}(T,x) = h(x), \quad \forall (x,a,b) \in \mathbb{R}^n \times \mathbb{M}_1 \times \mathbb{M}_2.
\end{cases}
\tag{7.42}
$$

Proof. For the simplicity of presentation, let us rewrite (7.41) as follows, using simplified notations:

$$
\begin{cases}
M_2 - V \leqslant 0, \quad M_1 - V \geqslant 0, \\
V_t + H \geqslant 0, \qquad \text{if } M_2 - V < 0, \\
V_t + H \leqslant 0, \qquad \text{if } M_1 - V > 0,
\end{cases}
\tag{7.43}
$$

and rewrite (7.42) as follows:

$$
\begin{cases}
\big[(V_t + H) \wedge (M_1 - V)\big] \vee \big[M_2 - V\big] = 0, \\
\big[(V_t + H) \vee (M_2 - V)\big] \wedge \big[M_1 - V\big] = 0.
\end{cases}
\tag{7.44}
$$

The meaning of the notations is obvious.

Now, suppose (7.43) holds. Then

$$
[0,T] \times \mathbb{R}^n = (M_2 \leqslant V \leqslant M_1)
$$
$$
= (M_2 = V = M_1) \bigcup (M_2 = V < M_1)
$$
$$
\bigcup (M_2 < V = M_1) \bigcup (M_2 < V < M_1).
$$

Consequently, we have the following:

On $(M_2 = V = M_1)$, one has

$$[(V_t + H) \wedge (M_1 - V)] \vee [M_2 - V] = [(V_t + H) \wedge 0] \vee 0 = 0,$$

and

$$[(V_t + H) \vee (M_2 - V)] \wedge [M_1 - V] = [(V_t + H) \vee 0] \wedge 0 = 0.$$

On $(M_2 = V < M_1)$, one has $V_t + H \leqslant 0$. Thus,

$$[(V_t + H) \wedge (M_1 - V)] \vee [M_2 - V] = [(V_t + H) \wedge (M_1 - V)] \vee 0$$
$$= (V_t + H) \vee 0 = 0,$$

and

$$[(V_t + H) \vee (M_2 - V)] \wedge [M_1 - V] = [(V_t + H) \vee 0] \wedge [M_1 - V]$$
$$= 0 \wedge [M_1 - V] = 0.$$

On $(M_2 < V = M_1)$, one has $V_t + H \geqslant 0$. Thus,

$$[(V_t + H) \wedge (M_1 - V)] \vee [M_2 - V] = [(V_t + H) \wedge 0] \vee [M_2 - V]$$
$$= 0 \vee [M_2 - V] = 0,$$

and

$$[(V_t + H) \vee (M_2 - V)] \wedge [M_1 - V] = (V_t + H) \wedge 0 = 0.$$

On $(M_2 < V < M_1)$, one has $V_t + H = 0$. Thus,

$$[(V_t + H) \wedge (M_1 - V)] \vee [M_2 - V] = [0 \wedge (M_1 - V)] \vee [M_2 - V]$$
$$= 0 \vee [M_2 - V] = 0,$$

and

$$[(V_t + H) \vee (M_2 - V)] \wedge [M_1 - V] = [0 \vee (M_2 - V)] \wedge [M_1 - V]$$
$$= 0 \wedge [M_1 - V] = 0.$$

This proves (7.44).

Conversely, if (7.44) holds, then

$$\begin{cases} (V_t + H) \wedge (M_1 - V) \leqslant 0, & M_2 - V \leqslant 0, \\ (V_t + H) \vee (M_2 - V) \geqslant 0, & M_1 - V \geqslant 0. \end{cases}$$

This leads to

$$M_2 \leqslant V \leqslant M_1,$$

and $(V_t + H) \wedge (M_1 - V) \leqslant 0$ implies

$$V < M_1 \quad \Rightarrow \quad V_t + H \leqslant 0,$$

and $(V_t + H) \vee (M_2 - V) \geqslant 0$ implies

$$V > M_2 \quad \Rightarrow \quad V_t + H \geqslant 0.$$

This means that (7.43) holds. □

For the upper value function $W(\cdot, \cdot)$, one has the same results as the lower value function. We state the following result and leave the details to the readers.

Theorem 7.2.6. *The upper Elliott–Kalton value function $W(\cdot, \cdot)$ satisfies the following:*

(i) *For any $(t, x, a, b) \in [0, T] \times \mathbb{R}^n \times \mathbb{M}_1 \times \mathbb{M}_2$,*

$$M_2^{a,b}[W](t, x) \leqslant W^{a,b}(t, x) \leqslant M_1^{a,b}[W](t, x).$$

(ii) *Suppose at $(t, x, a, b) \in [0, T] \times \mathbb{R}^n \times \mathbb{M}_1 \times \mathbb{M}_2$*

$$W^{a,b}(t, x) < M_1^{a,b}[W](t, x).$$

Then, there exists a $\bar{\tau} \in (t, T)$, such that for all $\tau \in (t, \bar{\tau})$,

$$W^{a,b}(t, x) \geqslant \int_t^\tau g(r, X(r), a, b) ds + W^{a,b}(\tau, X(\tau)).$$

(iii) *Suppose at $(t, x, a, b) \in [0, T] \times \mathbb{R}^n \times \mathbb{M}_1 \times \mathbb{M}_2$*

$$W^{a,b}(t, x) > M_2^{a,b}[W](t, x).$$

Then there exists a $\bar{\tau} \in (t, T)$, such that for all $\tau \in (t, \bar{\tau})$,

$$W^{a,b}(t, x) \leqslant \int_t^\tau g(r, X(r), a, b) ds + W^{a,b}(\tau, X(\tau)).$$

Suppose the upper value function $W(\cdot, \cdot)$ is C^1. Then $W(\cdot, \cdot)$ satisfies bilateral obstacle quasi-variational inequality system (7.42).

From the above, we have the following simple corollary.

Corollary 7.2.7. *If (7.42) admits at most one C^1 solution and $V(\cdot, \cdot)$ and $W(\cdot, \cdot)$ are C^1. Then,*

$$V(\cdot, \cdot) = W(\cdot, \cdot).$$

I.e., the game has an Elliott–Kalton value function.

Unfortunately, the upper and the lower value functions are not necessarily C^1 and similar to the usual first order Hamilton-Jacobi-Bellman equation, the problem (7.42) may have no C^1 solutions. Thus, we need some more investigations.

By the way, we point out that the two equations in (7.44) are not equivalent. In fact, if, say,

$$M_1 < V, \quad M_2 = V,$$

then the first equation in (7.44) holds, but the second equation in (7.44) fails. Likewise, if

$$M_1 = V, \quad M_2 > V,$$

then the second equation in (7.44) holds, but the first equation in (7.44) fails.

7.2.2 *Existence of the value function*

We now introduce the following notion.

Definition 7.2.8. (i) Function $V(\cdot, \cdot) \in C([0, T] \times \mathbb{R}^n; \mathbb{R}^{m_1 \times m_2})$ is called a *viscosity sub-solution* of (7.42) if

$$V^{a,b}(T, x) \leqslant h(x), \qquad \forall (a, b, x) \in \mathbb{M}_1 \times \mathbb{M}_2 \times \mathbb{R}^n, \qquad (7.45)$$

and as long as $\varphi \in C^1$ with $V^{a,b}(\cdot, \cdot) - \varphi(\cdot, \cdot)$ attains a local maximum at $(t_0, x_0) \in [0, T) \times \mathbb{R}^n$, the following holds:

$$
\max \Big\{ \min \big\{ \varphi_t(t_0, x_0) + H^{a,b}(t_0, x_0, \varphi_x(t_0, x_0)),
$$
$$
M_1^{a,b}[V](t_0, x_0) - V^{a,b}(t_0, x_0) \big\},
$$
$$
M_2^{a,b}[V](t_0, x_0) - V^{a,b}(t_0, x_0) \Big\} \geqslant 0,
$$
$$
\min \Big\{ \max \big\{ \varphi_t(t_0, x_0) + H^{a,b}(t_0, x_0, \varphi_x(t_0, x_0)),
$$
$$
M_2^{a,b}V](t_0, x_0) - V^{a,b}(t_0, x_0) \big\},
$$
$$
M_1^{a,b}[V](t_0, x_0) - V^{a,b}(t_0, x_0) \Big\} \geqslant 0.
$$

Here, if $t_0 = 0$, then, $\varphi_t(0, x_0)$ is understood as the right-derivative.

(ii) Function $V(\cdot, \cdot) \in C([0, T] \times \mathbb{R}^n; \mathbb{R}^{m_1 \times m_2})$ is called a *viscosity super-solution* of (7.42) if

$$V^{a,b}(T, x) \geqslant h(x), \qquad \forall (a, b, x) \in \mathbb{M}_1 \times \mathbb{M}_2 \times \mathbb{R}^n, \qquad (7.46)$$

and as long as $\varphi \in C^1$ with $V^{a,b}(\cdot, \cdot) - \varphi(\cdot, \cdot)$ attains a local minimum at

$(t_0, x_0) \in [0, T) \times \mathbb{R}^n$, the following holds:

$$\max \Big\{ \min \big\{ \varphi_t(t_0, x_0) + H^{a,b}(t_0, x_0, \varphi_x(t_0, x_0)),$$
$$M_1^{a,b}[V](t_0, x_0) - V^{a,b}(t_0, x_0) \big\},$$
$$M_2^{a,b}[V](t_0, x_0) - V^{a,b}(t_0, x_0) \Big\} \leqslant 0,$$

$$\min \Big\{ \max \big\{ \varphi_t(t_0, x_0) + H^{a,b}(t_0, x_0, \varphi_x(t_0, x_0)),$$
$$M_2^{a,b}[V](t_0, x_0) - V^{a,b}(t_0, x_0) \big\},$$
$$M_1^{a,b}[V](t_0, x_0) - V^{a,b}(t_0, x_0) \Big\} \leqslant 0.$$

(iii) If $V(\cdot)$ is both a viscosity sub- and super-solution of (7.42), then it is called a *viscosity solution* of (7.42).

Theorem 7.2.9. *The lower value function $V(\cdot, \cdot)$ and the upper value function $W(\cdot)$ of Problem (SG) are viscosity solutions of (7.42).*

Proof. We prove the conclusion for the lower value function $V(\cdot, \cdot)$. The conclusion for the upper value function is almost the same.

First of all, by Theorem 7.2.3, $V(\cdot, \cdot)$ satisfies (7.38). Now, let $V^{a,b}(\cdot, \cdot) - \varphi(\cdot, \cdot)$ attain a local maximum at $(t_0, x_0) \in [0, T) \times \mathbb{R}^n$. There will be the following two cases:

(a) If
$$M_2^{a,b}[V](t_0, x_0) = V^{a,b}(t_0, x_0) \leqslant M_1^{a,b}[V](t_0, x_0),$$
then regardless of the sign for $\varphi_t(t_0, x_0) + H^{a,b}(t_0, x_0, \varphi_x(t_0, x_0))$, one has

$$\max \Big\{ \min \big\{ \varphi_t(t_0, x_0) + H^{a,b}(t_0, x_0, \varphi_x(t_0, x_0)),$$
$$M_1^{a,b}[V](t_0, x_0) - V^{a,b}(t_0, x_0) \big\},$$
$$M_2^{a,b}[V](t_0, x_0) - V^{a,b}(t_0, x_0) \Big\}$$
$$= \max \Big\{ \min \big\{ \varphi_t(t_0, x_0) + H^{a,b}(t_0, x_0, \varphi_x(t_0, x_0)),$$
$$M_1^{a,b}[V](t_0, x_0) - V^{a,b}(t_0, x_0) \big\}, 0 \Big\} \geqslant 0,$$

and

$$\min \Big\{ \max \big\{ \varphi_t(t_0, x_0) + H^{a,b}(t_0, x_0, \varphi_x(t_0, x_0)),$$
$$M_2^{a,b}[V](t_0, x_0) - V^{a,b}(t_0, x_0) \big\},$$
$$M_1^{a,b}[V](t_0, x_0) - V^{a,b}(t_0, x_0) \Big\}$$
$$= \min \Big\{ \max \big\{ \varphi_t(t_0, x_0) + H^{a,b}(t_0, x_0, \varphi_x(t_0, x_0)), 0 \big\},$$
$$M_1^{a,b}[V](t_0, x_0) - V^{a,b}(t_0, x_0) \Big\} \geqslant 0.$$

(b) If

$$M_2^{a,b}[V](t_0, x_0) < V^{a,b}(t_0, x_0](t_0, x_0) \leqslant M_1^{a,b}[V](t_0, x_0),$$

then there exists a $\bar{\tau} \in (t_0, T)$ such that

$$V^{a,b}(t_0, x_0) \leqslant \int_{t_0}^{\tau} g(r, X(r), a, b) dr + V^{a,b}(\tau, X(\tau)), \quad \tau \in (t_0, \bar{\tau}).$$

Since $V^{a,b}(\cdot, \cdot) - \varphi(\cdot, \cdot)$ attains a local maximum at (t_0, x_0), we have

$$0 \leqslant V^{a,b}(\tau, X(\tau)) - V^{a,b}(t_0, x_0) + \int_{t_0}^{\tau} g(r, X(r)) dr$$

$$\leqslant \varphi(\tau, X(\tau)) - \varphi(t_0, x_0) + \int_{t_0}^{\tau} g(r, X(r), a, b) dr.$$

Dividing $\tau - t_0$ and sending $\tau \downarrow t_0$, we obtain

$$0 \leqslant \varphi_t(t_0, x_0) + H^{a,b}(t_0, x_0, \varphi_x(t_0, x_0)).$$

Therefore,

$$\max \Big\{ \min \big\{ \varphi_t(t_0, x_0) + H^{a,b}(t_0, x_0, \varphi_x(t_0, x_0)),$$
$$M_1^{a,b}[V](t_0, x_0) - V^{a,b}(t_0, x_0) \big\},$$
$$M_2^{a,b}[V](t_0, x_0) - V^{a,b}(t_0, x_0) \Big\} \geqslant 0$$

and

$$\min \Big\{ \max \big\{ \varphi_t(t_0, x_0) + H^{a,b}(t_0, x_0, \varphi_x(t_0, x_0)),$$
$$M_2^{a,b}[V](t_0, x_0) - V^{a,b}(t_0, x_0) \big\},$$
$$M_1^{a,b}[V](t_0, x_0) - V^{a,b}(t_0, x_0) \Big\} \geqslant 0.$$

This means that $V(\cdot, \cdot)$ is a viscosity sub-solution of (7.42).

In a similar manner, we are able to show that $V(\cdot, \cdot)$ is also a viscosity super-solution to (7.42).   □

The rest of this section is devoted to the uniqueness of viscosity solution to (7.42) which will lead to the existence of the value function for our differential game. To this end, similar to Proposition 7.1.9, we first state the following result whose proof is the same as Proposition 7.1.9.

Proposition 7.2.10. *Let (SG1)–(SG3) hold. Then function $V(\cdot, \cdot) \in C([0, T] \times \mathbb{R}^n; \mathbb{R}^{m_1 \times m_2})$ is a viscosity sub-solution (resp. viscosity super-solution) of (7.42) if and only if (7.45) (resp. (7.46)) is satisfied and the* following holds: $\forall (t, x, a, b) \in [0, T) \times \mathbb{R}^n \times \mathbb{M}_1 \times \mathbb{M}_2$,

$$
\begin{cases}
\max \Big\{ \min\{q + H^{a,b}(t, x, p), M_1^{a,b}[V](t, x) - V^{a,b}(t, x)\}, \\
\qquad M_2^{a,b}[V](t, x) - V^{a,b}(t, x) \Big\} \geqslant 0 \ (\text{resp.} \leqslant 0), \\
\min \Big\{ \max\{q + H^{a,b}(t, x, p), M_2^{a,b}[V](t, x) - V^{a,b}(t, x)\}, \\
\qquad M_1^{a,b}[V](t, x) - V^{a,b}(t, x) \Big\} \geqslant 0 \ (\text{resp.} \leqslant 0), \\
\forall (q, p) \in \overline{D}_{t,x}^{1,+} V^{a,b}(t, x) \quad (\text{resp.} \ \overline{D}_{t,x}^{1,-} V^{a,b}(t, x)).
\end{cases}
$$

Next, similar to Lemma 7.1.11, we have the following result.

Lemma 7.2.11. *Let $V(\cdot, \cdot)$ and $\widehat{V}(\cdot, \cdot)$ be a viscosity sub-solution and a viscosity super-solution of (7.42), respectively. Then, for all $(t, x, a, b) \in [0, T] \times \mathbb{R}^n \times \mathbb{M}_1 \times \mathbb{M}_2$,*

$$V^{a,b}(t, x) \leqslant M_1^{a,b}[V](t, x), \tag{7.47}$$

$$\widehat{V}^{a,b}(t, x) \geqslant M_2^{a,b}[\widehat{V}](t, x). \tag{7.48}$$

Proof. We carry out a proof for (7.47). The other can be proved similarly. Suppose at some point $(t_0, x_0, a, b) \in [0, T) \times \mathbb{R}^n \times \mathbb{M}_1 \times \mathbb{M}_2$, it holds

$$V^{a,b}(t_0, x_0) > M_1^{a,b}[V](t_0, x_0).$$

By continuity, we can find a $\delta > 0$ such that

$$
\begin{aligned}
V^{a,b}(t, x) &> M_1^{a,b}[V](t, x) + \delta, \\
&(t, x) \in [0, T) \times \mathbb{R}^n, \ |t - t_0| + |x - x_0| < \delta.
\end{aligned}
\tag{7.49}
$$

Let $\zeta(\cdot, \cdot)$ be smooth satisfying

$$
\begin{cases}
\operatorname{supp} \zeta \subseteq \{(t, x) \in [0, T] \times \mathbb{R}^n \mid |t - t_0| + |x - x_0| \leq \delta\}, \\
0 \leqslant \zeta(t, x) \leqslant 1, \qquad \forall (t, x) \in [0, T] \times \mathbb{R}^n, \\
\zeta(t_0, x_0) = 1, \qquad 0 \leqslant \zeta(t, x) < 1, \quad \forall (t, x) \neq (t_0, x_0).
\end{cases}
$$

Let

$$\Phi^{a,b}(t, x) = V^{a,b}(t, x) + 2R\zeta(t, x), \qquad (t, x) \in [0, T] \times \mathbb{R}^n,$$

with

$$R > \max_{|t - t_0| + |x - x_0| \leqslant \delta} |V^a(t, x)|.$$

Then for any (t,x) with $|t - t_0| + |x - x_0| = \delta$,

$$\Phi^{a,b}(t,x) = V^{a,b}(t,x) < R \leqslant \Phi^{a,b}(t_0,x_0).$$

Hence, there exists a point (t_1,x_1) with $|t_1 - t_0| + |x_1 - x_0| < \delta$ at which $\Phi^{a,b}(\cdot,\cdot)$ attains its local maximum. Then by the definition of viscosity solution, one has

$$\min\Big\{ \max\big\{ -2R\zeta_t(t_1,x_1) + H^{a,b}(t_1,x_1,-2R\zeta_x(t_1,x_1)),$$
$$M_2^{a,b}[V](t_1,x_1) - V^{a,b}(t_1,x_1)\big\},$$
$$M_1^{a,b}[V](t_0,x_0) - V^{a,b}(t_0,x_0)\Big\} \geqslant 0.$$

This implies

$$M_1^{a,b}[V](t_1,x_1) \geqslant V^{a,b}(t_1,x_1),$$

which contradicts (7.49). Hence, (7.47) holds. □

Next, we introduce the following further assumption.

(SG4) For any finite sequence

$$\{(a_i,b_i)\}_{1\leqslant i\leqslant j} \subseteq \mathbb{M}_1 \times \mathbb{M}_2,$$

it holds

$$\sum_{i=1}^{j+1} \kappa_1(t,a_i,a_{i+1}) - \sum_{i=1}^{j+1} \kappa_2(t,b_i,b_{i+1}) \neq 0, \quad \forall t \in [0,T],$$

where $a_{j+1} = a_1$, $b_{j+1} = b_1$.

Now, we are ready to state the following comparison result.

Theorem 7.2.12. *Let* (SG1)–(SG4) *hold. Let* $V(\cdot,\cdot)$ *and* $\widehat{V}(\cdot,\cdot)$ *be a viscosity sub-solution and a viscosity super-solution of* (7.42), *respectively. Then*

$$V^{a,b}(t,x) \leqslant \widehat{V}^{a,b}(t,x), \quad \forall(t,x,a,b) \in [0,T] \times \mathbb{R}^n \times \mathbb{M}_1 \times \mathbb{M}_2. \quad (7.50)$$

The proof is a proper modification of that for Theorem 7.1.7. For readers' convenience, we sketch the proof here, carefully pointing out the difference.

Proof. We split the proof into several steps.

Step 1. A reduction. The same as that in the proof of Theorem 7.1.7, it suffices to prove that for any $x_0 \in \mathbb{R}^n$, the following holds:

$$\sup_{(t,x)\in\Delta(x_0)} \max_{(a,b)\in\mathbb{M}_1\times\mathbb{M}_2} \big[V^{a,b}(t,x) - \widehat{V}^{a,b}(t,x)\big] \leqslant 0, \quad (7.51)$$

with

$$\begin{cases} \Delta(x_0) = \Big\{(t,x) \in [T_0, T] \times \mathbb{R}^n \mid |x - x_0| < L_0(t - T_0)\Big\}, \\ T_0 = \Big(T - \dfrac{1}{2L}\Big)^+, \qquad L_0 = 2L(1 + |x_0|). \end{cases}$$

Step 2. Construction of an auxiliary function.

Suppose (7.51) fails. Then we may suppose

$$\sup_{(t,x)\in\Delta(x_0)} \max_{(a,b)\in\mathbb{M}_1\times\mathbb{M}_2} \big[V^{a,b}(t,x) - \widehat{V}^{a,b}(t,x)\big] = \bar{\sigma} > 0. \tag{7.52}$$

Note that for any $(t, x_1), (t, x_2) \in \Delta(x_0)$, $(a, b) \in \mathbb{M}_1 \times \mathbb{M}_2$, and $p_1, p_2 \in \mathbb{R}^n$,

$$\begin{aligned} &|H^{a,b}(t, x_1, p_1) - H^{a,b}(t, x_2, p_2)| \\ &\leqslant L\big(1 + |p_1|\big)|x_1 - x_2| + L\big(1 + |x_2|\big)|p_1 - p_2| \\ &\leqslant L\big(1 + |p_1|\big)|x_1 - x_2| + L_0|p_1 - p_2|. \end{aligned}$$

Take small $\varepsilon, \delta > 0$ satisfying

$$\varepsilon + 2\delta < L_0(T - T_0) = L_0\Big(T \wedge \frac{1}{2L}\Big) = \Big(1 \wedge (2LT)\Big)(1 + |x_0|).$$

Let $K > 0$ be large so that

$$K > \sup_{(t,x,y)\in\Gamma(x_0)} \max_{(a,b)\in\mathbb{M}} \big[V^{a,b}(t,x) - \widehat{V}^{a,b}(t,y)\big],$$

where

$$\Gamma(x_0) = \big\{(t, x, y) \in [0, T] \times \mathbb{R}^{2n} \mid (t, x), (t, y) \in \Delta(x_0)\big\}.$$

Introduce $\zeta(\cdot) \in C^\infty(\mathbb{R})$ satisfying

$$\zeta_\delta(r) = \begin{cases} 0, & r \leqslant -2\delta, \\ -K, & r \geqslant -\delta, \end{cases} \qquad \zeta_\delta'(r) \leqslant 0, \qquad \forall r \in \mathbb{R}.$$

Define

$$\begin{aligned} \Phi^{a,b}(t, x, y) = {}&V^{a,b}(t,x) - \widehat{V}^{a,b}(t,y) - \frac{1}{\beta}|x - y|^2 \\ &+ \zeta_\delta\big(\langle x\rangle_\varepsilon - L_0(t - T_0)\big) \\ &+ \zeta_\delta\big(\langle y\rangle_\varepsilon - L_0(t - T_0)\big) + \sigma(t - T), \quad (t, x, y) \in \overline{\Gamma(x_0)}, \end{aligned}$$

where

$$\langle x\rangle_\varepsilon = \sqrt{|x - x_0|^2 + \varepsilon^2}, \qquad \langle y\rangle_\varepsilon = \sqrt{|y - x_0|^2 + \varepsilon^2}.$$

Let

$$\Phi^{\bar{a},\bar{b}}(\bar{t},\bar{x},\bar{y}) = \max_{(t,x,y)\in\Gamma_{\varepsilon,\delta}(x_0)} \max_{(a,b)\in\mathbb{M}_1\times\mathbb{M}_2} \Phi^{a,b}(t,x,y), \qquad (7.53)$$

where

$$\Gamma_{\varepsilon,\delta}(x_0) = \big\{ (t,x,y) \mid (t,x),(t,y) \in \Delta_{\varepsilon,\delta}(x_0) \big\},$$
$$\Delta_{\varepsilon,\delta}(x_0) = \big\{ (t,x) \in \Delta(x_0) \mid \langle x \rangle_\varepsilon < L_0(t-T_0) - \delta \big\}.$$

Note that $(\bar{t},\bar{x},\bar{y},\bar{a},\bar{b})$ depends on β, as well as other parameters $\varepsilon,\delta,\sigma$.

Step 3. We may assume that $(\bar{a},\bar{b})$ is independent of the parameters β,ε,δ, etc. and

$$\begin{cases} M_2^{\bar{a},\bar{b}}[V](\bar{t},\bar{x}) < V^{\bar{a},\bar{b}}(\bar{t},\bar{x}), \\ \widehat{V}^{\bar{a},\bar{b}}(\bar{t},\bar{y}) < M_1^{\bar{a},\bar{b}}[\widehat{V}](\bar{t},\bar{y}). \end{cases} \qquad (7.54)$$

In fact, if, say,

$$V^{\bar{a},\bar{b}}(\bar{t},\bar{x}) = M_2^{\bar{a},\bar{b}}[V](\bar{t},\bar{y}) = \max_{b\neq\bar{b}}\Big[V^{\bar{a},b}(\bar{t},\bar{x}) - \kappa_2(\bar{t},\bar{b},b) \Big]$$
$$= V^{\bar{a},\hat{b}}(\bar{t},\bar{x}) - \kappa_2(\bar{t},\bar{b},\hat{b}), \qquad (7.55)$$

for some $\hat{b} \in \mathbb{M}_2 \setminus \{\bar{b}\}$, then

$$V^{\bar{a},\hat{b}}(\bar{t},\bar{x}) - \widehat{V}^{\bar{a},\hat{b}}(\bar{t},\bar{y}) = V^{\bar{a},\bar{b}}(\bar{t},\bar{x}) + \kappa_2(\bar{t},\bar{b},\hat{b}) - \widehat{V}^{\bar{a},\hat{b}}(\bar{t},\bar{y})$$
$$= V^{\bar{a},\bar{b}}(\bar{t},\bar{x}) - \Big[\widehat{V}^{\bar{a},\hat{b}}(\bar{t},\bar{y}) - \kappa_2(\bar{t},\bar{b},\hat{b}) \Big]$$
$$\geqslant V^{\bar{a},\bar{b}}(\bar{t},\bar{x}) - M_2^{\bar{a},\bar{b}}[\widehat{V}](\bar{t},\bar{y}) \geqslant V^{\bar{a},\bar{b}}(\bar{t},\bar{x}) - \widehat{V}^{\bar{a},\bar{b}}(\bar{t},\bar{y}),$$

which implies

$$\Phi^{\bar{a},\hat{b}}(\bar{t},\bar{x},\bar{y}) \geqslant \Phi^{\bar{a},\bar{b}}(\bar{t},\bar{x},\bar{y}).$$

By the definition of $(\bar{t},\bar{x},\bar{y})$ and $(\bar{a},\bar{b})$, it is necessary that

$$\Phi^{\hat{a},\bar{b}}(\bar{t},\bar{x},\bar{y}) = \Phi^{\bar{a},\bar{b}}(\bar{t},\bar{x},\bar{y}) = \max_{(t,x,y)\in\overline{\Gamma}(x_0)} \max_{a\in\mathbb{M}_1,b\in\mathbb{M}_2} \Phi^{a,b}(t,x,y),$$

and

$$\widehat{V}^{\bar{a},\bar{b}}(\bar{t},\bar{y}) = M_2^{\bar{a},\bar{b}}[\widehat{V}](\bar{t},\bar{y}) = \widehat{V}^{\bar{a},\hat{b}}(\bar{t},\bar{y}) - \kappa_2(\bar{t},\bar{b},\hat{b}).$$

Further, similar to that in the proof of Theorem 7.1.7, we must have

$$V^{\bar{a},\hat{b}}(\bar{t},\bar{y}) > M_2^{\bar{a},\hat{b}}[V](\bar{t},\bar{y}).$$

Now, if

$$\widehat{V}^{\bar{a},\hat{b}}(\bar{t},\bar{y}) < M_1^{\bar{a},\hat{b}}[\widehat{V}](\bar{t},\bar{y}),$$

then (7.54) holds for $(\bar{a}, \hat{b})$. Otherwise, we will have

$$\widehat{V}^{\bar{a},\hat{b}}(\bar{t}, \bar{y}) = M_1^{\bar{a},\hat{b}}[\widehat{V}](\bar{t}, \bar{y}) = \min_{a \neq \bar{a}} \left[\widehat{V}^{a,\hat{b}}(\bar{t}, \bar{y}) + \kappa_1(\bar{t}, \bar{a}, a) \right]$$

$$= \widehat{V}^{\hat{a},\hat{b}}(\bar{t}, \bar{y}) + \kappa_1(\bar{t}, \bar{a}, \hat{a}),$$

for some $\hat{a} \in \mathbb{M}_1 \setminus \{\bar{a}\}$. Then

$$V^{\bar{a},\hat{b}}(\bar{t}, \bar{x}) - \widehat{V}^{\bar{a},\hat{b}}(\bar{t}, \bar{y}) \leqslant M_1^{\bar{a},\hat{b}}[V](\bar{t}, \bar{x}) - \widehat{V}^{\hat{a},\hat{b}}(\bar{t}, \bar{y}) - \kappa_1(\bar{t}, \bar{a}, \hat{a})$$

$$\leqslant V^{\hat{a},\hat{b}}(\bar{t}, \bar{x}) - \widehat{V}^{\hat{a},\hat{b}}(\bar{t}, \bar{y}),$$

which implies

$$\Phi^{\bar{a},\bar{b}}(\bar{t}, \bar{x}, \bar{y}) \leqslant \Phi^{\bar{a},\hat{b}}(\bar{t}, \bar{x}, \bar{y}) \leqslant \Phi^{\hat{a},\hat{b}}(\bar{t}, \bar{x}, \bar{y}).$$

By the definition of $(\bar{t}, \bar{x}, \bar{y})$ and $(\bar{a}, \bar{b})$, it is necessary that

$$\Phi^{\hat{a},\hat{b}}(\bar{t}, \bar{x}, \bar{y}) = \Phi^{\bar{a},\bar{b}}(\bar{t}, \bar{x}, \bar{y}) = \max_{(t,x,y) \in \overline{\Gamma(x_0)}} \max_{a \in \mathbb{M}_1, b \in \mathbb{M}_2} \Phi^{a,b}(t, x, y),$$

and

$$V^{\bar{a},\hat{b}}(\bar{t}, \bar{x}) = M_1^{\bar{a},\hat{b}}[V](\bar{t}, \bar{x}) = V^{\hat{a},\hat{b}}(\bar{t}, \bar{x}) + \kappa_1(\bar{t}, \bar{a}, \hat{a}). \qquad (7.56)$$

Further, the same as that in the proof of Theorem 7.1.7, we must have

$$\widehat{V}^{\hat{a},\hat{b}}(\bar{t}, \bar{y}) < M_1^{\hat{a},\hat{b}}[\widehat{V}](\bar{t}, \bar{y}).$$

Now, if

$$V^{\hat{a},\hat{b}}(\bar{t}, \bar{x}) > M_2^{\hat{a},\hat{b}}[V](\bar{t}, \bar{x}),$$

we obtain (7.54). Otherwise, we can continue the above procedure. Then, either we stop at a finite step and (7.54) holds for some $(\bar{a}, \bar{b})$, or one can continue indefinitely. In this case, we end up with a sequence $\{(a_i, b_i)\}_{i \geqslant 1} \subseteq \mathbb{M}_1 \times \mathbb{M}_2$ such that (see (7.55) and (7.56))

$$V^{a_1,b_1}(\bar{t}, \bar{x}) = V^{a_1,b_2}(\bar{t}, \bar{x}) - \kappa_2(\bar{t}, b_1, b_2)$$

$$= V^{a_2,b_2}(\bar{t}, \bar{x}) + \kappa_1(\bar{t}, a_1, a_2) - \kappa_2(\bar{t}, b_1, b_2)$$

$$= V^{a_2,b_3}(\bar{t}, \bar{x}) + \kappa_1(\bar{t}, a_1, a_2) - \kappa_2(\bar{t}, b_1, b_2) - \kappa_2(\bar{t}, b_2, b_3) = \cdots.$$

Since $\mathbb{M}_1 \times \mathbb{M}_2$ is a finite set, there exists a $j \geqslant 1$ such that

$$a_{j+1} = a_1, \quad b_{j+1} = b_1.$$

Then the above leads to a contradiction to (SG4). Hence, for any parameters $\beta, \varepsilon, \delta, \sigma$, we can find a $(\bar{a}, \bar{b}) \in \mathbb{M}_1 \times \mathbb{M}_2$ such that for some $(\bar{t}, \bar{x}, \bar{y}) \in \overline{\Gamma(x_0)}$, (7.53) and (7.54) hold. Again, by the finiteness of $\mathbb{M}_1 \times \mathbb{M}_2$, there must be one $(\bar{a}, \bar{b})$ appearing infinitely many times that (7.53)–(7.54) hold. By choosing such a pair $(\bar{a}, \bar{b})$ (corresponding to a sequence $\beta \downarrow 0$), we have the independence of $\bar{a}$ on the parameters.

Step 4. It holds

$$\frac{1}{\beta}|\bar{x} - \bar{y}|^2 \leqslant \omega_0\left(\sqrt{\beta\bar{\omega}_0}\right) \to 0, \qquad \beta \to 0,$$

where

$$\omega_0(r) = \frac{1}{2} \sup_{\substack{|x-y| \leqslant r \\ (t,x,y) \in \Gamma(x_0)}} \left(|V^{\bar{a},\bar{b}}(t,x) - V^{\bar{a},\bar{b}}(t,y)| + |\widehat{V}^{\bar{a},\bar{b}}(t,x) - \widehat{V}^{\bar{a},\bar{b}}(t,y)|\right),$$

and

$$\bar{\omega}_0 \equiv \sup_{r \geq 0} \omega_0(r).$$

The proof is the same as that of Theorem 7.1.7.

Step 5. It holds that

$$\langle \bar{x} \rangle_\varepsilon < L_0(\bar{t} - T_0) - \delta, \quad \langle \bar{y} \rangle_\varepsilon < L_0(\bar{t} - T_0) - \delta,$$

and when $\beta, \sigma > 0$ are small,

$$\bar{t} < T.$$

The proof is the same as that of Theorem 7.1.7.

Step 6. Completion of the proof.

Now, let us denote

$$\varphi(t,x,y) = \frac{1}{\beta}|x - y|^2 - \zeta_\delta\left(\langle x \rangle_\varepsilon - L_0(t - T_0)\right)$$
$$-\zeta_\delta\left(\langle y \rangle_\varepsilon - L_0(t - T_0)\right) + \sigma(T - t).$$

Then

$$\begin{cases} \varphi_t(t,x,y) = -\sigma + L_0\left[\zeta'_\delta(X_\varepsilon) + \zeta'_\delta(Y_\varepsilon)\right], \\[2mm] \varphi_x(t,x,y) = \dfrac{2}{\beta}(x - y) - \zeta'_\delta(X_\varepsilon)\dfrac{x - x_0}{\langle x \rangle_\varepsilon}, \\[2mm] \varphi_y(t,x,y) = \dfrac{2}{\beta}(y - x) - \zeta'_\delta(Y_\varepsilon)\dfrac{y - x_0}{\langle y \rangle_\varepsilon}, \end{cases}$$

where

$$X_\varepsilon = \langle x \rangle_\varepsilon - L_0(t - T_0), \quad Y_\varepsilon = \langle \bar{y} \rangle_\varepsilon - L_0(\bar{t} - T_0).$$

Applying Lemma 7.1.10 to the function

$$V^{\bar{a},\bar{b}}(t,x) + (-\widehat{V}^{\bar{a},\bar{b}})(t,y) - \varphi(t,x,y)$$

at point $(\bar{t}, \bar{x}, \bar{y})$, we can find $q_1, q_2 \in \mathbb{R}$ such that

$$\begin{cases} (q_1, \varphi_x(\bar{t}, \bar{x}, \bar{y})) \in \overline{D}_{t,x}^{1,+} V^{\bar{a},\bar{b}}(\bar{t}, \bar{x}), \\ (q_2, \varphi_y(\bar{t}, \bar{x}, \bar{y})) \in \overline{D}_{t,x}^{1,+} \big(-\widehat{V}^{\bar{a},\bar{b}} \big)(\bar{t}, \bar{y}), \\ q_1 + q_2 = \varphi_t(\bar{t}, \bar{x}, \bar{y}). \end{cases}$$

By Proposition 7.2.10, we have

$$\max \Big\{ \min \Big[q_1 + H^{\bar{a},\bar{b}}\Big(\bar{t}, \bar{x}, \frac{2}{\beta}(\bar{x} - \bar{y}) - \zeta'_\delta(X_\varepsilon)\frac{\bar{x} - x_0}{\langle \bar{x} \rangle_\varepsilon}\Big), $$
$$ M_1^{\bar{a},\bar{b}}[V](\bar{t}, \bar{x}) - V^{\bar{a}}(\bar{t}, \bar{x}) \Big], $$
$$ M_2^{\bar{a},\bar{b}}[V](\bar{t}, \bar{x}) - V^{\bar{a},\bar{b}}(\bar{t}, \bar{x}) \Big\} \geqslant 0$$

and

$$\min \Big\{ \max \Big[-q_2 + H^{\bar{a},\bar{b}}\Big(\bar{t}, \bar{x}, \frac{2}{\beta}(\bar{x} - \bar{y}) + \zeta'_\delta(Y_\varepsilon)\frac{\bar{y} - x_0}{\langle \bar{y} \rangle_\varepsilon}\Big), $$
$$ M_1^{\bar{a},\bar{b}}[\widehat{V}](\bar{t}, \bar{y}) - \widehat{V}^{\bar{a}}(\bar{t}, \bar{y}) \Big], $$
$$ M_2^{\bar{a},\bar{b}}[\widehat{V}](\bar{t}, \bar{y}) - \widehat{V}^{\bar{a},\bar{b}}(\bar{t}, \bar{y}) \Big\} \leqslant 0.$$

Thus, noting (7.54), we obtain

$$\begin{cases} \min \Big[q_1 + H^{\bar{a},\bar{b}}\Big(\bar{t}, \bar{x}, \frac{2}{\beta}(\bar{x} - \bar{y}) - \zeta'_\delta(X_\varepsilon)\frac{\bar{x} - x_0}{\langle \bar{x} \rangle_\varepsilon}\Big), \\ \qquad\qquad M_1^{\bar{a},\bar{b}}[V](\bar{t}, \bar{x}) - V^{\bar{a}}(\bar{t}, \bar{x}) \Big] \geqslant 0, \\ \max \Big[-q_2 + H^{\bar{a},\bar{b}}\Big(\bar{t}, \bar{x}, \frac{2}{\beta}(\bar{x} - \bar{y}) + \zeta'_\delta(Y_\varepsilon)\frac{\bar{y} - x_0}{\langle \bar{y} \rangle_\varepsilon}\Big), \\ \qquad\qquad M_1^{\bar{a},\bar{b}}[\widehat{V}](\bar{t}, \bar{y}) - \widehat{V}^{\bar{a}}(\bar{t}, \bar{y}) \Big] \leqslant 0. \end{cases}$$

This further leads to the following:

$$\begin{cases} q_1 + H^{\bar{a},\bar{b}}\Big(\bar{t}, \bar{x}, \frac{2}{\beta}(\bar{x} - \bar{y}) - \zeta'_\delta(X_\varepsilon)\frac{\bar{x} - x_0}{\langle \bar{x} \rangle_\varepsilon}\Big) \geqslant 0, \\ -q_2 + H^{\bar{a},\bar{b}}\Big(\bar{t}, \bar{y}, \frac{2}{\beta}(\bar{x} - \bar{y}) + \zeta'_\delta(Y_\varepsilon)\frac{\bar{y} - x_0}{\langle \bar{y} \rangle_\varepsilon}\Big) \leqslant 0. \end{cases}$$

The rest of the proof is the same as that of Theorem 7.1.7. □

Now, combining Theorems 7.2.9 and 7.2.12, we obtain the following result.

Theorem 7.2.13. *Let* (SG1)–(SG4) *hold. Then the Elliott–Kalton value of Problem* (SG) *exists.*

7.2.3 A limiting case

In this subsection, we investigate what happens if the switching costs $\kappa_1(\cdot,\cdot,\cdot,\cdot)$ and $\kappa_2(\cdot,\cdot,\cdot)$ approach to zero. More precisely, we let $\kappa_1^\varepsilon(\cdot,\cdot,\cdot)$ and $\kappa_2^\varepsilon(\cdot,\cdot,\cdot)$ be the switching costs for the two players, depending on a parameter $\varepsilon > 0$ such that

$$\begin{cases} \lim\limits_{\varepsilon\to0} \kappa_1^\varepsilon(t,a,\widehat{a}) = 0, & \forall t \in [0,T],\ a,\widehat{a} \in M_1, \\ \lim\limits_{\varepsilon\to0} \kappa_2^\varepsilon(t,b,\widehat{b}) = 0, & \forall t \in [0,T],\ b,\widehat{b} \in M_2. \end{cases}$$

We let (SG1)–(SG3) hold. From Proposition 7.2.2, we see that the family of the lower value functions (denoted by) $V_\varepsilon(\cdot,\cdot)$ corresponding to switching costs $(\kappa_1^\varepsilon, \kappa_2^\varepsilon)$, $\varepsilon > 0$ is uniformly bounded and equi-continuous in bounded sets. Thus, by Arzela–Ascoli Theorem, we can find a subsequence (still denoted by) $V_\varepsilon(\cdot,\cdot)$, such that

$$\lim_{\varepsilon\to0} V_\varepsilon^{a,b}(t,x) = \bar{V}^{a,b}(t,x), \tag{7.57}$$

uniformly for $t \in [0,T]$ and x in bounded sets. It is clear that $\bar{V}^{a,b}(\cdot,\cdot)$ also satisfies

$$\begin{cases} |\bar{V}(t,x_1) - \bar{V}(t,x_2)| \leqslant K|x_1 - x_2|, & \forall t \in [0,T],\ x_1,x_2 \in \mathbb{R}^n, \\ |\bar{V}(t_1,x) - \bar{V}(t_2,x)| \leqslant K(1+|x|)|t_1 - t_2|, & \forall t_1,t_2 \in [0,T],\ x \in \mathbb{R}^n. \end{cases}$$

Next, we define the following maps

$$\begin{cases} H^+(t,x,p) = \min\limits_{a\in M_1} \max\limits_{b\in M_2}\{\langle p, f(t,x,a,b)\rangle + g(t,x,a,b)\}, \\ H^-(t,x,p) = \max\limits_{b\in M_2} \min\limits_{a\in M_1}\{\langle p, f(t,x,a,b)\rangle + g(t,x,a,b)\}, \\ \qquad\qquad\qquad\qquad \forall(t,x,p) \in [0,T] \times \mathbb{R}^n \times \mathbb{R}^n. \end{cases}$$

Then, we have the following result:

Theorem 7.2.14. *Let (SG1)–(SG3) hold. Let $\{\bar{V}^{a,b}(\cdot,\cdot)\,|\,(a,b) \in M_1 \times M_2\}$ be any functions obtained through (7.57). Then, the following conclusions hold:*

(i) *There exists a scalar function $v(\cdot)$ such that*

$$\begin{cases} |v(t,x_1) - v(t,x_2)| \leqslant K|x_1 - x_2|, \\ \qquad\qquad \forall t \in [0,T],\ x_1,x_2 \in \mathbb{R}^n, \\ |v(t_1,x) - v(t_2,x)| \leqslant K(1+|x|)|t_1 - t_2|, \\ \qquad\qquad \forall t_1,t_2 \in [0,T],\ x \in \mathbb{R}^n, \end{cases} \tag{7.58}$$

and

$$\bar{V}^{a,b}(t,x) = v(t,x), \qquad \forall (t,x,a,b) \in [0,T] \times \mathbb{R}^n \times \mathbb{M}_1 \times \mathbb{M}_2. \qquad (7.59)$$

(ii) Function $v(\cdot,\cdot)$ is a viscosity sub-solution of the upper Isaacs equation

$$\begin{cases} v_t(t,x) + H^+(t,x,v_x(t,x)) = 0, & (t,x) \in [0,T) \times \mathbb{R}^n, \\ v(T,x) = h(x), & x \in \mathbb{R}^n. \end{cases}$$

(iii) Function $v(\cdot,\cdot)$ is a viscosity super-solution of the lower Isaacs equation

$$\begin{cases} v_t(t,x) + H^-(t,x,v_x(t,x)) = 0, & (t,x) \in [0,T) \times \mathbb{R}^n, \\ v(T,x) = h(x), & x \in \mathbb{R}^n. \end{cases}$$

Proof. (i) From

$$M_2^{a,b}[V_\varepsilon](t,x) \leqslant V_\varepsilon^{a,b}(t,x) \leqslant M_1^{a,b}[V_\varepsilon](t,x),$$

by letting $\varepsilon \to 0$ along subsequences in (7.57), we obtain (7.58)–(7.59).

(ii) Let $\varphi \in C^1([0,T) \times X)$ such that $v(\cdot,\cdot) - \varphi(\cdot,\cdot)$ attains a strict local maximum at $(t_0,x_0) \in [0,T) \times X$. Since the convergence in (7.57) is uniformly in $t \in [0,t]$ and x in bounded sets, we see that for any $a \in \mathbb{M}_1$, there exist $t_\varepsilon \to t_0$ and $x_\varepsilon \to x_0$ such that

$$\max_{b \in \mathbb{M}_2} V_\varepsilon^{a,b}(t_\varepsilon, x_\varepsilon) - \varphi(t_\varepsilon, x_\varepsilon) > \max_{b \in \mathbb{M}_2} V_\varepsilon^{a,b}(t,x) - \varphi(t,x),$$

$$\text{for } (t,x) \text{ near } (t_\varepsilon, x_\varepsilon).$$

We let $b_\varepsilon^a \in \mathbb{M}_2$ such that

$$V_\varepsilon^{a,b_\varepsilon^a}(t_\varepsilon, x_\varepsilon) = \max_{b \in \mathbb{M}_2} V_\varepsilon^{a,b}(t_\varepsilon, x_\varepsilon).$$

Since for any $b \in \mathbb{M}_2 \setminus \{b_\varepsilon^a\}$, $\kappa_2^\varepsilon(b_\varepsilon^a, b) > 0$, we must have

$$V_\varepsilon^{a,b_\varepsilon^a}(t_\varepsilon, x_\varepsilon) > M_2^{a,b_\varepsilon^a}[V_\varepsilon](t_\varepsilon, x_\varepsilon). \qquad (7.60)$$

Thus, by Definition 7.2.8 and (7.60), we obtain

$$\varphi_t(t_\varepsilon, x_\varepsilon) + H^{a,b_\varepsilon^a}(t_\varepsilon, x_\varepsilon, \varphi_x(t_\varepsilon, x_\varepsilon)) \geqslant 0.$$

Consequently, by choosing a subsequence if necessary, and taking the limits, we obtain

$$\varphi_t(t_0, x_0) + H^{a,\bar{b}}(t_0, x_0, \varphi_x(t_0, x_0)) \geqslant 0,$$

for some $\bar{b} \in \mathbb{M}_2$ (depending on a, in general). Therefore,

$$\min_{a \in \mathbb{M}_1} \max_{b \in \mathbb{M}_2} \{\varphi_t(t_0, x_0) + H^{a,b}(t_0, x_0, \varphi_x(t_0, x_0))\} \geqslant 0,$$

i.e.,

$$\varphi_t(t_0, x_0) + H^+(t_0, x_0, \varphi_x(t_0, x_0)) \geqslant 0.$$

Finally, it is easy to see that

$$v(T, x) = h(x), \qquad \forall x \in \mathbb{R}^n.$$

This proves (ii). The proof of (iii) is similar. ☐

From the above theorem we can obtain the following interesting result.

Corollary 7.2.15. *Let* (SG1)–(SG3) *hold. Let the Isaacs condition hold:*

$$H^+(t, x, p) = H^-(t, x, p) \equiv H(t, x, p), \quad \forall (t, x, p) \in [0, T] \times \mathbb{R}^n \times \mathbb{R}^n.$$

Then, there exists a function $v(\cdot, \cdot)$ satisfying (7.58)–(7.59) *such that for any $(a, b) \in \mathbb{M}_1 \times \mathbb{M}_2$,*

$$\lim_{\varepsilon \to 0} V_\varepsilon^{a,b}(t, x) = v(t, x), \tag{7.61}$$

uniformly for $t \in [0, T]$ and x in any bounded sets. Moreover, the function $v(\cdot, \cdot)$ is the unique viscosity solution of the following Isaacs equation:

$$\begin{cases} v_t(t, x) + H(t, x, v_x(t, x)) = 0, & (t, x) \in [0, T) \times \mathbb{R}^n, \\ v(T, x) = h(x), & x \in \mathbb{R}^n. \end{cases} \tag{7.62}$$

Proof. We only need to notice that the uniqueness of the viscosity solutions of (7.62) implies the whole sequence $V_\varepsilon^{a,b}(\cdot, \cdot)$ converges. ☐

It is not hard to see that $v(\cdot, \cdot)$ obtained in (7.61) is exactly the Elliott–Kalton value function of the classical two-player zero-sum differential game of fixed duration with control sets $\mathbb{M}_1$ and $\mathbb{M}_2$. We also see that the same result as Corollary 7.2.15 holds for the upper value functions $W_\varepsilon^{a,b}(\cdot)$. Finally, as far as the above convergence is concerned, the condition (SG4) is irrelevant.

7.3 Brief Historic Remarks

Optimal switching problems were firstly studied by Capuzzo Dolcetta–Evans for ordinary differential equation [25] in 1984, which is an extension of the so-called optimal stopping time problems. See [111] for some extension to infinite-dimensional systems, [124] for the case of systems with continuous, switching and impulse controls, and [114], [74] for stochastic cases. Two-person zero-sum differential games with switching strategies were studied by Yong [125, 126] in 1990, and later was extended to the case of switching and impulse strategies in [127]. The material presented in this chapter for the optimal switching control is essentially based on [25], with some modification and that for the differential games with switching strategies is based on [126]. Lemma 7.1.10 is due to Crandall–Ishii ([29]), which played a very subtle role in the presentation.

Bibliography

[1] M. Bardi and I. Capuzzo-Dolcetta, *Optimal Control and Viscosity Solutions of Hamilton-Jacobi-Bellman Equations*, Birkhäuser, Boston, 1997.

[2] M. Bardi and F. Da Lio, *On the Bellman equation for some unbounded control problems*, NoDEA, 4 (1997), 491–510.

[3] T. Basar and P. Bernhard, H^∞-*Optimal Control and Related Minimax Design Problems: A Dynamic Game Approach*, Birkhäuser, Boston, 1991.

[4] A. Beck, *Uniquess of Flow Solutions of Differential Equations*, Lecture Notes in Math. vol. 318, Springer-Verlag, Berlin, 1973.

[5] R. Bellman, *On the theory of dynamic programming*, Proc. Nat. Acad. Sci. USA, 38 (1952), 716–719.

[6] R. Bellman, *Dynamic Programming*, Princeton Univ. Press, Princeton, NJ, 1957.

[7] R. Bellman, I. Glicksberg, and O. Gross, *Some Aspects of the Mathematical Theory of Control Processes*, Rand Corporation, Santa Monica, 1958.

[8] L. D. Berkovitz, *A variational approach to differential games*, Advances in Game Theory, Princeton Univ. Press, Princeton, N.J, (1964), 127–174.

[9] L. D. Berkovitz, *Lectures on differential games*, Differential Games and Related Topics, H. W. Kuhn and G. P. Szego, eds., North-Holland, Amsterdam, 1971, 3–45.

[10] L. D. Berkovitz, *Optimal Control Theory*, Springer-Verlag, New York, 1974.

[11] L. D. Berkovitz, *The existence of value and saddle point in games of fixed duration*, SIAM J. Control Optim., 23 (1985), 172–196.

[12] L. D. Berkovitz, *Differential games of generalized pursuit and evasion*, SIAM J. Control Optim., 24 (1986), 361–373.

[13] L. D. Berkovitz, *Characterization of the values of differential games*, Appl. Math. Optim., 17 (1988), 177–183.

[14] L. D. Berkovitz and W. H. Fleming, *On differential games with integral payoff*, Contributions to the theory of games, vol. 3, 413-435; Annals of Mathematics Studies, no. 39. Princeton University Press, Princeton, N. J., 1957.

[15] P. Bernhard, *Linear-quadratic, two-person, zero-sum differential games: Ne- cessary and sufficient conditions*, J. Optim. Theory Appl.,27 (1979),

51–69.

[16] S. Biton, *Nonlinear monotone semigroups and viscosity solutions*, Ann. I. H. Poincaré Anal. Non Linéaire, 18 (2001), 383–402.

[17] V. G. Boltyanski, *The maximum principle in the theory of optimal processes*, Dokl. Akad. Nauk SSSR, 119 (1958), 1070–1073 (Russian).

[18] V. G. Boltyanski, R. V. Gamkrelidze, and L. S. Pontryagin, *On the theory of optimal processes*, Doklady Akad. Nauk SSSR, 110 (1956), 7–10 (Russian).

[19] V. G. Boltyanski, R. V. Gamkrelidze, and L. S. Pontryagin, *On the theory of optimal processes I. The maximum principle*, Izvest Akad. Nauk SSSR, Ser. Mat. 24 (1960), 3–42 (Russian). English transl. in Amer. Math. Soc. Transl. (2) 18 (1961), 341–382.

[20] M. H. Breitner, *The genesis of differential games in light of Isaacs' contributions*, J. Optim. Theory & Appl., 124 (2005), 523–559.

[21] E. Borel, *The theory of play and integral equations with skew symmetric kernels*, Comptes Rendus Academie des Sciences, 173 (1921), 1304–1308 (French). English transl. in Econometrica, 21 (1953), 97–100.

[22] E. Borel, *On the games that involve chance and the skill of the players*, Theorie des Probabilites. Paris: Librairie Scientifique, J. Hermann, (1924), 204–224 (French). English transl. in Econometrica, 21 (1953), 101–115.

[23] E. Borel, *On systems of linear forms of skew symmetric determinant and the general theory of paly*, Comptes Rendus Academie des Sciences, 184 (1927), 52–53 (French). English transl. in Econometrica, 21 (1953), 116–117.

[24] R. Buckdahn, S. Peng, M. Quincampoix, and C. Rainer, *Existence of stochastic control under state constraints*, C. R. Acad. Sci. Paris, Sér. I Math., 327 (1988), 17–22.

[25] I. Capuzzo-Dolcetta and L. C. Evans, *Optimal switching for ordinary differential equations*, SIAM J. Control Optim., 22 (1984), 1133–1148.

[26] C. Carathéodory, *Calculus of variations and partial differential equations of the first order. Part I: Partial differential equations of the first order; Part II: Calculus of variations*, B. G. Teubner, Leipzig, Germany, 1935 (German). English transl. Holden-Day, Inc., San Francisco, Calif.-London-Amsterdam 1965/1967.

[27] S. Chen, *Matrix Riccati equations and linear Fredholm integral equations*, Zhejiang Daxue Xuebao, 19 (1985), no. 2, 137–145. (Chinese)

[28] A. A. Cournot, *Recherches sur les Principes Mathematiquesde la Theorie des Richesses*. Paris: Hachette, 1838. English translation: *Researches into the Mathematical Principles of the Theory of Wealth*, Macmillian, New York, 1897; Reprinted, Augustus M. Kelley, New York, 1971.

[29] M. G. Crandall and H. Ishii, *The maximum principle for the semicontinuous functions*, Diff. Int. Eqs., 3 (1990), 1001–1014.

[30] M. G. Crandall and P. L. Lions, *Viscosity solutions of Hamilton-Jacobi equations*, Trans. AMS, 277 (1983), 1–42.

[31] M. G. Crandall and P. L. Lions, *On existence and uniqueness of solutions of Hamilton-Jacobi equations*, Nonlinear Anal., 10 (1986), 353–370.

[32] M. G. Crandall and P. L. Lions, *Remarks on the existence and uniqueness*

of unbounded viscosity solutions of Hamilton-Jacobi equations, Illinois J. Math., 31 (1987), 665–688.

[33] F. Da Lio, *On the Bellman equation for infinite horizon problems with unblounded cost functional*, Appl. Math. Optim., 41 (2000), 171–197.

[34] F. Da Lio and O. Ley, *Convex Hamilton-Jacobi equations under superlinear growth conditions on data*, Appl. Math. Optim., 63 (2011), 309–339.

[35] M. C. Delfour, *Linear quadratic differential games: saddle point and Riccati differential equations*, SIAM J. Control Optim., 46 (2007), 750–774.

[36] M. C. Delfour and O. D. Sbarba, *Linear quadratic differential games: closed loop saddle points*, SIAM J. Control Optim., 47 (2009), 3138–3166.

[37] S. Dreyfus, *Richard Bellman on the birth of dynamic programming*, Operations Reserch, 50 (2002), 48–51.

[38] I. Ekeland, *On the variational principle*, J. Math. Anal. Appl., 47 (1974), 324–353.

[39] I. Ekeland, Nonconvex minimization problems, *Bull. Amer. Math. Soc.* (New Serise), 1 (1979), 443–474.

[40] R. J. Elliott and N. J. Kalton, *The Existence of Value in Differential Games*, Memoirs of AMS, No. 126, Amer. Math. Soc., Providence, R.I., 1972.

[41] L. C. Evans and P. E. Souganidis, *Differential games and representation formulas for solutions of Hamilton-Jacobi-Isaacs equations*, Indiana Univ. Math. J., 5 (1984), 773–797.

[42] H. O. Fattorini, *The maximum principle for nonlinear nonconvex systems in infinite dimensional spaces*, Lecture Notes in Control & Inform. Sci., Vol. 75, Springer-Verlag, 1985, 162–178.

[43] H. O. Fattorini, *A unified theory of necessary conditinos for nonlinear nonconvex control systems*, Appl. Math. Optim., 15 (1987), 141–185.

[44] W. H. Fleming, *On a class of games over function space and related variational problems*, Ann. of Math. (2), 60 (1954), 578–594.

[45] W. H. Fleming, *A note on differential games of prescribed duration*, Contributions to the theory of games, vol. 3, 407–412, Annals of Mathematics Studies, no. 39 (1957), Princeton University Press, Princeton, N.J.

[46] W. H. Fleming, *The convergence problem for differential games*, J. Math. Anal. Appl., 3 (1961), 102–116.

[47] W. H. Fleming, *The convergence problem for differential games, II*, Advances in Game Theory, Princeton Univ. Press, Princeton, N.J, 1964, 195–210.

[48] M. Fréchet, *Emile Borel, initiator of the theory of psychological games and its application*, Econometrica, 21 (1953), 95–96.

[49] M. Fréchet, *Commentary on the three notes of Emile Borel*, Econometrica, 21 (1953), 118–124.

[50] A. Friedman, *On the definition of differential games and the existence of value and of saddle points*, J. Diff. Eqs, 7 (1970), 69–91.

[51] A. Friedman, *Differential Games*, Wiley-Interscience, New York, 1971.

[52] R. V. Gamkrelidze, *Discovery of the maximum principle in optimal control, Mathematics and War*, B. Boob-Bavnbek, J. Hoyrup eds., Springer, 2003, 160–173.

[53] M. Garavello and P. Soravia, *Optimality principles and uniqueness for Bellman equations of unbounded control problems with discontinuous running cost*, NoDEA, 11 (2004), 271–298.

[54] M. R. Hestenes, *Numerical Methods for Obtaining Solutions of Fixed End Point Problems in the Calculus of Variations*, Research Mem. No. 102, RAND Corporation, 1949.

[55] M. R. Hestenes, *A General Problem in the Calculus of Variations with Applications to the Paths of Least Time*, Research Mem. No. 100, RAND Corporation, 1950.

[56] Y. C. Ho, A. E. Bryson, and S. Baron, *Differential games and optimal pursuit-evasion strategies*, IEEE Trans, AC, 10 (1965), 385–389.

[57] R. Isaacs, *Games of pursuit*, Rand Corporation Report, P-257, 17 Nov., 1951.

[58] R. Isaacs, *Differential Games*, Wiley, New York, 1965.

[59] R. Isaacs, *Differential games: their scope, nature, and future*, J. Optim. Theory & Appl., 3 (1969), 283–295.

[60] H. Ishii, *Uniqueness of unbounded viscosity solutions of Hamilton-Jacobi equations*, Indiana Univ. Math. J., 33 (1984), 721–748.

[61] H. Ishii, *Representation of solutions of Hamilton-Jacobi equations*, Nonlinear Anal., 12 (1988), 121–146.

[62] R. E. Kalman, *Contributions to the theory of optimal control*, Bol. Soc. Math. Mexicana, 5 (1960), 102–119.

[63] B. Káskosz, *On a nonlinear evasion problem*, SIAM J. Control Optim., 15 (1977), 661–673.

[64] N. N. Krasovskii and A. I. Subbotin, *Optimal deviation in a differential game*, Differencial'nye Uravenija, 4 (1968), 2159–2165.

[65] N. N. Krasovskii and A. I. Subbotin, *Game-theoretical control problems. Translated from the Russian by Samuel Kotz. Springer Series in Soviet Mathematics*, Springer-Verlag, New York, 1988.

[66] V. N. Lagunov, *A nonlinear differential game of evasion*, Dokl. Akad. Nauk. SSSR, 202 (1972), 522–525 (Russian). English transl. in Soviet Math. Dokl., 13 (1972), 131–135.

[67] A. M. Letov, *Analytical design of regulator*, Avtomat. i Telemekh., (1960), 436–446, 561–571, 661–669 (in Russian); English transl. in Automat. Remote Control, 21 (1960).

[68] X. Li and Y. Yao, *On optimal control for distributed parameter systems*, Proc. IFAC 8th Triennial World Congress, Kyto, Japan, 1981, 207–212.

[69] X. Li and J. Yong, *Necessary conditions of optimal control for distributed parameter systems*, SIAM J. Control Optim., 29 (1991), 985–908.

[70] X. Li and J. Yong, *Optimal Control Theory for Infinite Dimensional Systems*, Birkhäuser, Boston, 1995.

[71] P. L. Lions and P. E. Souganidis, *Differential games, optimal control and directional derivatives of viscosity solutions of Bellman's and Isaacs' equations*, SIAM J. Control Optim., 23 (1985), 566–583.

[72] P. L. Lions and P. E. Souganidis, *Differential games, optimal control and directional derivatives of viscosity solutions of Bellman's and Isaacs' equa-*

tions II, SIAM J. Control Optim., 24 (1986), 1086–1089.

[73] H. Lou and J. Yong, *A Concise Course of Optimal Control Theory*, High Education Press, Beijing, 2006, (in Chinese).

[74] J. Ma and J. Yong, *Dynamic programming for multidimensional stochastic control problems*, Acta Math. Sinica, 15 (1999), 485–506.

[75] W. McEneaney, *A uniqueness result for the Isaacs equation corresponding to nonlinear H_∞ control*, Math. Control Signals Systems, 11 (1998), 303–334.

[76] E. F. Mishchenko, *On the problem of evading the encounter in differential games*, SIAM J. Control Optim., 12 (1974), 300–310.

[77] E. F. Mishchenko and L. S. Pontryagin, *Linear differential games*, Dokl. Akad. Nauk. SSSR, 174 (1967), 27–29 (Russian). English transl. in *Soviet Math. Dokl.*, 8 (1967), 585–588.

[78] L. Mou and J. Yong, *Two-person zero-sum linear quadratic stochastic differential games by a Hilbert space method*, J. Industrial & Management Optim., 2 (2006), 95–117.

[79] J. F. Nash, *Equilibrium Points in N-Person Games*, Proc. Nat. Acad. Sci. USA, 36 (1951), 48–49.

[80] J. F. Nash, *The Bargaining Problem*, Econometrica, 18 (1950), 155–162.

[81] J. F. Nash, *Non-cooperative games*, Ann. of Math. (2), 54 (1951), 286–295.

[82] J. F. Nash, *Two Person Cooperative Games*, Econometrica, 21 (1953), 128–140.

[83] V. V. Ostapenko, *A nonlinear escape problem*, Kibernetika (Kiev), (1978), No.4, 106–112 (Russian). English transl. in *Cybernetics*, 14 (1978), 594–601.

[84] V. V. Ostapenko, *A nonautonomous evasion problem*, Avtomatika i Telemekhanika, 43 (1982), No.6, 81–86 (Russian). English transl. in *Automation & Remote Control*, 43 (1882), 768–773.

[85] S. Peng and J. Yong, *Determination of controllable set for a controlled dynamic system*, J. Austral. Math. Soc. Ser. B, 33 (1991), 164–179.

[86] R. Penrose, *A general inverse of matrices*, Proc. Cambridge Philos. Soc., 52 (1955), 17–19.

[87] H. J. Pesch and R. Bulirsch, *The maximum principle, Bellman's equation, and Carathéodory's work*, J. Optim. Theory Appl., 80 (1994), 199–225.

[88] H. J. Pesch and M. Plail, *The maximum principle of optimal control: a history of ingenious ideas and missed opportunities*, Control & Cybernetics, 38 (2009), 973–995.

[89] H. J. Pesch, *Carathéodory's royal road of the calculus of variations: missed exits to the maximum principle of optimal control theory*, Numer. Algebra Control Optim., 3 (2013), 161–173.

[90] L. S. Pontryagin, *Optimal processes of regulation*, Proc. Internat. Congr. Math. (Edinburgh, 1958), Cambridge Univ. Press, 1960, 182–202 (in Russian).

[91] L. S. Pontryagin, *Optimal process of regulation*, Uspekhi Mat. Nauk, 14 (1959), No.1(85), 3–20 (in Russian); English transl. in *Amer. Math. Soc. Transl.*, 18 (1961), No.2.

[92]   L. S. Pontryagin, *Linear differential games 1,2*, Dokl. Akad. Nauk. SSSR, 174 (1967), 1278–1280; 175 (1967), 764–766 (Russian). English transl. in *Soviet Math Dokl.*, 8 (1967), 769–771; 8 (1967), 910–912.

[93]   L. S. Pontryagin, *A linear differential escape game*, Trudy Mat. Inst. Steklov, 112 (1971), 30–63 (Russian). English transl. in *Proc. Steklov Inst. Math.*, 112 (1971), 27–60.

[94]   L. S. Pontryagin, *On the evasion process in differential games*, Appl. Math. Optim., 1 (1974), 5–19.

[95]   L. S. Pontryagin, *Linear differential games of pursuit*, Mat. Sb., 112(154) (1980), 307–330 (Russian). English transl. in *Math. USSR Sb.*, 40 (1981), 285–303.

[96]   L. S. Pontryagin, *The mathematical theory of optimal control processes and differential games*, Proc. Steklov Inst. Math., (1986), No.4, 123–159.

[97]   L. S. Pontryagin, V. G. Boltyanski, R. V. Gamkrelidze, and E. F. Mishchenko, *Mathematical Theory of Optimal Processes*, Wiley, New York, 1962.

[98]   L. S. Pontryagin and E. F. Mishchenko, *A problem on the escape of one controlled object from another*, Dokl. Akad. Nauk. SSSR, 189 (1969), 721–723 (Russian); English transl. in *Soviet Math. Dokl.*, 10 (1969), 1488–1490.

[99]   L. S. Pontryagin and E. F. Mishchenko, *The contact avoidance problem in linear differential games*, Differencial'nye Uravenija, 7 (1971), 436–445 (Russian); English transl. in *Diff. Eqs.*, 7 (1971), 335–352.

[100] B. N. Pshenichnyi, *Linear differential games*, Avtomatika i Telemekhanika, (1968), No.5, 46–54 (Russian). English transl. in *Automation & Remote Control*, (1968), No.1, 55–67.

[101] B. N. Pshenichnyi, *The flight problem*, Kibernetika (Kiev), (1975), No.4, 120–127 (Russian). English transl. in *Cybernetics*, 11 (1975), 642–651.

[102] H. Qiu and J. Yong, *Hamilton-Jacobi equations and two-person zero-sum differential games with unbounded control*, ESIAM COCV, 19 (2013), 404–437.

[103] F. Rampazzo, *Differential games with unbounded versus bounded controls*, SIAM J. Control Optim., 36 (1998), 814–839.

[104] N. Satimov, *On a way to avoid contact in differential games*, Mat. Sb., 99(141) (1976), 380–393 (Russian). English transl. in *Math. USSR Sb.*, 28 (1976), 339–352.

[105] W. E. Schmitendorf, *Differential games with open-loop saddle point conditions*, IEEE Trans. Auto. Control, 15 (1970), 320–325.

[106] W. E. Schmitendorf, *Existence of optimal open-loop strategies for a class of differential games*, J. Optim. Theory Appl., 5 (1970), 363–375.

[107] W. E. Schmitendorf, *Differential games without pure strategy saddle-point solutions*, J. Optim. Theory Appl., 18 (1976), 81–92.

[108] U. Schwalbe and P. Walker, *Zermelo and the early history of game theory*, Games and Economic Behavior, 34 (2001), 123–137.

[109] P. Soravia, *Equivalence between nonlinear $\mathcal{H}_\infty$ control problems and existence of viscosity solutions of Hamilton-Jacobi-Isaacs equations*, Appl. Math. Optim., 39 (1999), 17–32.

[110] P. Soravia, *Pursuit-evasion problems and viscosity solutions of Isaacs equation*, SIAM J. Control Optim., 31 (1993), 604–623.

[111] S. Stojanovic and J. Yong, *Optimal switching for partial differential equations I, II*, J. Math. Anal. Appl., 138 (1989), 418–438; 439–460.

[112] J. Sun and J. Yong, *Linear quadratic stochasitc differential games: open-loop and closed-loop saddle pints*, submitted.

[113] H. J. Sussmann and J. C. Willems, *300 years of optimal control: from the brachystochrone to the maximum principle*, IEEE Control Systems, 17 (1997), No.3, 32–44.

[114] S. Tang and J. Yong, *Finite horizon stochastic optimal switching and impulse controls with a viscosity solution approach*, Stochastics & Stochastics Reports, 45 (1993), 145–176.

[115] J. von Neumann, *On the Theory of Games of Strategy*, Mathematische Annalen, 100 (1928), 295–320 (German). English transl. in *Contributions to the Theory of Games, Volume IV (Annals of Mathematics Studies, 40) (A. W. Tucker and R. D. Luce, eds.)*, Princeton University Press, Princeton, 1959, 13–42.

[116] J. von Neumann, *Communication on the Borel notes*, Econometrica, 21 (1953), 124–127.

[117] J. von Neumann and O. Morgenstern, *Theory of Games and Economic Behavior*, Princeton Univ. Press, New York, 1944.

[118] J. Yong, *On Differential Games of Evasion and Pursuit*, Ph.D. Dissertation, Purdue University, 1986.

[119] J. Yong, *On differential evasion games*, SIAM J. Control & Optim., 26 (1988), 1–22.

[120] J. Yong, *On differential pursuit games*, SIAM J. Control & Optim., 26 (1988), 478–495.

[121] J. Yong, *On the evadable sets of differential evasion games*, J. Math. Anal. Appl., 133 (1988), 249–271.

[122] J. Yong, *Evasion with weak superiority*, J. Math. Anal. Appl., 134 (1988), 116–124.

[123] J. Yong, *A sufficient condition for the evadability of differential evasion games*, J. Optim. Theory & Appl., 57 (1988), 501–509.

[124] J. Yong, *Systems governed by ordinary differential equations with continuous, switching and impulse controls*, Appl. Math. Optim., 20 (1989), 223–236.

[125] J. Yong, *Differential games with swithcing strategies*, J. Math. Anal. Appl., 145 (1990), 455–469.

[126] J. Yong, *A zero-sum differential game in a finite duration with switching strategies*, SIAM J. Control & Optim., 28 (1990), 1234–1250.

[127] J. Yong, *Zero-sum differential games involving impusle controls*, Appl. Math. Optim., 29 (1994), 243–261.

[128] J. Yong and X. Y. Zhou, *Stochastic Control: Hamiltonian Systems and HJB Equations*, Springer-Verlag, New York, 1999.

[129] Y. You, *Syntheses of differential games and pseudo-Riccati equations*, Abstr. Appl. Anal., 7 (2002), 61–83.

[130] E. Zermelo, *On an application of set theory to the theory of the game of chess*, Proc. Fifth Congress Mathematicians, Cambridge Univ. Press, 1913, 501–504 (in German).

[131] P. Zhang, *Some results on two-person zero-sum linear quadratic differential games*, SIAM J. Control Optim., 43 (2005), 2157–2165.

Index

Printed in the United States
By Bookmasters